CRC Series in Naturally Occurring Pesticides
Series Editor-in-Chief
N. Bhushan Mandava

Handbook of Natural Pesticides: Methods
Volume I: Theory, Practice, and Detection
Volume II: Isolation and Identification
Editor
N. Bhushan Mandava

Handbook of Natural Pesticides
Volume III: Insect Growth Regulators
Volume IV: Pheromones
Editors
E. David Morgan
N. Bhushan Mandava

Future Volumes
Handbook of Natural Pesticides

Insect Attractants, Deterrents, and Defensive Secretions
Editors
E. David Morgan
N. Bhushan Mandava

Plant Growth Regulators
Editor
N. Bhushan Mandava

Microbial Insecticides
Editors
Carl M. Ignoffo

CRC
Handbook
of
Natural Pesticides

Volume IV
Pheromones

Part A

Editors

E. David Morgan, D.Phil.
Reader
Department of Chemistry
University of Keele
Staffordshire, England

N. Bhushan Mandava, Ph.D.
Senior Partner
Todhunter, Mandava and Associates
Washington, D.C.

CRC Series in Naturally Occurring Pesticides
Series Editor-in-Chief
N. Bhushan Mandava, Ph.D.

CRC Press, Inc.
Boca Raton, Florida

Library of Congress Cataloging-in-Publication Data
(Revised for vol. 4)

CRC handbook of natural pesticides.

(CRC series in naturally occurring pesticides)
Includes bibliographies and index.
Contents: v. 1. Theory, practice and detection —
v. 2. Isolation and identification —   — v. 4. Pher-
omones [2 vols.]
   1. Natural pesticides — Collected works. I. Mandava,
N. Bhushan, 1934-   . II. Handbook of natural pesti-
cides — methods. III. Series.
SB951.145.N37C73   1985   632'.95   84-12092
ISBN 0-8493-3651-1 (v. 1) √
ISBN 0-8493-3652-X (v. 2)

Direct all inquiries to CRC Press, Inc., 2000 Corporate Blvd., N.W., Boca Raton, Florida, 33431.

© 1988 by CRC Press, Inc.

International Standard Book Number 0-8493-3656-2 (set)
International Standard Book Number 0-8493-3657-0 (v. 4A)
International Standard Book Number 0-8493-3658-9 (v. 4B)

Library of Congress Card Number 84-12092
Printed in the United States

INTRODUCTION

The United States has been blessed with high quality, dependable supplies of low cost food and fiber, but few people are aware of the never-ending battle that makes this possible. There are at present approximately 1,100,000 species of animals, many of them very simple forms, and 350,000 species of plants that currently inhabit the planet earth. In the U.S. there are an estimated 10,000 species of insects and related acarinids which at sometime or other cause significant agricultural damage. Of these, about 200 species are serious pests which require control or suppression every year. World-wide, the total number of insect pests is about ten times greater. The annual losses of crops, livestock, agricultural products, and forests caused by insect pests in the U.S. have been estimated to aggregate about 12% of the total crop production and to represent a value of about $4 billion (1984 dollars). On a world-wide basis, the insect pests annually damage or destroy about 15% of total potential crop production, with a value of more than $35 billion, enough food to feed more than the population of a country like India. Thus, both the losses caused by pests and the costs of their control are considerably high. Insect control is a complex problem for there are more than 200 insects that are or have been subsisting on our main crops, livestock, forests, and aquatic resources. Today, in the U.S., conventional insecticides are needed to control more than half of the insect problems affecting agriculture and public health. If the use of pesticides were to be completely banned, crop losses would soar and food prices would also increase dramatically.

About 1 billion pounds of pesticides are used annually in the U.S. for pest control. The benefits of pesticides have been estimated at about $4/$1 cost. In other words, chemical pest control in U.S. crop production costs an estimated $2.2 billion and yields a gross return of $8.7 billion annually.

Another contributing factor for increased crop production is the effective control of weeds, nematodes, and plant diseases. Crop losses due to unwanted weed species are very high. Of the total losses caused by pests, weeds alone count for about 10% of the agricultural production losses valued at more than $12 billion annually. Farmers spend more than $6.2 billion each year to control weeds. Today, nearly all major crops grown in the U.S. are treated with herbicides. As in insect pest and weed control programs, several chemicals are used in the disease programs. Chemical compounds (e.g., fungicides, bactericides, nematicides, and viracides) that are toxic to pathogens are used for controlling plant diseases. Several million dollars are spent annually by American farmers to control the diseases of major crops such as cotton and soybeans.

Another aspect for improved crop efficiency and production is the use of plant growth regulators. These chemicals that regulate the growth and development of plants are used by farmers in the U.S. on a modest scale. The annual sale of growth regulators is about $130 million. The plant growth regulator market is made up of two distinct entities — growth regulators and harvest aids. Growth regulators are used to increase crop yield or quality. Harvest aids are used at the end of the crop cycle. For instance, harvest aids defoliate cotton before picking or desiccate potatoes before digging.

The use of modern pesticides has accounted for astonishing gains in agricultural production as the pesticides have reduced the hidden toll exacted by the aggregate attack of insect pests, weeds, and diseases, and also improved the health of humans and livestock as they control parasites and other microorganisms. However, the same chemicals have allegedly posed some serious problems to health and environmental safety, because of their high toxicity and severe persistence, and have become a grave public concern in the last 2 decades. Since the general public is very much concerned about their hazards, the U.S. Environmental

Protection Agency enforced strong regulations for use, application, and handling of the pesticides. Moreover, such toxic pesticides as DDT, 2,4,5-T and toxaphene were either completely banned or approved for limited use. They were, however, replaced with less dangerous chemicals for insect control. Newer approaches for pest control are continuously sought, and several of them look very promising.

According to a recent study by the National Academy of Sciences, pesticides of several kinds will be widely used in the foreseeable future. However, newer selective and biodegradable compounds must replace older highly toxic persistent chemicals. The pest control methods that are being tested or used on different insects and weeds include: (1) use of natural predators, parasites, and pathogens, (2) breeding of resistant varieties of species, (3) genetic sterilization techniques, (4) use of mating and feeding attractants, (5) use of traps, (6) development of hormones to interfere with life cycles, (7) improvement of cultural practices, and (8) development of better biodegradable insecticides and growth regulators that will effectively combat the target species without doing damage to beneficial insects, wildlife, or man. Many leads are now available, such as the hormone mimics of the insect juvenile and molting hormones. Synthetic pyretheroids are now replacing the conventional insecticides. These insecticides, which are a synthesized version of the extract of the pyrethrum flower, are much more attractive biologically than the traditional insecticides. Thus, the application rates are much lower in some cases, one tenth the rates of more traditional insecticides such as organophosphorus pesticides. The pyrethroids are found to be very specific for killing insects and apparently exhibit no negative effects on plants, livestock, or humans. The use of these compounds is now widely accepted for use on cotton, field corn, soybean, and vegetable crops.

For the long term, integrated pest management (IPM) will have tremendous impact on pest control for crop improvement and efficiency. Under this concept, all types of pest control — cultural, chemical, inbred, and biological — are integrated to control all types of pests and weeds. The chemical control includes all of the traditional pesticides. Cultural controls consist of cultivation, crop rotation, optimum planting dates, and sanitation. Inbred plant resistance involves the use of varieties and hybrids that are resistant to certain pests. Finally, the biological control involves encouraging natural predators, parasites, and microbials. Under this system, pest-detection scouts measure pest populations and determine the best time for applying pesticides. If properly practiced, IPM could reduce pesticide use up to 75% on some crops.

The naturally occurring pesticides appear to have a prominent role for the development of future commercial pesticides not only for agricultural crop productivity but also for the safety of the environment and public health. They are produced by plants, insects, and several microorganisms, which utilize them for survival and maintenance of defense mechanisms, as well as for growth and development. They are easily biodegradable, often times species-specific and also sometimes less toxic (or nontoxic) on other non-target organisms or species, an important consideration for alternate approaches of pest control. Several of the compounds, especially those produced by crop plants and other organisms, are consumed by humans and livestock, and yet appear to have no detrimental effects. They appear to be safe and will not contaminate the environment. Hence, they will be readily accepted for use in pest control by the public and the regulatory agencies. These natural compounds occur in nature only in trace amounts and require very low dosage for pesticide use. It is hoped that the knowledge gained by studying these compounds is helpful for the development of new pest control methods such as their use for interference with hormonal life cycles and trapping insects with pheromones, and also for the development of safe and biodegradable chemicals (e.g., pyrethroid insecticides). Undoubtedly, the costs are very high as compared to the presently used pesticides. But hopefully, these costs would be compensated for by the benefits derived through these natural pesticides from the lower volume of pesticide use and reduction of risks. Furthermore, the indirect or external costs resulting from pesticide poisoning, fatalities, livestock losses, and increased control expenses (due to the destruction of natural enemies and beneficial insects as well as the environmental contamination and pollution from chlorinated, organophosphorus, and carbamate pesticides) could be assessed against benefits vs. risks. The development and use of such naturally occurring chemicals could become an integral part of IPM strategies.

As long as they remain endogenously, several of the natural products presented in this handbook series serve as hormones, growth regulators, and sensory compounds for growth, development, and reproduction of insects, plants, and microorganisms. Others are useful for defense or attack against other species or organisms. Once these chemicals or their analogs and derivatives are applied by external means to the same (where produced) or different species, they come under the label "pesticides" because they contaminate the environment. Therefore, they are subject to regulatory requirements, in the same way the other pesticides are handled before they are used commercially. However, it is anticipated that the naturally occurring pesticides would easily meet the regulatory and environmental requirements for their safe and effective use in pest control programs.

A vast body of literature has been accumulated on natural pesticides during the last 2 or 3 decades; we have been assembling this information in these handbooks. We have limited our attempts to chemical and a few biological aspects concerned with biochemistry and physiology. Wherever possible, we tried to focus attention on the application of these compounds for pesticidal use. We hope that the first two volumes which dealt with theory and practice served as introductory volumes and will be useful to everyone interested in learning about the current technology that is being adapted from compound identification to the field trials. The subsequent volumes deal with the chemical, biochemical, and physiological aspects of naturally occurring compounds, grouped under such titles as insect growth regulators, plant growth regulators, etc.

In a handbook series of this type with diversified subjects dealing with plant, insect, and microbial compounds, it is very difficult to achieve either uniformity or complete coverage while putting the subject matter together. This goal was achieved to a large extent with the understanding and full cooperation of chapter contributors who deserve my sincere appreciation.

The editors of the individual handbooks relentlessly sought to meet the deadlines and, more importantly, to bring a balanced coverage of the subject matter, but, however, that seems to be an unattainable goal. Therefore, they bear full responsibility for any pitfalls and deficiencies. We invite comments and criticisms from readers and users as they will greatly help to update future editions. It is hoped that these handbooks will serve as a source book for chemists, biochemists, physiologists, and other biologists alike — those engaged in active research as well as those interested in different areas of natural products that affect the growth and development of plants, insects, and other organisms.

The editors wish to acknowledge their sincere thanks to the members of the Advisory Board for their helpful suggestions and comments. Their appreciation is extended to the publishing staff, especially Amy Skallerup, Melanie Mortellaro, and Sandy Pearlman for their ready cooperation and unlimited support from the initiation to the completion of this project.

N. Bhushan Mandava
Editor-in-Chief

FOREWORD

Pests of crops and livestock annually account for multi-billion dollar losses in agricultural productivity and costs of control. Insects alone are responsible for more than 50% of these losses.

For the past 40 years the principal weapons used against these troublesome insects have been chemical insecticides. The majority of such materials used during this period have been synthetic organic chemicals discovered, synthesized, developed, and marketed by commercial industry. In recent years, environmental concerns, regulatory restraints, and problems of pest resistance to insecticides have combined to reduce the number of materials available for use in agriculture. Replacement materials reaching the marketplace have been relatively few due to increased costs of development and the general lack of knowledge about new classes of chemicals having selective insecticidal activity.

In response to these trends, it is gratifying to note that scientists in both the public and private sectors have given significant attention to the discovery and evaluation of natural products as fertile sources of new insecticidal agents. Not only are these materials directly useful as insect control agents, but they also serve as models for new classes of chemicals with novel modes of action to attack selective target sites in pest species. Such new control agents may also be less susceptible to the cross resistance difficulties encountered with most classes of currently used synthetic pesticide chemicals to which insects have developed immunity.

Natural products originating in plants, animals, and microorganisms are providing a vast source of bioactive substances. The rapid development and application of powerful analytical instrumentation, such as mass spectrometry, nuclear magnetic resonance spectroscopy, gas chromatography, high performance liquid chromatography, immuno- and other bioassays, have greatly facilitated the identification of miniscule amounts of active biological chemicals isolated from natural sources. These new scientific approaches and tools are addressed and reviewed extensively in these volumes.

Some excellent examples of success in this research involve the discovery of insect growth regulators, especially the so-called juvenoids, which are responsible for control of insect metamorphosis, reproduction, and behavior. Pheromones which play essential roles in insect communication, feeding, and sexual behavior represent another important class of natural products holding great promise for new pest insect control technology. All of these are discussed in detail in Volumes dealing with insects.

It is hoped that the scientific information provided in these volumes will serve researchers in industry, government, and academia, and stimulate them to continue to seek even more useful natural materials that produce effective, safe, and environmentally acceptable materials for use against insect pests affecting agriculture and mankind.

Orville G. Bentley
Assistant Secretary
Science and Education
U.S. Department of Agriculture

PREFACE

The end of the Second World War was a time of great hopes; hopes of a new organization of governments that would settle disputes between nations peacefully; new nations with high intentions of feeding their growing populations adequately, and new insecticides like DDT that would make those promises of abundant food possible. There were great hopes of eliminating the insect pests that destroyed or damaged so much food, and there was promise of removing the scourges of malaria and other insect-transmitted diseases. Today these hopes and many others of that time seem tarnished by reality. The goals were not so easily achieved as we then thought, and we see now that more thought and effort and more strategy must be put into their achievement.

The idea too of a pancratic insecticide is less prominent. No one would seriously suggest today a single pharmaceutical product to treat all infectious diseases. We must recognize also that insect control will be most efficacious if it is directed towards a specific pest or group of pests.

One of our best hopes in finding such means of control is to look again at how nature controls insect populations, and the natural substances of insects themselves. Volume III in this series has dealt with the natural substances affecting insect growth and development. This volume concerns itself with pheromones, another apsect of insect regulation.

Insects make use of all their senses, to varying degrees, for their communication, mating, food-seeking, regulation of maturity, and survival. Research in recent decades has shown they make particular use of the chemical sense of "smell" if that is an appropriate term to use by extension from our human sense. The term pheromone is now widely accepted for those substances emitted by one member of a species, conveying some message to another member of the species. The appeal of breaking into this message system to block or disrupt it, to send false messages, is immediate and obvious. The use of pheromones has the advantages of specificity, economy in the use of material, the air to transmit them, and few problems of persistence and residues.

Pheromone strategies need not be confined to the natural substances. Sometimes an unnatural isomer or enantiomer of a pheromone can effectively suppress the message of the natural substance. This is frequently found for geometric isomers of lepidopteran phero-mones. We know very little as yet of the avoidance by one species of an area marked with the pheromone of another species. Although considerable experience has been gained in the use of pheromones for attraction and disruption, and there are now a number of effective uses for pheromones in agriculture and silviculture, the possibilities of the use of pheromones and their mimics or antagonists are far from being thoroughly examined. Sometimes, a too limited view of insects in agriculture is uppermost, and we lose sight of the damage done by insects in stored products and in packaging, or the spoilage of food by insect filth. There, is, also, the importance of insects as parasites and vectors of pathogens in medical and veterinary practice. The curious and powerful attraction of the copulatory pheromone for the male tsetse fly is just one example that should be capable of exploitation.

In this volume, the first chapter introduces the subject of perception of odor and the function of molecular structure. The current theories linking molecular structure and odor are considered, to help in the understanding of odor perception. In the following chapters, the major insect orders are considered in turn, by international experts in each speciality. Our knowledge is certainly greatest at present for the Lepidoptera, but there it is limited largely to the subject of sexual attraction. Among Coleoptera the use of pheromones is perhaps more varied and still more so with Diptera. The greatest and most varied use of pheromones is among the social insects, so that the Hymenoptera and Isoptera are considered together in one chapter. Finally the other insect orders where our knowledge, as yet, is still but fragmentary are gathered together by B. S. Fletcher and T. E. Bellas.

The reader is commended to the wealth of information collected and tabulated here, both to extend further our knowledge of insects and their communication and to find those more effective, selective, and acceptable methods of control that a hard-pressed natural world requires.

The contributors must have our deepest thanks for their labors with a difficult and ever-growing task. Our thanks also to Mrs. Margaret Furnival and Mrs. Christine Owen for their considerable secretarial help.

E. D. M.
N. B. M.

THE EDITORS

E. David Morgan, D.Phil., is a Chartered Chemist, a Fellow of the Royal Society of Chemistry, and a Fellow of the Royal Entomological Society of London. He received his scientific training in Canada and England, and has worked for the National Research Council of Canada, Ottawa, The National Institute for Medical Research, London, the Shell Group of Companies and is now Reader in Chemistry at the University of Keele, Staffordshire, England. He is co-author of a textbook on aliphatic chemistry with the Nobel prizewinner, Sir Robert Robinson, and with him is a co-inventor of a number of patents. Dr. Morgan has contributed to over 130 papers, most of them on aspects of insect chemistry and has written a number of reviews on insect hormones and pheromones.

N. Bhushan Mandava, holds B.S., M.S., and Ph.D. degrees in chemistry and has published over 140 papers including two patents, several monographs and reviews, and books in the areas of pesticides and plant growth regulators and other natural products. As editorial advisor, he has edited three special issues on countercurrent chromatography for the *Journal of Liquid Chromatography*. He is now a consultant in pesticides and drugs. Formerly, he was associated with the U.S. Department of Agriculture and the Environmental Protection Agency as Senior Chemist. He has been active in several professional organizations, was President of the Chemical Society of Washington, and serves as Councilor of the American Chemical Society.

CONTRIBUTORS

Thomas E. Bellas, Ph.D.
Principal Research Scientist
Division of Entomology
CSIRO
Canberra, Australia

Hans Jürgen Bestmann, Ph.D.
Professor and Head
Institute of Organic Chemistry
University of Erlangen-Nuremburg
Erlangen, W. Germany

Richard Duffield, Ph.D.
Department of Zoology
Howard University
Washington, D.C.

Richard P. Evershed, Ph.D.
Senior Experimental Officer
Department of Biochemistry
University of Liverpool
Liverpool, England

Brian S. Fletcher, Ph.D.
Senior Principal Research Scientist
Division of Entomology
CSIRO
Canberra, Australia

Yoshio Tamaki, Ph.D.
Chief
Insect Biochemistry Laboratory
National Institute of Agricultural Sciences
Nannondai, Ibaraki, Japan

Otto Vostrowsky, Dr. Phil.
Institute of Organic Chemistry
University of Erlangen-Nuremburg
Erlangen, W. Germany

James W. Wheeler, Ph.D.
Professor
Department of Chemistry
Howard University
Washington, D.C.

TABLE OF CONTENTS

Part A

Part B

INSECT OLFACTION AND MOLECULAR STRUCTURE

Richard P. Evershed

INTRODUCTION

Olfaction is the perception of relatively low concentrations of airborne chemicals by an organism. These chemicals — termed olfactory stimulants or odorants — act on receptor cells to elicit a biological response. Olfaction is related to contact chemoreception (where direct contact is made between the chemical stimulant in solid/liquid or solution form), but distinct from chemosensitivity (the slow response of the entire body surface or certain internal organs to high concentrations of harmful substances).[1]

Of the sensory stimuli emanating from their natural habitat, odorant substances are believed to have by far the largest influence on insect behavior and physiology.[2] Such stimuli, termed semiochemicals (also include nonvolatile substances)[3] are known to be employed for a wide variety of functions including food location, avoidance of predators, location of an oviposition site, and signaling to a mate.[4]

Recent advances in microanalytical chemical techniques have led to the unambiguous structural elucidation of many hundreds of semiochemicals whose action is mediated via the insect olfactory system. Much of the incentive for this work has come from the realization that semiochemicals, such as the sex or aggregation pheromones of pest insect species, may be incorporated into control programs which avoid the use of ecologically damaging, so-called "hard", insecticides.[4,5]

From a purely practical point of view, insects are considered to be excellent subjects for olfactory research since they respond to certain specialized odorants, such as pheromones, with recognizable and measurable responses. Their olfactory organs, the antennae, are also readily accessible and morphologically well understood.[1,6,7]

DETECTION AND QUANTIFICATION OF OLFACTION

Man classifies his perception of odors on a psychological basis, in insects (and other animals) this can only be realized by monitoring their behavioral or physiological reactions. Both behavioral and electrophysiological bioassays are used as means of detecting and measuring olfactory response in insects. Testing different concentrations of odorants allows dose-response relationships to be established enabling threshold concentrations to be determined. Precise and reliable determination of such relationships for synthetic analogs and homologs of semiochemicals is valuable for understanding their mode of action and establishing the molecular properties essential for biological activity (see later, Structure-Activity Investigations). Bioassays are also used in monitoring the isolation and identification of semiochemicals.

Behavioral Bioassays

Laboratory and field bioassays based on behavioral or physiological responses of insects are employed as an indirect means of studying their olfactory processes. The most effective behavioral bioassays are based on measuring well-defined behavioral responses such as those elicited by recognized semiochemicals. The behavioral bioassays which have been employed are, necessarily, as varied as insect behavior itself. By far the most common form of behavioral bioassay is that which measures the reflex-type response of insects to releaser pheromones, e.g., the sex attraction of male moths or the alarm or trail following responses of many social Hymenoptera. In contrast, and much less used, are the measurements of physiological changes in response to primer pheromones, e.g., the effects of Hymenopteran queen substances on larval development.

As our knowledge of the identities and subtlety of mode of action of semiochemicals has increased so the sophistication of bioassays has developed to monitor these effects. For example, many of the early workers on moth sex attractants failed to distinguish between long range attraction and short range stimulatory responses. This resulted in the identification of stimulatory substances from females which were ineffective in attracting males in the field.[8] With these potential pitfalls in mind, bioassays are now designed accordingly. No matter what species of insect is being investigated or odorant tested, the bioassay must be carefully designed. The environmental conditions must be strictly controlled. In the laboratory this can be achieved by using specialized enclosures such as wind tunnels or mazes. When assaying a series of odorants or different concentrations of the same odorant it is important to avoid contamination. Cleaning the apparatus between assays is essential. Ideally, the air supply to the apparatus should be filtered and contaminated air vented well away from the test area. Problems are also avoided, when testing a range of odorant concentrations, if the lowest concentration is presented before proceeding to increasingly higher ones. In order to obtain reproducible results, care must also be taken to test insects of similar physiological state. The responsiveness of an insect may vary with age, circadian rhythm, mating, sensory adaption etc. As strict an attention must be paid to these latter factors as to the purity and concentration of test odorants, if quantifiable, reproducible, and comparable data are to be obtained. The problems associated with both field and laboratory behavioral bioassays have been comprehensively reviewed by Young and Silverstein.[8]

Electrophysiological Assays

Electrophysiological techniques provide a direct means of investigating the function and activity of the olfactory organs. The electroantennogram (EAG), the first electrophysiological technique to be applied to the study of insect olfaction, was pioneered by Schneider in Germany.[9] In his early work he tested the idea that the antennal receptors of the male silkworm moth *Bombyx mori* were highly sensitive and very specific for the perception of the female sex pheromone. The EAGs that he recorded corresponded to the summated slow generator potentials elicited simultaneously in a large number of olfactory receptor cells (analogous to the electro-olfactogram [EOG] response recorded from the olfactory epithelium of vertebrates[10]). The EAG is now widely used to study antennae possessing many similarly reacting olfactory cells. However, in antennae possessing several types of olfactory cells reacting to certain odorants in varying degrees, the EAG may give a complex picture. In such cases recordings from single sensilla may allow a greater insight into the function of their associated olfactory receptor cells. Single cell recordings were first made by Boeckh[11] in the sensilla basiconica of the *Necrophorus* beetle antenna by inserting tungsten microelectrodes into the base of individual sensilli. In a modified procedure the tip of a long sensillum is removed and the remainder inserted into a glass capillary microelectrode.[12] In addition to distal and proximal contacting electrodes, an amplifier and a means of recording the electrical responses (e.g., oscilloscope and camera) are required for both EAG and single unit techniques. The odorants are introduced into a stream of clean filtered air which blows continuously over the antenna at a constant rate.

Although in use for nearly 30 years EAG procedures have remained largely unchanged. Methodologies have been described which improve precision and sensitivities.[13] A differential adaption technique has been described[14] which allows a greater insight into the olfactory receptor system of an organism. Briefly, this involves exposing an antennal preparation to one compound until adapted (i.e., no more change in electrophysiological activity on restimulation within seconds or less). The adapted preparation is then exposed to an equal amount of a second compound. If no response occurs then the preparation is adapted to both compounds, hence, they must be acting through common receptors. If similar or even reduced response is observed the antennae must possess different specific receptors for the two compounds or a combination of specific and nonspecific receptors. This technique can also be applied to single cell investigations.

Another modification of the EAG method incorporates a gas chromatograph (GC) to allow separation of mixtures of volatile odorants, e.g., crude insect extracts.[15,16] The eluting volatile components are split between the GC detector and EAG recording apparatus to allow simultaneous monitoring of the GC detector response and electrophysiological activity. In this way it is possible to correlate the GC retention times of the eluting components with their EAG responses. A modification to this system has recently been described to allow GC detection and single unit recordings to be made.[17]

PERCEPTION OF OLFACTORY STIMULI

Olfactory Organs: Antennae and Sensilla

The olfactory receptors of insects are associated with sensilla (sensory hairs) located primarily on the antennae. Olfactory receptors have been located on the labial[18] and maxillary palps.[19,20] It has been suggested that these latter sensilla serve to detect carbon dioxide and water vapor. However, olfactory and gustatory receptors have been identified on various mouth parts in many insect species.[21] Also found on the antennae are sensilla serving as touch, taste, air movement, or temperature sensors.[6]

Although olfactory sensilla are known to occur in all the major orders of insects, no single morphological type of sensilla is concerned solely with olfaction. The form of olfactory sensilla is variable, ranging from the slender hairs (sensilla trichodea of up to 300 μm in length) on the antennae of many male moths, to short hairs or pegs (sensilla basiconica of up to 10 μm in length) in the locust, or flat plates on the antennae of the honey bee and aphids.[2] One aspect of the function of these sensilla, when odorant concentration is low, is to adsorb as many of the odorant molecules from the surrounding air as possible. The efficiency of this process depends largely on the surface area of the antenna (related to the number and dimensions of the sensilla) and the arrangement of sensilla. The number of olfactory sensilla present on the antennae is variable even between members of the same insect order. The plumose antenna of *B. mori* male is estimated to support some 16,000 sex pheromone-sensitive sensilla trichodea[22] while the small filamentous antenna of *Choristoneura fumiferana* (the eastern spruce budworm) possess 2,300 of which only 300 are sex pheromone-sensitive.[22]

Olfactory sensilla, characteristically, comprise a cuticular structure sheathing a lymph space containing one or more sensory dendrite branches (Figure 1). The cuticle is penetrated by numerous tiny pores of at least 10 to 15 nm diameter (in sensilla trichodea and basiconica).[2] These pores play a central role in the olfactory process, providing the stimulant molecules adsorbed in the sensilla cuticle with access to the dendrite in order that stimulation may occur (Figure 1). The fine structure of insect sensilla has been the subject of a number of reviews.[2,7,24-30]

Receptor Sites

It is generally accepted that the olfactory receptors (or acceptors as they are also known) of insects are three-dimensional, proteinaceous structures located on or in the dendrite membrane (Figure 1). Receptor sites are probably normal structural features of proteinaceous cell membranes and they are only receptor sites because they happen to be associated with a cell equipped both to generate and transmit information to other components of the system to which it belongs.[31] Odorants reach the receptors via the pore tubule system (Figure 1) and bind with the receptor site to bring about an increase in the electrical conductance of

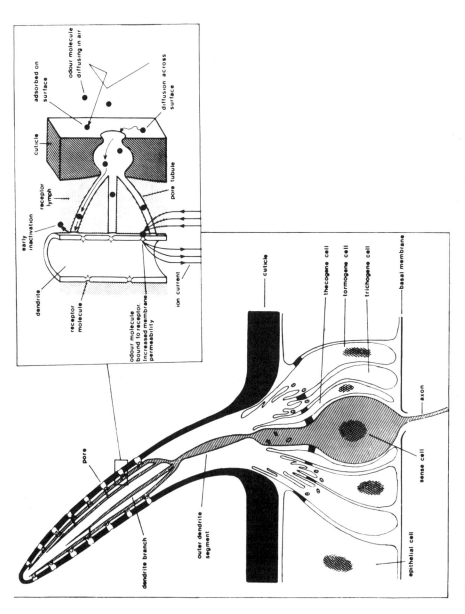

FIGURE 1. Diagram showing structure of an olfactory sensillum displaying formative and sensory cells. Inset shows the fine structure of the pore tubule system and depicts, schematically, the capture and conveyance of an odor molecule to an olfactory receptor.[12,28]

the dendrite membrane by an unknown mechanism. Also unknown is the precise means by which the olfactory organs are able to distinguish a huge variety of odorants at a wide range of thresholds while retaining a discriminating power sufficient to distinguish chemically and structurally very similar substances, e.g., enantiomer pairs. Such a wide dynamic range concomitant with this resolving power is rarely achieved by the most up to date chemical analytical instruments, e.g., computerized gas chromatography-mass spectrometry. An insight into the means by which this is achieved has come through electrophysiological investigations. Single unit recordings have demonstrated the existence of odor "generalist" and "specialist" receptor cells.[2,32] The odor "generalist" receptors are stimulated by a broad spectrum of substances and so enable the insect to recognize and respond to a variety of odors. "Specialist" receptors on the other hand are highly sensitive and selective to a single odorant, e.g., pheromone. The odor "generalist" and "specialist" receptors are extreme cases and receptors have been observed which respond to slightly differing groups of odors while displaying some overlap in responsiveness. Results from work on pheromones suggest "specialists" are by far the most common receptor type on insect antennae, and there is very often more than one group of "specialist" receptors present.

Sensory Transduction

Sensory transduction is defined as the conversion of a physical or chemical stimulus into the excitation of a sensory (or receptor) cell. In olfactory transduction the information carried by an airborne three-dimensional molecule is translated into a bioelectric response in the dendrite by an interaction at a receptor site. In insects the electrical changes which occur during olfactory transduction are relatively easily recorded using electrophysiological techniques. The transduction of the odor stimulus to the receptor potential must involve at least six steps[12] (represented diagramatically in Figure 1). Although olfactory transduction is a complex process and one that is incompletely understood, elements of this scheme have found experimental support.

The primary step in olfactory transduction is the adsorption of the stimulant molecules. Obviously these molecules are adsorbed by all parts of the insect's body, however, only those adsorbed by the antennae can become involved in olfactory transduction. Calculations[33] predict that, in certain moths, at least 80% of the odorant molecules trapped by the antennae become adsorbed on the sensilla. This has been confirmed experimentally in the case of *B. mori,* even though the sensilla trichodea comprise only 25% of the total antennal surface area.[34] The results provided by studies employing radiolabeled bombykol, *(E,Z)*-10,12-hexadecadien-1-ol, (**1**) and *n*-hexadecan-1-ol (**2**) show that the adsorption process is relatively nonspecific, i.e., it does not contribute to the very high level of specificity often observed in the olfactory process.

1

2

It is unlikely that only those stimulant molecules making "direct hits" on the sensilla pores become involved in transduction. Presumably, therefore, molecules adsorbed on other parts of the sensilla diffuse rapidly through the surface coating (composed of "stabilized lipid" or protein)[35,36] and pore tubule system to the olfactory receptors (assumed to be located on the dendrite surface).[1,12] Electron micrographs of thin sections recorded after treating the sensilla with penetrating agents, such as dyes or colloidal silver, have demonstrated the presence of such a diffusive pathway.[1,6] Whether or not the pore tubules make direct contact with the dendrite membranes has been a point of contention. However, electron micrographs provide good evidence for an intimate connection.[2,34,37]

An odor molecule reaching the dendrite surface via adsorption on the antenna and diffusion through the surface coating and pore tubule system, is believed to interact with a receptor site. The nature of this interaction is unknown, as is the mechanism by which this leads to an increase in electrical conductance of the dendrite membrane. The electrical response at a single receptor site is referred to as an elementary receptor potential, and results from stimulation of the dendrite surface by several odor molecules (or in some cases a single molecule) that gives rise to the generator potential. Generator potentials above a given magnitude produce action potentials which are propagated along the neural axon to the brain to elicit behavioral or physiological responses.[1]

The binding of odor molecules at receptor sites appears to be an irreversible process, as there is evidence that pheromones at least are converted to inactive compounds by enzymes present in the antennae (see later Enzyme Theories). This latter mechanism, established through studies on radiolabeled pheromone components, ensures that stimulant molecules are rapidly removed from the receptor area to prevent them from triggering further action potentials or blocking the receptors to incoming odorant molecules carrying important environmental information.

OLFACTORY THEORIES

Perhaps the most intriguing, and indeed vexing, aspect of olfactory transduction is the nature of the interaction between the odorant molecules and the receptor at the dendrite surface. The interaction is believed to proceed through an "activated complex" to produce the observed bioelectric response. The mechanism of this interaction has long been the subject of discussion and speculation, and has led to the proposal of numerous "olfactory theories". The majority of these more correctly warrant the title "hypotheses" as they are virtually unsupported by experimental evidence. The many theories and hypotheses can be grouped into three discernible classes: (1) chemical, (2) vibrational, and (3) enzymatic. Advantage has been taken of the high sensitivity and stereotyped responses of insects to specific semiochemicals (e.g., pheromones) in assessing olfactory theories. Rather than reviewing the many theories and hypotheses, as this has been done by several authors,[38-40] the most prominent theories will be discussed with special reference to supporting (and nonsupporting) studies, specifically those performed on insects.

Chemical Theories

The vast majority of olfactory theorists favor a chemical interaction between the odor molecule and receptor site as providing the necessary activating olfactory stimulus. More than 30 theories have been proposed which invoke either chemical reactivity or, more popularly, a physicochemical property as accounting for the properties of odorants.

The idea that molecular shape is the dominant factor in determining the properties of odorants is generally accepted. This concept is based on the ideas of Moncrieff[38,41] who postulated that for a substance to be odorous, it must satisfy three criteria: (1) it must be volatile, (2) it must be absorbed and possess a molecular configuration complementary to that of certain sites on the olfactory receptor system, and (3) the substance must be one which is not already present in the olfactory epithelium.

These ideas and that invoking "fundamental" or "primary" odors and odor receptors in human olfaction have been extensively tested by Amoore and his co-workers[42-44] in developing the "stereochemical theory" of odor. Substantial experimental support for the role of stereochemical factors and electronic status in olfactory discrimination in man was deduced from statistical correlation of dimensions and shapes of molecular models with the odors of more than 100 substances.[45]

The relationship between molecular configuration and behavioral activity has been thoroughly tested on insects. One of the earliest studies performed by Blum et al. tested the abilities of 49 ketones to produce typical alarm behavior in the dolichoderine ant *Iridomyrmex pruinosus*.[46] Quantification of the ants' responses enabled relationships between chemical structure and alarm releasing activity to be established. The data supported a stereochemical theory indicating that a relatively flat molecule bearing no more than nine carbon atoms and a carbonyl moiety towards the chain end were required for eliciting a response comparable to that of the natural alarm pheromone, 2-heptanone (**3**).[46,47] Blum et al. also assessed the molecular requirements for alarm activity of 4-methyl-3-heptanone (**4**) in the ant *Pogonomyrmex badius*[48] and sex attractancy of 9-oxo-(*E*)-2-decenoic acid (**5**) in the honey bee *Apis melifera*.[49] Some 99 ketones were tested on *P. badius* and 19 alkenoic acids and derivatives (**6**—**24**) on *A. melifera*; the results of these studies were also in agreement with the theory that geometry and size were the determinants of odor quality. The investigation of the activity of the 9-oxo-(*E*)-2-decenoic acid showed that all modifications of this structure resulted in complete loss of activity. This was explained as resulting from a highly specific receptor site for this substance and drew attention to the importance of functional groups in conferring rigid geometric requirements on odorants. The importance of functional groups in addition to molecular shape and size in conferring behavioral activity has been observed in studies of representatives of several insect orders (see later, Functional Groups).

These latter observations relating functional groups and molecular dimensions to odor quality, are consistent with the "profile-functional group theory" proposed by Beets[50] based on the ideas of Ruzicka and others.[51] The theory proposes that the form and bulk of the molecule, and the nature and disposition of its functional group(s) are especially important for odor. Beets argues that it is these factors which irrevocably and fully define all the chemical, physical, and physiological, including odor, properties of a given compound, and hence, the only question which must be answered is whether the relationship between odor and structure is simple enough to be recognized. Beets views odor as arising from molecules interacting with an olfactory receptor to produce a "transition state" complex. The molecules impinging on a given receptor site may be arranged in a chaotic or organized way depending upon the tendency of orientation.[50,52] Rigid or polar molecules, possessing sterically accessible functional groups, will tend to interact with a common receptor in the same orientation. As the polarity or accessibility to the functional group decreases, the orientation of the absorbed molecules becomes more random. The theory does not attempt to explain the nature of the receptor but defines the molecular properties which confer odor or behavioral activity on particular compounds. The theory was propounded over many years mainly on the basis of Beets' own work on the odor similarities and differences in human olfaction of various synthetic musks, and in its most refined form odor is regarded not as the shape of a molecule, but the oriented profile of a population of molecules as a statistical concept.[53] While no investigation has set out to test rigorously and systematically the applicability of this theory to insect olfaction, many of the structural activity studies that have been performed conclude that molecular dimensions and the presence of specific functional groups are inseparable in determining maximum behavioral activity in insect semiochemicals.

The "induced fit"[54] and "cooperative multi-point interaction"[55] models warrant particular mention as these have been developed through studies on insects and represent attractive explanations of the perception of behaviorally and/or electrophysiologically active chemicals.

Both models assume the most effective compounds have functional groups positioned on the molecule complementary to binding positions on antennal receptor proteins. Binding is thought to occur through an initial single-point attachment followed by conformational rearrangements in the unbound portions of the molecule to allow complete binding to complementary acceptor subsites.[56,57] Cooperative interaction of these binding positions is presumed to initiate the ionic mechanism of cell excitation via conformational changes in the receptor proteins. A quantitative model has been developed based on fitting Boltzmann statistics to electrophysiologically determined efficacies to predict electron polarizabilities and dipole moments at the receptor subsites.[55,58,59] These models do not fully explain odorant discrimination in insects. This may be achieved in the central nervous system through evaluation of the integrated responses from many hundreds or thousands of receptor cells. By this means insects might be able to detect in signal quality minute differences otherwise undetectable experimentally through receptor recordings.[57]

The "penetration and puncturing theory" put forward by Davies examines the mechanism by which the increased membrane conductance is brought about through the interaction of an odor molecule with an olfactory nerve cell wall. Davies' theory[60] was inspired by Hodgkin and Kratz'[61] ideas explaining nervous conduction by proposing a local disorientation of the lipid membrane, allowing movement of potassium ions in, and sodium ions out of the cell. The movement of ions causes a wave of reversed polarization (100 mV action potential) constituting a nerve impulse along the fiber. Davies extended this theory to olfaction by proposing a model of olfactory nerve membranes comprising a biomolecular layer of oriented lipid molecules. Bulky odorant molecules were viewed as "puncturing" the cell wall and before the uniform lipid membrane is re-established the movement of K^+ and Na^+ ions occurs initiating the nerve impulse. Experimental support for this theory was derived from the correlation of human olfactory thresholds with the hemolysis accelerating power of various odorants and the ability of the theory to predict olfactory thresholds based on the cross-sectional area of odorants and their adsorption at an oil-water interface. There was also good correlation between the estimated number of cells lining the human nose and the number of molecules required for threshold stimulation.[40,62] Davies did not address himself to the mechanism of olfactory discrimination in detail but attributed the odor properties (equivalent to behavioral activity in lower organisms such as insects) of molecules to a spatial and temporal pattern which he likened to a visual picture.

The above theory was proposed before the physical chemistry of cell membranes and phospholipids was fully elucidated. It is now thought that odorants do not "puncture" the cell membranes and leave a transient hole. Rather, it is envisaged that odorants interact with the lipid regions of cell membranes causing phase transitions or altered packing of lipid molecules, the result being changes in lipid-protein interactions, producing a receptor potential through conformational changes in ion gating proteins.[63]

Vibrational Theories

Many of the early theories of olfaction likened the mechanism of the transmission and perception of odors to the transduction processes involved in sight or hearing. Such theories, based on the distant perception of odors and activation ("tele-activation") of receptors by transmitted electromagnetic radiation (e.g., ultraviolet [UV] or infrared [IR]), have now been abandoned.[38-40]

In contrast, the theories of Dyson,[64,65] and more recently Wright[66,67] proposed that odor arises from differences in the characteristic vibrations of molecules at the receptor surface. Dyson correlated certain odors with specific Raman frequencies in the 1500 to 3000 cm^{-1} range.[64,65] As the vibrational frequencies observed in this range ($>$1000 cm^{-1}) are characteristic of particular functional groups, Wright questioned this relationship, reasoning that there would be no need for an olfactory theory if Dyson's hypothesis was rigorous.[66] Instead he postulated a link between odor and Raman frequencies in the range 100 to 700 cm^{-1}, as these are more characteristic of whole molecules.[66-68] Olfactory nerve cells are known to contain pigments[69] possessing weakly bound electrons which may readily alter their energy states under the influence of molecular vibrations. Wright proposed that the "near-synchronous 'throbbing' of the odorous molecule and the receptor site might, however, permit an unusually close approach and allow greater-than-normal electronic (Van der Waals) interactions to establish themselves".[68] This interaction is conceived as triggering the return of an electronically excited pigment molecule to its ground state with a concomitant increase in polarity. The simultaneous de-excitation of a number of pigment molecules in an olfactory nerve cell membrane may be sufficient to bring about a localized breakdown in the membrane potential to initiate the observed action potential.[70]

Wright has produced substantial experimental evidence to support his theory. One approach he adopted was to compare the positions of absorption maxima in the low frequency vibrational spectra of a range of substances with their odors as expressed by a panel of judges.[67,71] He also introduced the idea of osmically "active" and "inactive" vibrations.[68] Wright has also tested his theory extensively using insects and found a significant correlation with the presence or absence of certain characteristic IR frequencies to the behavioral activity of odorant substances.[72-77] In one test of the vibrational theory, Wright and Brand[75] extended the investigations of Blum et al.[46] and Amoore et al.[47] aimed at determining the characteristics of the natural alarm pheromone, 2-heptanone (3), of the ant *Iridomyrmex pruinosus* necessary for conferring behavioral activity. In order to test the vibrational theory, the IR spectra of 19 of the active and 26 of the inactive substances previously tested, were compared. The "favorable frequencies" for behavioral activity were thus determined, and used to test the theory. Nine compounds were selected for behavioral testing purely on the basis of their IR spectra. Behavioral testing showed that seven of the nine compounds displayed "distinct responses" and Wright et al. draw attention to the fact that comparable activity would not have been predicted in substances such as triethylamine (**25**), heptyl butyrate (**26**), and terpineol acetate (**27**) on the basis of the "stereochemical theory". Wright et al. claim that this and the similar predictive studies in man[71] and insects[74,77] "lend confidence to a belief that the relationship between alarm pheromone activity and molecular vibrational theory is able to go further in accounting for the stimulus specificity necessary to explain the variety of olfactory sensations".[78] Wright also takes account of "specialist" and "generalist" receptors in insects.[79]

25   **26**   **27**

Under normal laboratory conditions, the IR spectra of optical isomers are identical, while the odors of certain enantiomer pairs have been found to be dissimilar in man[80-82] and readily distinguishable by insects. Wright incorporated the ideas of Hayward[83,84] in extending his theory to account for these odor differences.[85] He pointed out that enantiomeric molecules have the same vibrational frequencies only if they are not perturbed by an external agency and proposed that such effects may come into play when an odorant molecule comes into close proximity to a chiral receptor site.[85] The differences in the odor properties of chiral molecules are explained as resulting from the differing twistings of enantiomers by short-range intermolecular forces akin to solvent-induced circular dichroism.[86]

In spite of the successes of Wright's theory, his method of comparing spectra has been criticized by Amoore as being too qualitative. Amoore argues that Wright's theory takes no account of the relative intensities of absorption bands, believing that if the vibrational theory is valid then there should be a relationship between the absorption band intensity and the intensity of the sensation of smell.[87]

Davies has also questioned the supporting nature of the correlation of molecular vibration frequencies with those in the mammalian olfactory bulb, claiming that a similar correlation could be achieved with randomly chosen figures. Davies does however acknowledge the predictive ability of Wright's theory.[40] Another doubt cast on this theory was the ineffectiveness of deuteration in extinguishing behavioral activity of the melon fly attractant[88] and alarm releasing ketones in the ant *Pogonomyrmex badius*.[48] Wright argues against this, saying that the absorption maxima remained within the favorable bands.

Enzyme Theories

Because many of the chemical transformations which occur in living cells are mediated by enzymes it is perhaps not surprising that olfactory theorists should propose that enzymes, well known for their high substrate specificity, catalytic properties, and proteinaceous nature, might be responsible for olfactory discrimination. While initially an attractive idea, the fact that an organism would have to biosynthesize an enzyme for every odorant it perceives renders the theory impractical.[38-40,89] Although odorants might not interact through enzymes, enzymes do appear to play an important part in the latter stages of olfactory transduction. A number of studies have demonstrated pheromone-enzyme or protein-binding reactions. Among the first was that of Riddiford[90] who reversibly blocked the reception of the female sex pheromone by male silk moths (*Antheraea pernyi*) and the perception of (*E*)-2-hexenal by female *A. polyphemus* moths by exposing the moths to formaldehyde vapors and bathing the antennae with Ringer's solution. Formaldehyde is known to react with imidazole and sulfydryl groups of proteins and the α-amino groups in lysine and arginine.[91,92] Hence, she concluded that the action of formaldehyde in blocking insect olfaction was due to its combination with protein(s) involved in the reception of the sex pheromone. The blocking of the sex pheromone activity following elution of the antennae with Ringer's solution was presumed to result from removal of the receptor protein. Restoration of reception was concomitant with the reappearance of protein, as revealed by electrophoretograms. The proteins are concluded to be sourced in the pore tubule system of the sensilli trichodea and sensilli basiconica, externalized and subject to constant resynthesis.[90]

A number of studies have demonstrated the presence of pheromone degrading enzymes on the antennae and other parts of the insect's body. Kassang[93-96] observed the conversion of tritium (^{3}H)-labeled bombykol into ester and acidic fractions by thin layer chromatography (TLC). It was concluded that this conversion was enzymatic as it was blocked by specific enzyme inhibiting agents. This process, however, was thought not to involve the specific pheromone receptors as similar metabolites were observed in male and female antennae and legs. Furthermore, dihydrobombykol was also metabolized, albeit at a higher threshold concentration. Further evidence for this process *not* forming the basis of olfactory transduction was the fact that metabolism occurred in minutes, whereas the response of the olfactory cells to odorants, as recorded by EAG, decays with a half-time of 1 sec.[93] Further investigations demonstrated an enzymatic process and protein binding reaction for the sex pheromone, (*Z*)-7-dodecen-1-ol acetate of proteins isolated from the antennal sensilli of the cabbage looper moth (*Trichoplusia ni*).[97,98] The enzymes may be located on the cuticular surface or in or on the dendritic endings of the primary olfactory receptor cells.[99,100] Substrate specificity of the antennal enzyme(s) of *T. ni* for the pheromone over closely related isomers and analogs has been demonstrated in vivo at short exposure times (4 sec); with longer times (1 min) no pattern of specificity was observed.[101] Investigations into enzyme specificity led to the conclusion that transduction is not primarily enzymatic, however, it has been proposed that pheromone-protein binding is an integral part of the mechanism.[100] More recently, enzymes isolated from the chemosensory sensilli on the antennae have been shown, by use of enzyme blocking agents, to be acetylesterases, and their close association with olfactory membranes is taken to be indicative of the functional part in the olfactory process.[102] The discovery of esterases on the legs and wings of *T. ni* as well as the antennae probably means that the enzymes of the antennae are involved in clearing pheromone from the vicinity of the acceptor to prevent repeated stimulation and prepare it for interaction with subsequent incoming molecules. The esterases present on other parts of the body may serve to prevent surface accumulation of pheromone.[103]

In summary, no one theory so far proposed has been irrefutably proven to account for all aspects of olfaction. It does not appear to be possible readily to correlate odor (behavioral or electrophysiological activity in organisms other than humans) to any single molecular property. Odor and molecular structure are inseparable, as it is the arrangement of the atoms in the molecule which must fully define all its properties.[52] It is therefore likely that the current, apparently competitive, olfactory theories may eventually be shown to be emphasizing different aspects of the same effect.[86,88] A complete understanding of olfaction, particularly of odor discrimination and the nature of the interaction between an odorant molecule and the receptor may prove intractable until more is learned of the molecular structures of receptors.

PROBING RECEPTOR STRUCTURE AND SPECIFICITY

The small amount of receptor material available from a definite olfactory receptor currently excludes the direct molecular analysis of what are almost certainly complex proteinaceous structures.[2] However, inferences concerning the nature and disposition of certain essential structural features of semiochemicals and receptors have been made on the basis of structure-activity investigations and through the effects of protein-specific reagents.

Structure-Activity Investigations

Detailed analysis of receptor specificity should afford a complementary copy of a receptor site with respect to steric requirements and position and nature of functional groups.[2] Receptor specificity may be deduced from the behavioral or electrophysiological activities of a series of structural analogs, compared to the activity of a structurally related known semiochemical. The most informative structure-activity studies are those in which a series of compounds is selected which represent systematic stepwise variations in the molecular structural features of a known semiochemical. Many such investigations have been performed on insects, some more rigorous than others. They range from those in which the activities of pairs of stereoisomers have been compared, to studies involving many tens or even hundreds of different compounds.

The structural features of molecules most commonly addressed and varied in structure-activity studies are discussed below. Consideration is also given to certain cases in which blends of structural analogs e.g. isomers, homologs etc. occur naturally in multi-component insect pheromones and confer special properties on them.

Functional Groups

The presence of a particular atom or group of atoms i.e., the functional group, in an organic compound determines much of the chemistry and properties of that substance. The presence of a given functional group does not, however, mean a compound will possess comparable electrophysiological or behavioral activity to a semiochemical bearing the same functional groups, but otherwise different molecular structure. Among the most striking demonstrations of this came from assays of 19 alkenoic acids and derivatives (**6—24**), related to the queen substance (**5**) as attractants to honey bee drones.[49] In spite of 17 of the substances possessing $-CO_2H$ moiety all structural modifications resulted in complete loss of attractant activity. It was suggested that the absence of activity in the analogs resulted from a highly specific receptor on the drones' antennae which complements the rigid spatial conformation of the pheromone molecule.[49] Although few, if any, other studies have revealed such extreme receptor specificity, it is generally the case that deviations in structure from that of a key molecule result in reduced biological activity.

Almost every imaginable physicochemical molecular property e.g., shape, size, electrical (dipole moment, polarizability, electron donor-acceptor), chemical reactivity, hydrogen-bonding etc., has been implicated in the interaction of odorants with receptors. Polar groups are probably especially important in the initial binding of odorants to receptor sites. One means of testing which specific property or properties of a functional group are crucial for its optimum interaction with the receptor involves assaying synthetic analogs in which the functional group of interest has been replaced by another group possessing significant physical or chemical differences (or similarities) to the group in the key molecule.

The most subtle change in a functional group which can be achieved is to replace an atom(s) with one of its heavier isotopes. This was done in investigations of attractancy of various analogs of cue-lure (4-(*p*-hydroxyphenyl)-2-butanone acetate; **28**) to the male melon fly.[88] The structures of the analogs of this compound were systematically altered by replacing hydrogen atoms (^1H) at various positions in the molecule with deuterium (^2H). These sub-stitutions did not reduce the attractiveness of the analogs nor did similar isotopic substitutions in the alarm substance (4-methyl-3-heptanone; **3**) of *P. badius* workers.[48] Clearly, the ol-factory receptors were unable to distinguish hydrogen from deuterium atoms. The fact that deuteration shifted the IR absorption bands significantly without affecting behavioral activity was believed to conflict with the suggestion that the osmic properties of molecules are related to far IR absorptions.[66,68]

28

Owing to its somewhat similar size, fluorine (Van der Waals radius 1.38 Å) has also been substituted for hydrogen (Van der Waals radius 1.04 Å), however, their disparate electrical properties may be important.[104,105] The similar reactions of subterranean termites to fluor-inated analogs of their trail following pheromone, albeit at higher threshold concentration, seemed to indicate that steric rather than electronic properties dominated.[105] In a detailed structure-activity study of analogs of an aggregation pheromone component (4-methyl-3-heptanol; **29**) of the European elm bark beetle (*Scolytus multistriatus*), increasing fluorination (**30—32**) produced a gradual decrease in behavioral and electrophysiological activity.[104]

29	30	31	32

Steric rather than electronic properties also appeared to dominate the reactions of workers of the leaf-cutting ant *Atta texana* to analogs of its trail substance, methyl 4-methylpyrrole-2-carboxylate (**33**). Substitution of the β-pyrrole methyl by chlorine (**34**) or bromine (**35**) had little or no detectable effect on trail-following activity. Even substitution of this methyl group by iodine, -CHO, -CO₂CH₃, -CN, or -CH₂CH₃ (**36—40**) did not entirely eliminate the trail-following response.[106,107] Although these observations were rationalized largely on stereochemical ground,[106,107] extended Hückel calculations subsequently showed a significant correlation between the charge on the pyrrolic nitrogen atom and trail following activity of methyl 4-methylpyrrole-2-carboxylate and its analogs.[108] This charge is determined by the electron donating and withdrawing properties of the substituents.

33	34	35	36

37	38	39	40

From a practical point of view interest has been shown in formate derivatives as behaviorally active substitutes for aldehydes as Lepidopteran mating disruptants.[109,110] The advantage of formates is their relatively higher chemical stability compared to aldehydes, which would lead to their activity being maintained in the field over a longer period.[99] The sex pheromone of the tobacco budworm, *Heliothis virescens,* was identified as a 16:1 mixture of (Z)-11-hexadecenal (**41**) and *(Z)-9-tetradecenal (**42**).[111,112] It has been found, through field tests, that (Z)-9-tetradecen-1-ol formate (**43**) can be substituted for (Z)-11-hexadecenal (**41**) and is highly attractive to males.[113] Presumably, the steric similarity (compare **41** to **43**) of the formate mimics the sensory input of the naturally occurring aldehyde and, hence, elicits strong attraction. Surprisingly, substitution of (Z)-7-dodecen-1-ol (**44**) formate for (Z)-9-tetradecenal (**42**) was unsuccessful in mimicking the mixture of aldehydes (**41 and 42**). The formate also failed to reduce the male response when evaporated in the atmosphere around *H. virescens* females. Substitution of (Z)-9-tetradecen-1-ol formate (**43**) for (Z)-11-hexadecenal (**41**) in an attractant composed of (Z)-11-hexadecenal (**41**) and (Z)-9-hexadecenal (**45**) (92:2) has also been effective in laboratory tests on *H. zea.*[114]

41	**42**

43

44

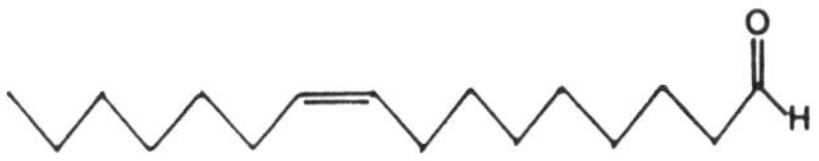

45

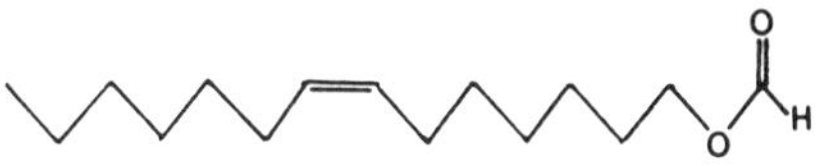

46

(Z)-9-hexadecenal (**45**) has also been shown to be a component of the trail pheromone of the Argentine ant *Iridomyrmex humilis*.[115] In this case, however, the formate analog (**46**) showed no significant trail activity compared to solvent controls.[116] The inability of the formate to mimic the corresponding aldehyde, in this latter example, might indicate a differing mode of receptor-odorant interaction. The exchange of an aldehyde for a formate moiety effectively corresponds to substitution of the heteroatom, oxygen, for a methylene group. Presumably, if electronic rather than purely steric factors are important, then the difference between these groups is significant. These molecular properties were considered in assessing the results of field tests of the effectiveness of analogs, including **48** and **49** of the housefly (*Musca domestica*) attractant,[117] (Z)-9-tricosene (**47**). The structure-activity of these analogs can only be adequately correlated on the basis of the Van der Waals radius of the heteroatom compared to the methylene group.

47

48

49

Enantiomers

Molecules that are not superimposable on their mirror image are said to be chiral. A chiral molecule and its non-superimposable mirror image constitute a pair of enantiomers (optical isomers or antipodes).

Although it has long been known that organisms generally biosynthesize chiral molecules stereospecifically (i.e., one enantiomer in preference to the other) it was not until the early 1970s that it was proven that enantiomers of chiral odorants could possess differing odors.[80-82] These findings from tests on man were thought to indicate chirality in the odor receptors themselves and have been likened to taste discrimination to sugars, amino acids, and pharmaceuticals, or enzyme-substrate or receptor-substrate interactions. These studies prompted investigations the results of which suggested insects may also possess chiral olfactory receptors. Stereoenantiomorphism is to be expected if insects are to attain maximum selectivity to chiral semiochemicals functioning as attractants, feeding stimulants etc.[118]

The first of these investigations demonstrated the ability of insects to discriminate enantiomers through the electrophysiological responses of locusts and worker honeybees and behavioral (proboscis reflex) response of the latter, to very pure (99.5%) enantiomers of 4-methyl-hexanoic acid (**50** and **51**).[119] The differing responses they obtained to the two enantiomers led to the conclusion that a minimum of two variations of one highly specialized receptor structure were present in different numbers on the membranes of individual cell types. Although the discrimination of these enantiomers is biologically meaningless since neither of the compounds is a known semiochemical, the findings indicate the olfactory discriminatory power of insects is considerable, and not necessarily restricted to semiochemicals with high olfactory activity.

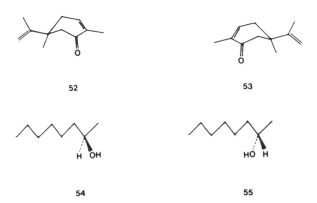

Honeybees were also used by Lensky and Blum in their behavioral experiments. They used the proboscis reflex as a measure of conditioning to the enantiomers of carvone (**52** and **53**) and 2-octanol (**54** and **55**).[118] Their findings were in agreement with those of Kafka et al.,[119] as the bees were readily able to discriminate between the enantiomer pairs. It is perhaps not surprising the bees could distinguish between the carvone enantiomers since these are readily distinguished by man. However, it was significant that the enantiomers of 2-octanol, which are indistinguishable by man, are readily distinguished by honeybees. This latter discovery suggests a greater discriminatory power in insects than in man in this case and provided further evidence for the existence of chiral chemoreceptors in insects. It was reasoned that as most biochemical reactions are stereospecific these chiral receptor sites might be associated with receptor proteins.[118]

There are indications that chiral receptors are also involved in the perception of achiral pheromone components. These indications come from the differential precopulatory responses of male moths of the European corn borer (*Ostrinia nubilalis*) and red-banded leaf roller (*Angyrotaemia velitinana*) to the racemate and enantiomers (**56** and **57**) of the chiral mimic, 9-(2'-cyclopenten-1'-yl)nonyl acetate, and of (Z)-11-tetradecenyl acetate (**58**), an achiral sex pheromone component. The authors argue that the enantiomers mimic two conformers of (Z)-11-tetradecen-yl acetate which "simply coils differently in the two chiral receptors".[120,121]

CH₃CO₂(CH₂)₉

56   57

58

The first evidence of the insects biosynthesizing and responding preferentially to a chiral pheromone was reported for *Atta texana*.[122,123] The principal alarm pheromone component of workers was shown to be (S)-(+)-4-methyl-3-heptanone. Both the (R) and (S) enantiomers (**59** and **60**, respectively) were synthesized in high optical purity and tested for alarm activity. The (S) enantiomer, the naturally occurring compound, was about 100 × more active than the (R) form. It was also noted that the (R) enantiomer did not inhibit the response to the (S) form (mutual inhibition of stereoisomers is discussed later in this section). A close relative of *A. texana, Atta cephalotes* also produces (S)-(+)-4-methyl-3-heptanone (**60**) as its principal alarm pheromone component and responds similarly to the (R) form and mixtures of the two enantiomers. It was reasoned that as enantiomers have identical chemical and physical properties in an achiral medium, then discrimination at receptor sites will only be possible if those receptors are also chiral. Hence, the workers of both *A. cephalotes* and *A. texana* were presumed to have chiral receptors that preferentially discriminate the spatial disposition of (S)-(+)-4-methyl-3-heptanone (**60**).[123]

H   H

59   60

There are many recorded examples of insects producing chiral pheromones,[124] however, not all display the preferential response to a single enantiomer shown by the leaf-cutting ants.[123] In reviewing chirality in insect pheromones Silverstein defined nine possible response categories. Not all of these have been, nor are expected to be, observed in nature (Table 1).[124] Of the categories which have been observed most examples fall into category 1. These include the responses of myrmicine ants,[122,123,125] *Scolytus* spp.,[126-128] and others. *Dendroctonus frontalis* gave a category 7 response to frontalin in a laboratory walking assay.[129] In a recent paper[130] (−)-frontalin is reported to elicit a similar response, in a field bioassy, to the racemic mixture and naturally produced enantiomer ratio (85% (−): 15% (+)), which is greater than the response to (+)-frontalin; this latter response is not strictly category 7. A number of similar examples are also known but in which the natural enantiomeric composition has yet to be determined. Several category 5 type responses have been recorded. Among the most notable are those cases in which differential interspecific responses to enantiomers

Table 1
**CATEGORIES OF RESPONSE OF INSECTS TO
ENANTIOMERS**[124]

Category	Enantiomer(s) biosynthesized	Activity	Occurrence[a]
1	Single	Natural enantiomer more active than the other	+
2	Single	Both enantiomers equally active	−
3	Single	Unnatural enantiomers more active than natural	−
4	Single	Unnatural synergizes response of natural enantiomer	−
5	Single	Unnatural blocks response to natural enantiomer	+
6	Both	Strongest response to natural ratio	+
7	Both	Strongest response to one rather than other or natural ratio	+
8	Both	Equal response to both enantiomers in all ratios	−
9	Both	One inhibits response to other	−

^a +, Known to occur in nature, −, yet to be observed.

serve as a species isolating mechanism between sympatric species utilizing a common host material. Such a mechanism is believed to ensure reproductive isolation among *Ips* bark beetles in California.[131] *Ips pini* produces (R)-(−)-ipsdienol (**61**) as the principal component of its aggregation pheromone. The aggregating action of this compound is inhibited by the presence of (S)-(+)-ipsdienol (**62**). Its close relation *I. paraconfusus,* however, employs the (S) enantiomer in addition to (−)-ispsenol and *cis*-verbenol as their aggregation pheromone. Although both species share the same habitat they are rarely found infesting the same piece of host material. This behavior is attributed to the composition of their aggregation pheromone. The presence of (S)-(+)-ipsdienol (**62**) in the pheromone of *I. paraconfusus* inhibits the attraction of *I. pini*. The inhibitory effect of this isomer on the aggregative effect of (R)-(−)-ispdienol (**61**) on *I. pini* is very powerful, the presence of more than 2 to 3% of the (+), relative to the (−) enantiomer results in complete inhibition. Conversely the presence of the (−) enantiomer causes inhibition in *I. paraconfusus.* Consequently, when both species do infest the same host log few individuals of either species are attracted.

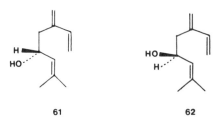

61   62

The response of the ambrosia beetle (*Gnathotrichus sulcatus*) to a mixture of enantiomers of 6-methyl-5-hepten-2-ol (sulcatol; **63** and **64**) provided the first recognized example of a synergistic (category 6) response to a mixture of enantiomers.[132]

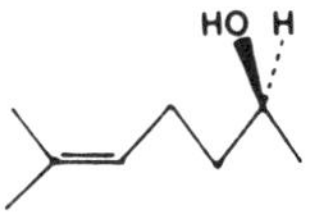

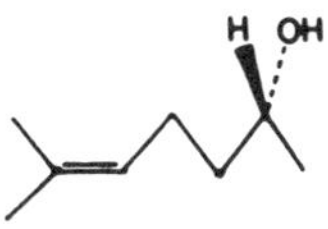

63   **64**

Geometric Isomers

Geometric isomerism is perhaps the most commonly observed form of stereoisomerism observed in insect semiochemicals, particularly among Lepidopteran sex pheromones. Consequently, considerable effort has been made to measure the responses of insects to geometric isomers, whether or not all possible isomers are semiochemicals. Geometric isomers are *not* related as mirror images, but owe their existence to hindered rotation about double bonds, e.g., 11-tetradecenyl acetate may exist in two different forms (**58** and **65**). The two configurations are assigned with prefixes (Z) (German, zusammen, meaning together) and (E) (entgegen, meaning opposite). Geometrical isomerism is clearly impossible if either of the carbon atoms attached by the double bond bears identical groups. It has long been known that the odors of geometric isomers are readily distinguished by man, and this is perhaps not surprising as the (Z) and (E) isomers, unlike (R) and (S) enantiomers, possess measureably different physical properties e.g., boiling and melting points, spectroscopic properties etc. It was not, however, until comparatively recently that the ability of insects to distinguish geometric isomers was revealed.

65

One of the earliest demonstrations of the effect of varying geometric configurations on biological activity was reported for *Bombyx mori*. The activities of the four possible geometric isomers of its principal sex attractant, bombykol, (*E,Z*)-10,12-hexadecadienol (**1**), structure shown in Table 2, together with those of its geometric isomers were compared by means of behavioral and electrophysiological bioassays. As can be seen from Table 2 the natural isomer was very much more active than any of its isomers in both assays.[133,134]

While *B. mori* appears to produce only a single geometric isomer of its principal attractant component other species may produce a mixture of isomers. (*Z*)-11-tetradecenyl acetate (**58**) has been found to be the sex pheromone of two economically important pest species of moth, the European corn borer *Ostrinia nubilalis*[135,136] and the redbanded leafroller *Angyrotaemia velitinana*.[137] However, the pure synthetic material was only weakly attractive to male moths in field tests. Further investigation showed that when small amounts of (*E*)-11-tetradecenyl acetate (**65**) were added to the (*Z*) isomer the attraction for males of both species increased dramatically. As little as 0.5% of the (*E*) isomer carried a substantial increase in attraction of European corn borer males, although above 5% of the (*E*) isomer resulted in decreased attraction. The synergistic effect of the (*E*) isomer was apparent over a wider range in the redbanded leafroller; the (*E*) isomer only became inhibitory when present at greater than the 8% level. The olfactory system of these moths was clearly capable of distinguishing between the geometric isomer compositions of the attractant.[138] The presence of the important minor synergistic geometric isomer (and additional minor components) has now been confirmed in both species using more advanced separation techniques.[139,140]

Table 2
RELATIVE RESPONSES OF *BOMBYX MORI* TO SEX
ATTRACTANT, *(E,Z)*-10,12-HEXADECADIENOL (1) AND
GEOMETRIC ISOMERS[133,134]

Structure number	Structure	Relative response[a]	
		Behavioral	Electrophysiological
1		10^{-12}	10^{-3}
66		10^{-3}	10^{-1}
67		1	1
68		1	1

[a]   Constitutes relative amounts (in µg) required to elicit detectable response.

These studies emphasized the importance of specific blends of geometric isomers in producing optimal activity in certain Lepidopteran sex pheromones. This has been established as a common occurrence and now researchers are careful to determine and state isomeric composition in reporting the identity of such sex pheromones.

Geometric isomers are also produced stereospecifically and responded to selectively by members of other insect orders. Among the Coleoptera, *Trogoderma* beetles have been found to produce (Z) and (E)-14-methyl-8-hexadecenal (**69** and **70**) as sex pheromone components.[141] In common with many Lepidoptera, *Trogoderma granarium* (khapra beetle) produces a mixture of the two geometric isomers (92.8 Z:E). In laboratory behavioral bioassays, males of four *Trogoderma* species are able to discriminate between the geometric isomers.

Workers of the Argentine ant *Iridomyrmex humilis* were readily able to distinguish between their major trail pheromone component, (Z)-9-hexadecenal (**45**) and its (E) isomer in laboratory trail-following bioassays.[116] The choice test employed in this study revealed little difference in trail-following response even between the unnatural (E) isomer and solvent control. This receptor indicated again specificity for the natural geometric isomer. Other

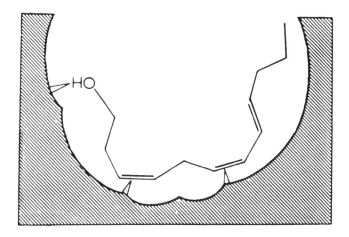

FIGURE 2. Schematic representation of the interactions of the trail substance, (Z,Z,E)-3,6,8-dodecatrienol, of the southern subterranean termite
(*Reticulitermes virginicus*) at its hypothetical ∪-shaped receptor site. The
configuration of the receptor site was proposed on the basis of the least
strained molecular conformation of the trail substance.[143]

structural analogs were active, however, all structural modifications reduced activity compared to the natural compound. An even higher degree of receptor specificity was observed
for the honey bee queen substance 9-oxo-(E)-2-decenoic acid (**5**). The unnatural (Z) isomer
and all analogs (**6—24**) were completely inactive. Furthermore, when presented together,
the (Z) isomer had no blocking effect on the (E) isomer.[49] The geometrical configuration of
the double bond in the central position of the aphid alarm pheromone, (E)-α-farnesene (**71**)
and a number of structural analogs has been found to be critical in conferring alarm activity.
All analogs possessing the (Z) configuration were inactive. Such was the thoroughness of
this study that the authors were able to specify the structural requirements for activity very
precisely.[142]

71

Structure-activity investigations on the trail-following pheromone of the subterranean
termite led to one of the few speculations on receptor shape and dimensions. The pheromone
(Z,Z,E)-3,6,8-dodecatrien-1-ol was found to be considerably more active than its synthetic
(Z,Z,Z) and (Z,E,Z) isomers. The preferred conformation of the (Z,Z,E) isomer and several
other highly active structural analogs was explained on the basis of their being complementary
for a common, possibly ∪-shaped receptor site (Figure 2).[143]

Homologs

A series of compounds in which each member differs from the next by a constant amount
e.g., CH_4, CH_3CH_3, $CH_3CH_2CH_3$ etc., is called a homologous series and the members called
homologs. The various members of a particular homologous series have measurably different
properties e.g., boiling point, molecular dimensions etc. The fact that many of the volatile
secretions produced by insects comprise homologous components has aroused interest.

Ants are well known for their production of homologous exocrine gland components. Dumpert[144] examined the receptor system of the ant *Lasius fuliginosus* electrophysiologically. He exposed the antenna to a range of compounds including homologous alkanes. One of the receptors responded maximally to *n*-undecane (**72**) the major alarm pheromone component of this species' Dufour gland secretion. Subsequently, it was shown that this receptor cell was 10^3 times more sensitive to *n*-undecane (**72**) than to either its next lower or higher homologs *n*-decane (**73**) or *n*-dodecane (**74**).[145]

72

73   74

The variations in electrophysiological responses of noctuide and tortricide moths to 40 systematically structurally modified monoethylenic acetates (empirical formula, $CH_3(CH_2)_nCH=CH)(CH_2)_mOCOCH_3$, a common pheromone structure in these moths) have been studied in detail.[57] The results showed again that altering the molecules chain length (lengthening or shortening) relative to a key molecule gradually decreases the response (Table 3). This response pattern has been observed for many other insects. Behavioral bioassay of responses of *I. pruinosus* workers to homologs of its principal alarm pheromone component, 2-heptanone (**3**), revealed an analogous response spectrum.[46] Multicomponent pheromones may comprise homologous compounds. This is the case in the sex attractant of the greater wax moth, *Galleria mellonella,* produced in the wing gland of the male. Both *n*-nonanal (**75**) and *n*-undecanal (**76**) have been shown to be present and together are capable of eliciting typical sexual behavior.[146,147] An EAG study of the female using synthetic *n*-aldehydes in the C_7-C_{12} chain length range showed greatest response to the two pheromone components.[148] The analogs elicited the typically decreasing response the more they differed in structure (i.e., chain length) from the natural pheromone components.

75   76

Selectivity towards a particular homolog is generally very pronounced when that homolog is a key pheromone component and other members of the series are not. Although, and this has been demonstrated many times, in cases where none of the members of the series is a semiochemical, responses are generally weak (i.e., threshold concentrations high) and variations in activity between homologs are slight.

Positional Isomerism

The compounds of interest here have common molecular formulas but (unlike stereoisomers) differ with respect to which atoms are joined to which other atoms.

Many studies have considered the effects of moving a functional group in a key molecule

Table 3

MALE EAG RESPONSE SPECTRA OF 5 LEPIDOPTERAN SPP. TO 40 ACETATES, ALTERED IN CHAIN LENGTH (12 TO 16) AND IN POSITION (5 TO 14) OF A *(Z)* DOUBLE BOND

	Tricho-plusia ni	Syngrapha variabilis	Cucullia umbratica	Chersotis multangula	Spilonota laricana
Z5-12:Ac[a]	0.01	0.03	0.03	0.3	1
Z6-12:Ac	0.01	0.1	0.1	0.3	1
Z7-12:Ac	0.001	0.03	0.01	0.1	0.3
Z8-12:Ac	0.01	0.1	0.03	0.3	0.03
Z9-12:Ac	0.03	0.1	0.03	0.3	0.3
Z10-12:Ac	0.1		0.1		1
Z5-13:Ac	0.03	0.03	0.03	0.3	
Z6-13:Ac	0.03	0.03	0.03		0.3
Z7-13:Ac	0.01	0.01	0.01	0.1	0.1
Z8-13:Ac	0.01	0.03	0.01	0.03	0.003
Z9-13:Ac	0.03	0.03	0.01	0.1	0.1
Z10-13:Ac	0.1	0.1	0.03	0.3	0.1
Z11-13:Ac	0.1		0.1	0.3	0.3
Z5-14:Ac	0.1	0.01	0.03	0.3	0.3
Z6-14:Ac	0.1	0.01		0.3	0.3
Z7-14:Ac	0.03	0.001	0.01	0.1	0.03
Z8-14:Ac	0.03	0.01	0.01	0.03	0.001
Z9-14:Ac	0.01	0.01	0.001	0.01	0.03
Z10-14:Ac	0.1	0.1	0.01	0.03	0.1
Z11-14:Ac	0.1	0.1	0.03	0.1	0.1
Z12-14:Ac	0.3	0.3	0.1	0.3	0.1
Z5-15:Ac	0.3		0.3	0.3	1
Z6-15:Ac	0.3	0.03		0.3	1
Z7-15:Ac	0.1	0.01	0.1	0.1	0.3
Z8-15:Ac	0.1	0.01	0.03	0.1	0.01
Z9-15:Ac	0.03	0.01	0.01	0.03	0.03
Z10-15:Ac		0.03	0.01	0.01	0.1
Z11-15:Ac	0.1	0.1	0.03	0.01	0.3
Z12-15:Ac	0.3		0.1	0.1	1
Z13-15:Ac		0.3	0.3	0.3	1
Z5-16:Ac	1	0.3	0.3	0.3	1
Z6-16:Ac	1		1	0.3	
Z7-16:Ac	0.3	0.1	0.3	0.1	1
Z8-16:Ac	0.3	0.1	0.3	0.1	0.03
Z9-16:Ac	0.1	0.01	0.1	0.03	0.1
Z10-16:Ac	0.3	0.1	0.03	0.01	
Z11-16:Ac	0.1	0.03	0.01	0.001	0.3
Z12-16:Ac	0.3	0.1	0.1	0.03	0.3
Z13-16:Ac	1	0.3	0.1	0.1	0.3
Z14-16:Ac		1	0.3	0.3	1

Note: Values indicate equipotent stimulus amounts in a half log scale (0.01 represents a range from 0.0056 to 0.018 µg; 0.03, from 0.018 to 0.056 µg; etc.).

[a]   Z5-12:Ac = (Z)-5-dodecenyl acetate; Z6-12:Ac = (Z)-6-dodecenyl acetate etc.

From Priesner, E., in *Chemical Ecology: Odour Communication in Animals,* Ritter, F. J., Ed., Elsevier/North Holland, Amsterdam, 1979, 57. With permission.

while retaining other structural elements. For example, it was found that movement of the carbonyl group of 2-heptanone (the principal pheromone component) towards the center of the chain (compare **3** with **77** and **78**) reduced the alarm response of *I. pruinosus*.[46] A similar effect was noted in assays of alarm behavior in honey bee workers for the same series of compounds.[149] In addition to examining the influence of functional group position Moser et al.,[150] also investigated the importance of the position of substitution of the methyl group in conferring behavioral activity in 4-methyl-3-heptanone (racemic mixture of **59** and **60**) the principal alarm pheromone component of *Atta texana*. Forty-one compounds (ketones, aldehydes, alcohols, and an ester) in the C_2-C_{10} carbon number range, including the natural isomer (racemic mixture) were examined. None of the 16 positional isomers (Table 4) was as active as the natural compound, confirming that both structural features (i.e., functional group and methyl substitution position) are essential in conferring optimal behavioral activity.

77   78

The wide occurrence of positional isomers as active constituents of multicomponent sex pheromones of Lepidoptera has stimulated considerable investigation. An important observation is that of synergism between positional isomers (akin to synergism of geometric isomers). In the case of tortricid moths,[151] (*Z*)-9-tetradecenyl acetate (**95**) and (*Z*)-11-tetradecenyl acetate (**58**) were found to be highly active in EAG studies of *Adoxophyes orana* and *Clepsis spectrana,* although, in field test, both compounds were inactive when presented alone. However, when both compounds were presented together in a definite optimum ratio they acted synergistically, being highly attractive to males. This and similar observations further emphasized the importance of precisely determining and stating isomeric ratios when reporting pheromone compositions. Rather than acting synergistically to enhance attraction one positional isomer may also act to reduce the attractiveness of another. This rarely happens when both compounds are pheromone components (provided they are presented in the correct ratio). A notable exception to this is the European corn borer, *Ostrinia nubilalis* which employs (*E*) and (*Z*)-11-tetradecenyl acetate (**65** and **58**, respectively) as its sex attractant. However, (*E*)-9-tetradecenyl acetate (**96**) observed in female ovipositor washings apparently suppresses male attraction and precopulatory behavior in field and laboratory bioassays. It is not thought that this substance serves to maintain pheromonal specificity between the sympatric redbanded leafroller which also employs an attractant composed of (*Z*) and (*E*)-11-tetradecenyl acetates. In his studies of the acceptor specificity of pheromone receptors of noctuid and tortricid moths, Priesner recorded EAG and single unit responses to a wide range of (*Z*) alkenyl acetates. As part of this study the effect of shifting the double bond position from the 5-position to the penultimate carbon was measured (see Table 3). For all five species tested, their responses gradually decreased upon shifting from the optimum position.[57]

95   96

Table 4
RELATIVE ALARM RESPONSES OF *ATTA TEXANA* WORKERS TO C$_8$ KETONES[150]

Structure number	Structure	Relative response	Structure number	Structure	Relative response
79		0.1	88		1
80		0.001	89		1
81		0.1	90		1
82		1	4		0.000001[b]
83		0.1	91		0.001
84		1[a]	92		0.1
85		1	93		0.01
86		1[a]	94		0.01
87		0.1			

Note: Structures **79—94** are positional isomers of 4-methyl-3-heptanone (**4**) an alarm substance produced in the mandibular glands of the worker ants.

[a] Not alarm response.
[b] 2.7×10^8 molecules cm^{-3} or 5.7×10^{-12} g cm^{-3} of 4-methyl-3-heptanone required to elicit alarm response.

Hence, it would appear as a general principle that for a series of systematically altered positional isomers, the greater the deviation from the structure of the key molecule the more significant the reduction in behavioral or electrophysiological activity. Where one of the isomers is a pheromone component this generally elicits the largest response.

Protein-Specific Olfactory Inhibitors

A somewhat lesser used method of studying the structure of odorant receptors and the mechanism of olfactory transduction is to employ blocking or inhibitory chemicals. In this approach, behavioral or electrophysiological response to a specific odorant is measured. A reagent, which will react with a specific functional group, is then applied to the antennae and the biological activity in response to the same odorant is recorded again. If there is a significant decrease in biological activity following chemical treatment, then the reagent can be presumed to be reacting with a functional group involved in olfactory transduction in the insect species under test, and may even comprise part of the receptor site. Although reactions of this sort have been most extensively employed in the study of taste reception this approach has also been used to investigate insect olfaction. Examples of such investigations are given below.

The relatively nonspecific inhibitor, formaldehyde, was employed to inhibit the mating response of male *Antheraea pernyi* moths and the calling behavior of female *A. polyphemus* moths.[90] Formaldehyde is known to react with the imidazole and sulfhydryl groups of proteins and the β-amino group in lysine and arginine.[91,92] Hence, it was concluded that the blocking of olfaction was due to the combination of formaldehyde with protein(s).

Both *in situ*[152-160] (ultrastructural, electrophysiological, behavioral, and biochemical) and in vitro[161] (dropping mercury electrode polarography) techniques have been employed to study the mechanism of feeding inhibition in *Periplaneta americana* and *Scolytus multistriatus* by various naphthoquinones. Treatments with inhibitors such as the thiol specific reagent N-ethylmaleimide provided evidence for the energy transfer mechanisms in these insects for naphthaquinone-receptor interactions in chemosensory nerves, involving shifts in the -SH/S-S equilibrium in the receptor macromolecule. Such a mechanism is known to bring about alterations of inorganic ion flow, analogous to that characteristic of nervous excitation and impulse conduction.[161,162]

In contrast, Villet[163] showed that this was probably not the case in the silk moths *A. pernyi* and *B. mori*. The amputated male antennae were perfused with solutions of reagents selective to protein side chains. The amino blockers 2,3-dimethylmaleic anhydride and 2-methylmaleic anhydride[164] produced reversible inhibition, as did 2,4,6-nitrobenzenesulfonic acid.[165] The possibility of the involvement of sulfhydryl groups in olfaction was tested by perfusing the antennae separately with N-ethylmaleimide, *p*-chloromercuribenzoate, salyrganic acid, and iodoacetic acid. The reversible reaction obtained gave strong indication of the involvement of sulfhydryl groups in olfaction in these silkmoths. These results were in agreement with those obtained in other studies of taste and olfactory receptors in insects[152,166,167] and man.[168] The apparently nonspecific action of reduced glutathione, L-cysteine, dithioerythritol,[169] and β-mercaptoethanol in blocking olfaction in *A. pernyi* and *B. mori* led to the tentative conclusion that disulfide groups were probably not involved in silk moth chemosensory transduction.

A combined electrophysiological, behavioral, and biochemical study was performed to determine the effects of fluorescein mercuric acetate (FMA) on olfaction in the tobacco budworm moth *Heliothis virescens*.[170] FMA reacts specifically with sulfhydryl groups at neutral pH[171] and has been used as a fluorescent structural probe of enzyme active sites.[172,173] EAG and behavioral responses of the moth to *n*-pentyl acetate showed a time- and concentration-dependent reversible inhibition by FMA. The fluorescence of FMA in sodium phosphate buffer (pH 7.4) was found to decrease with successive additions of antennal sonicate. Partial recovery of the fluorescence by addition of cysteine was taken to indicate that FMA was reacting with relatively large antennal protein(s) probably by forming mercaptide linkages between the mercury atom of FMA and -SH groups of the receptor protein.

While it appears that an approach of this sort can afford information concerning the nature of certain protein functionalities involved in olfactory transduction, Villet has offered a cautionary note. He points out that the exact site of action of an inhibitor on a whole antenna is difficult to predict and tentatively suggests that rather than it acting at a receptor macromolecule, ion channels or an ion-pump mechanism might be impaired.[163] Resolving this question ironically takes us full circle, as unambiguous identification of the inhibitor target site and the modified group would require isolation of the receptor and inhibition products.[163,174]

SUMMARY

Insects have been found to be excellent subjects for olfactory research. Their well-defined behavioral responses have allowed the activities of key semiochemicals and structural analogs to be thoroughly assessed. The ready accessibility of their olfactory organs has led to the development of electrophysiological methodologies, so allowing direct investigation of the function and activity of the olfactory organs.[9,11]

Olfactory discrimination is thought to occur at the receptor level, however its mechanism is poorly understood. Olfactory theories correlating odor with a variety of molecular properties have been proposed and tested experimentally to widely varying degrees.[38-40] By far the most popular theories are those which correlate odor (= behavioral or electrophysiological activity in organisms other than man) to molecular shape.[41,42] However, neither these nor any other theories have been wholly successful in explaining all aspects of olfactory discrimination and transduction. A major stumbling block is the lack of knowledge concerning the nature of receptor sites. Structure-activity investigations have revealed a very high discriminatory power in insect olfaction. It is assumed that the structure of a semiochemical must complement a receptor site with respect to molecular dimensions and functional group positions. However, surprisingly few investigators have used structure-activity data to speculate on receptor characteristics e.g., see Reference 143. The possibilities of employing pharmacological techniques of drug evaluation although recognized,[104] are yet to be widely applied to the study of olfaction.

The differential responses of insects to chiral compounds are thought to indicate the chiral nature of receptors.[124] Evidence for the proteinaceous nature of receptors has come from the use of protein-specific olfactory blocking reagents.[90] This approach has also been employed in an attempt to identify the functional groups involved in transduction at the receptor level. However, this approach is also limited, as the exact site of action on a whole antenna is difficult to identify and olfaction may be blocked by interaction at some other site e.g., ion pumps or ion channels, rather than a receptor.[163]

There appears therefore to be no alternative to deducing the mechanism of olfactory detection and discrimination except to approach it through biochemistry and neurophysiology.[39] Assuming that olfactory discrimination is an example of cells and tissue detecting and recognizing information-carrying molecules, then perhaps one approach might be to examine analogous processes in simpler organisms such as bacteria or algae, where some progress has already been made.[175,178]

ACKNOWLEDGMENTS

In presenting this work I would like to express my thanks to Professor L. M. Schoonhoven for his critical reading of the manuscript and Miss J. M. McGreavey for her invaluable assistance in typing the manuscript.

REFERENCES

1. **Schneider, D.,** Insect olfaction: deciphering system for chemical messages, *Science,* 163, 1031, 1969.
2. **Kaissling, K.-E.,** Insect olfaction, in *Handbook of Sensory Physiology,* Vol. 4, Beidler, L. M., Ed., Springer-Verlag, Basel, 1971, 351.
3. **Law, J. H. and Regnier, F. W.,** Pheromones, *Ann. Rev. Biochem.,* 40, 533, 1971.
4. **Shorey, H. H. and McKelvey, J. J., Jr.,** Eds., *Chemical Control of Insect Behaviour,* John Wiley & Sons, New York, 1977.

5. **Birch, M. C., Ed.,** *Pheromones,* Frontiers of Biology No. 32, North-Holland, Amsterdam, 1974.

6. **Schneider, D.,** Insect antennae, *Ann. Rev. Entomol.,* 9, 103, 1964.

7. **Steinbrecht, R. A.,** Comparative morphology of olfactory receptors, in *Third Int. Symp. Olfaction and Taste,* Pfaffmann, C., Ed., Rockefeller University Press, New York, 1969, 3.

8. **Young, J. C. and Silverstein, R. M.,** Biology and chemical methodology in the study of insect communication, in *Methods of Olfactory Research,* Moulton, E. G., Turk, A., and Johnson, J. W., Jr., Eds., Academic Press, New York, 1975, 75.

9. **Schneider, D.,** Elektrophysiologische untersuchungen von chemo- und mechanorezeptoren der antenne des seidenspinners *Bombyx Mori L., Z. Vergl. Physiol.,* 40, 8, 1957.

10. **Ottoson, D.,** Analysis of the electrical activity of the olfactory epithelium, *Acta Physiol. Scand.,* 35(122), 1, 1956.

11. **Boeckh, J.,** Elektrophysiologische untersuchungen an einzelnen geruchsrezeptoren auf den antennen des totengräbers (Necrophorus, Coleoptera), *Z. Vergl. Physiol.,* 46, 212, 1962.

12. **Kaissling, K.-E.,** Sensory transduction in insect olfactory receptors, *Colloq. Gesellschaft Biol. Chem. Mosbachbaden,* 25, 243, 1974.

13. **Mayer, M. S., Mankin, R. W., and Lemire, G. F.,** Quantitation of the insect electroantennogram: measurement of sensillar contributions, elimination of background potentials, and relationship to olfactory sensation, *J. Insect Physiol.,* 30, 757, 1984.

14. **Payne, T. L. and Dickens, J. C.,** Adaption to determine receptor system specificity in insect olfactory communication, *J. Insect Physiol.,* 22, 1569, 1976.

15. **Moorehouse, J. E., Yeadon, R., Beevor, P. S., and Nesbitt, B. F.,** Method for use in studies of insect chemical communication, *Nature (London),* 223, 1174, 1969.

16. **Arn, H., Städler, E., and Rauscher, S.,** The electroantennographic detector — a selective and sensitive tool in the gas chromatographic analysis of insect pheromones, *Z. Naturforsch.,* 30, 722, 1975.

17. **Van Der Pers, J. N. C. and Löfstedt, C.,** Continuous single sensillum recording as a detection method for moth pheromone components in the effluent of a gas chromatograph, *Physiol. Entomol.,* 8, 203, 1983.

18. **White, R. A., Paim, U., and Seabrook, W. D.,** Maxillary and labial sites of carbon dioxide-sensitive receptors of larval *Orthosoma brunneum* (Forster), *J. Comp. Physiol.,* 88, 235, 1974.

19. **Bässler, U.,** Versuche zur Orientierung der Stechmücken: die Schwarmbildung und die Bedeutung des Johnstonschen Organs, *Z. Vergl. Physiol.,* 41, 300, 1958.

20. **Kellogg, F. E.,** Water vapour and carbon dioxide receptors in *Aedes aegypti, J. Insect Physiol.,* 16, 99, 1970.

21. **Schoonhoven, L. M.,** Insect chemosensory responses to plant and animal hosts, in *Chemical Control of Insect Behaviour,* Shorey, H. H. and McKelvey, J. J., Jr., Eds., John Wiley & Sons, New York, 1977, 7.

22. **Schneider, D.,** Olfactory receptors for the sexual attractant (Bombykol) of the silk moth, in *Neurosciences Second Study Program,* Schmitt, F. O., Ed., Rockefeller University Press, New York, 1970, 511.

23. **Albert, P. J., Seabrook, W. D., and Paim, U.,** Isolation of a sex pheromone receptor in males of the eastern spruce budworm, *Choristoneura fumiferana* (Clem), *J. Comp. Physiol.,* 91, 79, 1974.

24. **Schneider, D. and Steinbrecht, R. A.,** Checklist of insect olfactory sensilla, *Symp. Zool. Soc. (London),* 23, 279, 1968.

25. **Slifer, E. H.,** Thin walled olfactory sense organs on the insect antenna, in *Insects and Physiology,* Beamont, J. W. L. and Treherne, J. E., Eds., Oliver and Boyd, Edinburgh, 1967, 232.

26. **Slifer, E. H.,** The structure of arthropod chemoreceptors, *Ann. Rev. Entomol.,* 15, 121, 1970.

27. **Sinoir, Y.,** L'ultrastructure des organes sensoriels des insectes, *Ann. Zool. Ecol. Anim.,* 1, 339, 1969.

28. **Mustaparta, H.,** Olfaction, in *Chemical Ecology of Insects,* Bell, W. J. and Cardé, R. T., Eds., Chapman & Hall, London, 1984, 37.

29. **Zacharuk, R. Y.,** Ultrastructure and function of insect chemosensilla, *Ann. Rev. Entomol.,* 25, 27, 1980.

30. **Altner, H. and Prillinger, L.,** Ultrastructure of invertebrate chemo-, thermo- and hygroreceptors and its functional significance, *Int. Rev. Cytol.,* 67, 69, 1980.

31. **Beets, M. G. J.,** Odor and stimulant structure, in *Handbook of Perception VI A,* Carterette, E. C. and Friedman, M. P., Eds., Academic Press, New York, 1978, 245.

32. **Priesner, E.,** Progress in the analysis of pheromone receptor systems, *Ann. Zool. Ecol. Anim.,* 11, 533, 1979.

33. **Adam, G. and Delbrück, M.,** Reduction of dimensionality in biological diffusion processes, in *Structural Chemistry & Molecular Biology: A Volume Dedicated to Linus Pauling by His Students, Colleagues, and Friends,* Rich, A. and Davidson, N., Eds., W. C. Freeman, San Francisco, 1968, 198.

34. **Steinbrecht, R. A. and Kasang, G.,** Capture and conveyance of odour molecules in an insect olfactory receptor, in *Int. Symp. Olfaction and Taste IV,* Schneider, D., Ed., Wissenschaftliche Verlagsgesellschaft, Stuttgart, 1971, 193.

35. **Wiggleworth, V. B.,** Structural lipids in the insect cuticle and the function of the oenocytes, *Tissue Cell,* 2, 155, 1970.

36. **Seabrook, W. D.,** Insect chemosensory responses to other insects, in *Chemical Control of Insect Behaviour,* Shorey, H. H. and McKelvey, J. J., Jr., Eds., John Wiley & Sons, New York, 1977, 15.

37. **Steinbrecht, R. A.,** Cryofixation without cryoprotectants, freeze substitution and freeze etching of an insect olfactory receptor, *Tissue Cell,* 12, 73, 1980.

38. **Moncrieff, R. W.,** *The Chemical Senses,* 3rd ed., Leonard Hill, London, 1967.

39. **Amoore, J. E.,** Odor theory and odor classification, in *Fragrence Chemistry,* Theimer, E. T., Ed., Academic Press, New York, 1982, 27.

40. **Davies, J. T.,** Olfactory theories, in *Handbook of Sensory Physiology,* Vol. 4, Beidler, L. M., Ed., Springer-Verlag, Basel, 1971, 322.

41. **Moncrieff, R. W.,** A new theory of odour, *Perfum. Essent. Oil Rec.,* 40, 279, 1949.

42. **Amoore, J. E.,** The stereochemical specificities of human olfactory receptors, *Perfum. Essent. Oil Rec.,* 43, 321, 1952.

43. **Amoore, J. E.,** The stereochemical theory of olfaction, *Proc. Sci. Sect. Toilet Goods Assoc.,* 37(Suppl.), 1, 1962.

44. **Amoore, J. E., Johnston, J. W., and Rubin, M.,** The stereochemical theory of odor, *Sci. Am.,* 210, 42, 1964.

45. **Amoore, J. E. and Venstrom, D.,** Correlations between stereochemical assessments and organoleptic analysis of odorous compounds, in *Olfaction and Taste II,* Hayashi, T., Ed., Pergamon Press, Oxford, 1967, 3.

46. **Blum, M. S., Warter, S. L., and Traynham, J. G.,** Chemical releasers of social behaviour. VI. The relation of structure to activity of ketones as releasers of alarm for *Iridomyrmex pruinosus* (Roger), *J. Insect Physiol.,* 12, 419, 1966.

47. **Amoore, J. E., Palmeri, G., Wanke, E., and Blum, M. S.,** Ant alarm pheromone activity: correlation with molecular shape by scanning computer, *Science,* 165, 1256, 1969.

48. **Blum, M. S., Doolittle, R. E., and Beroza, M.,** Alarm pheromones: utilization in evaluation of olfactory theories, *J. Insect Physiol.,* 17, 2351, 1971.

49. **Blum, M. S., Boch, R., Doolittle, R. E., Tribble, M. T., and Traynham, J. G.,** Honey bee sex attractant: conformational analysis, structural specificity, and lack of masking activity of congeners, *J. Insect Physiol.,* 17, 349, 1971.

50. **Beets, M. G. J.,** Structure and odour, in *Molecular Structure and Organoleptic Quality,* Macmillan, New York, 1957, 54.

51. **Ruzicka, L.,** Die Grundlagen der Geruchschemie, *Chem. Ztg.,* 93, 1920.

52. **Beets, M. G. J.,** Olfactory response and molecular structure, in *Handbook of Sensory Physiology,* Vol. 4, Beidler, L. M., Ed., Springer-Verlag, Basel, 1971, 257.

53. **Beets, M. G. J.,** Odor and stimulant structure, in *Fragrance Chemistry,* Theimer, E. T., Ed., Academic Press, New York, 1982, 77.

54. **Roelofs, W. L. and Comeau, A.,** Sex pheromone perception: synergists and inhibitors for the red-banded leaf roller attractant, *J. Insect Physiol.,* 17, 435, 1971.

55. **Kafka, W. A.,** Physiochemical aspects of odour reception in insects, *Ann. N.Y. Acad. Sci.,* 237, 115, 1974.

56. **Burgen, A. S. V., Roberts, G. C. K., and Feeney, J.,** Binding of flexible ligands to macromolecules, *Nature (London),* 253, 753, 1975.

57. **Priesner, E.,** Specificity studies on pheromone receptors of noctuid and tortricid lepidoptera, in *Chemical Ecology: Odour Communication in Animals,* Ritter, F. J., Ed., Elsevier/North Holland, Amsterdam, 1979, 57.

58. **Kafka, W. A.,** Energy transfer and odour recognition, in *Structure-Activity Relationships in Chemoreception,* Benz, G., Ed., IRL, London, 1975, 123.

59. **Kafka, W. A. and Neuwirth, J.,** A model of pheromone molecule acceptor interaction, *Z. Naturforsch.,* 30c, 278, 1975.

60. **Davies, J. T.,** Olfactory stimulation, *Int. Perfum.,* 3, 17, 1953.

61. **Hodgkin, A. L. and Kratz, B.,** The effect of sodium ions on the electrical activity of the giant axon of the squid, *J. Physiol.,* 108, 37, 1949.

62. **Davies, J. T.,** The mechanism of olfaction, in *Symp. Soc. Exp. Biol.,* 16, 170, 1962.

63. **Dodd, G. and Persaud, K.,** Biochemical mechanisms in vertebrate primary olfactory neurons, in *Biochemistry of Taste and Olfaction,* Cagan, R. H. and Kare, M. R., Eds., Academic Press, New York, 1981, 333.

64. **Dyson, G. M.,** Raman effect and the concept of odor, *Perfum. Essent. Oil Rec.,* 28, 13, 1937.

65. **Dyson, G. M.,** The scientific basis of odor, *Chem. Ind.,* 16, 647, 1938.

66. **Wright, R. H.,** Odour and molecular vibrations. I. Quantum and thermodynamic considerations, *J. Appl. Chem.,* 4, 611, 1954.

67. **Wright, R. H. and Serenius, R. S. E.,** Odour and molecular vibration. II. Raman spectra of substances with the nitrobenzene odour, *J. Appl. Chem.,* 4, 611, 1954.

68. **Wright, R. H.,** Odour and molecular vibration, *Nature (London),* 209, 571, 1966.
69. **Briggs, M. H. and Duncan, R. B.,** Odour receptors, *Nature (London),* 191, 1310, 1961.
70. **Wright, R. H.,** Odour and molecular vibration, in *Molecular Structure and Organoleptic Quality,* Macmillan, New York, 1957, 91.
71. **Wright, R. H.,** Molecular vibration and the 'green' odour, *J. Appl. Chem. Biotechnol.,* 21, 10, 1971.
72. **Wright, R. H.,** Molecular vibration and insect sex attractants, *Nature (London),* 198, 455, 1963.
73. **Wright, R. H.,** Primary odors and insect attraction, *Can. Entomol.,* 98, 1083, 1966.
74. **Wright, R. H., Chambers, D. L., and Keiser, I.,** Insect attractants, antiattractants, and repellents, *Can. Entomol.,* 103, 627, 1971.
75. **Wright, R. H. and Brand, J. M.,** Correlation of ant alarm pheromone activity with molecular vibration, *Nature (London),* 239, 225, 1972.
76. **Wright, R. H.,** Correlation of far infrared spectra and mediterranean fruit fly (*Diptera Tephritidea*) attraction, *Can. Entomol.,* 103, 284, 1971.
77. **Wright, R. H.,** Insect attraction (wasps), *Israel J. Entomol.,* 4, 83, 1969.
78. **Wright, R. H.,** Stereochemical and vibrational theories of odour, *Nature (London),* 239, 226, 1972.
79. **Wright, R. H.,** Odor and molecular vibration: neural coding of olfactory information, *J. Theor. Biol.,* 64, 473, 1977.
80. **Friedman, L. and Miller, J. G.,** Odour incongruity and chirality, *Science,* 172, 1044, 1971.
81. **Russell, G. F. and Hills, J. I.,** Odour differences between enantiomeric isomers, *Science,* 172, 1043, 1971.
82. **Leiterey, T. J., Guadagni, D. G., Harris, J., Mon, T. R., and Teranishi, R.,** Evidence for the difference between the odours of the optical isomers (+) and (−)- carvone, *Nature (London),* 230, 455, 1971.
83. **Hayward, L. D.,** A new theory of olfaction based on dispersion-induced optical activity, *Nature (London),* 267, 554, 1977.
84. **Hayward, L. D.,** Quantitative correlation of biological activity and solvent induced circular dichroism, in *Chemical Ecology: Odour Communication in Animals,* Ritter, F. J., Elsevier/North-Holland, Amsterdam, 1979, 19.
85. **Wright, R. H.,** Odor and molecular vibration: optical isomers, *Chem. Senses Flavour,* 3, 35, 1978.
86. **Hayward, L. D. and Totty, R. N.,** Induced optical rotation and circular dichroism of symetric and racemic aliphatic carbonyl compounds, *Chem. Commun.,* p.676, 1969.
87. **Ritter, F. J., Ed.,** General introduction and overview, in *Chemical Ecology: Odour Communication in Animals,* Elsevier/North-Holland, Amsterdam, 1979, 1.
88. **Doolittle, R. E., Beroza, M., Keiser, I., and Schneider, E. L.,** Deuteration of the melon fly attractant, cue-lure, and its effect on olfactory response and infra-red absorption, *J. Insect Physiol.,* 14, 1697, 1968.
89. **Dravnieks, A.,** Properties of receptors through molecular parameters of odorivectors, in *Wenner-Gren Centre Int. Symp. Series 8,* Pergamon Press, London, 1967, 89.
90. **Riddiford, L. M.,** Antennal proteins of saturnid moths — their possible role in olfaction, *J. Insect Physiol.,* 16, 653, 1970.
91. **Fraenkal-Conrat, H.,** Methods for investigating the essential groups for enzyme activity, in *Methods in Enzymology,* Colwick, S. P. and Kaplan, N. O., Eds., Academic Press, New York, 1957, 247.
92. **Martin, C. J. and Marini, M. A.,** Spectral detection of the reaction of formaldehyde with the histidine residues of α-chymotrypsin, *J. Biol. Chem.,* 242, 5736, 1967.
93. **Kasang, G.,** Bombykol reception and metabolism on the antennae of the silk moth *Bombyx mori,* in *Gustation and Olfaction,* Ohloff, G. and Thomas, A. F., Eds., Academic Press, New York, 1971, 245.
94. **Kasang, G. and Kaissling, K.-E.,** Specificity of primary and secondary olfactory processes in *Bombyx* antennae, in *Olfaction and Taste IV,* Schneider, D., Ed., Wiss. Verl. Ges, Stuttgart, 1972, 200.
95. **Kasang, G.,** Physikochemische vorgänge beim riechen des seidenspinners, *Naturwissenschaften,* 60, 95, 1973.
96. **Kasang, G.,** Uptake of the sex pheromone ³H-Bombykol and related compounds by male and female *Bombyx* antennae, *J. Insect Physiol.,* 20, 2407, 1974.
97. **Ferkovich, S. M., Mayer, M. S., and Rutter, R. R.,** Conversion of the sex pheromone of the cabbage looper, *Nature (London),* 2, 53, 1973.
98. **Ferkovich, S. M., Mayer, M. S., and Rutter, R. R.,** Sex pheromone of the cabbage looper: reactions with antennal proteins *in vitro, J. Insect Physiol.,* 19, 2231, 1973.
99. **Mayer, M. S.,** Hydrolysis of sex pheromone by the antennae of *Trichoplusia ni, Experientia,* 31, 452, 1975.
100. **Mayer, M. S., Ferkovich, S. M., and Rutter, R. R.,** Localization and reactions of a pheromone degradative enzyme isolated from an insect antenna, *Chem. Senses Flavour,* 2, 51, 1976.
101. **Ferkovich, S. M. and Mayer, M. S.,** Localization and specificity of pheromone degrading enzyme(s) from antennae of *Trichoplusia ni,* in *Olfaction and Taste V,* Denton, D. A. and Coghan, J. P., Eds., Academic Press, New York, 1975, 337.

102. **Ferkovich, S. M., Van Essen, F., and Taylor, T. R.,** Hydrolysis of sex pheromone by antennal esterases of the cabbage looper, *Trichoplusia ni, Chem. Senses Flavour,* 5, 33, 1980.

103. **Ferkovich, S. M., Oliver, J. E., and Dillard, C.,** Pheromone hydrolysis by cuticular and interior esterases of the antennae, legs, and wings of the cabbage looper moth, *Trichoplusia ni* (Hubner), *J. Chem. Ecol.,* 8, 859, 1982.

104. **Pignatello, J. J. and Grant, A. J.,** Structure-activity correlations among analogs of 4-methyl-3-heptanol, a pheromone component of the European elm bark beetle *(Scolytus multistriatus), J. Chem. Ecol.,* 9, 615, 1983.

105. **Prestwich, G. D., Eng, W.-S., Deaton, E., and Wichern, D.,** Structure-activity relationships among aromatic analogs of trail-following pheromones of subterranean termites, *J. Chem. Ecol.,* 10, 1201, 1984.

106. **Sonnet, P. E. and Moser, J. C.,** Synthetic analogues of the trail pheromone of the leaf cutting ant, *Atta texana* (Buckley), *J. Agric. Food Chem.,* 20, 1191, 1972.

107. **Sonnet, P. E. and Moser, J. C.,** Trail pheromones: responses of the Texas leaf cutting ant, *Atta texana* to selected halo- and cyanopyrrole-2-aldehydes, ketones and esters, *Environ. Entomol.,* 2, 851, 1973.

108. **Caputo, J. F., Caputo, R. E., and Brand, J. M.,** Significance of the pyrrolic nitrogen atom in receptor recognition of *Atta texana* (Buckley) (Hymenoptera: Formicidae) trail pheromone and parapheromones, *J. Chem. Ecol.,* 5, 273, 1979.

109. **Tumlinson, J. H.,** The need for biological information in developing strategies for applying semiochemicals, in *Chemical Ecology: Odour Communication in Animals,* Ritter, F. J., Ed., Elsevier/North-Holland, Amsterdam, 1979, 301.

110. **Beevor, P. S. and Campion, D. G.,** The field use of 'inhibitory' components of lepidopterous sex pheromones and pheromone mimics, in *Chemical Ecology: Odour Communication in Animals,* Ritter, F. J., Ed., Elsevier/North-Holland, Amsterdam, 1979, 313.

111. **Roelofs, W. L., Hill, A. S., Cardé, R. T., and Baker, T. C.,** Two sex pheromone components of the tobacco budworm moth, *Heliothis virescens, Life Sci.,* 14, 1555, 1974.

112. **Tumlinson, J. H., Hendricks, D. E., Mitchell, E. R., Doolittle, R. E., and Brennan, M. M.,** Isolation, identification and synthesis of the sex pheromone of the tobacco hornworm, *J. Chem. Ecol.,* 2, 203, 1975.

113. **Mitchell, E. R., Tumlinson, J. H., and Baumhover, A. H.,** *Heliothis virescens:* attraction of males to blends of (Z)-9-tetradecen-1-ol formate and (Z)-9-tetradecenal, *J. Chem. Ecol.,* 4, 709, 1978.

114. **Mitchell, E. R. and Tumlinson, J. H.,** unpublished data cited in Reference 109.

115. **Cavill, G. W. K., Davies, N. W., and MacDonald, F. J.,** Characterization of aggregation factors and associated compounds from the Argentine ant, *Iridomyrmex humilis, J. Chem. Ecol.,* 6, 371, 1980.

116. **Van Vorhis Key, S. E. and Baker, T. C.,** Specificity of laboratory trail following by the Argentine ant *(Iridomyrmex humilis)* (Mayr.), to (Z)-9-hexadecenal, analogs and gaster extract, *J. Chem. Ecol.,* 8, 1057, 1982.

117. **Carroll, F. A., Boldrige, D. W., Lee, J. T., Martin, R. R., Turner, M. J., and Venable, T. L.,** Synthesis and field tests of analogues of the house fly pheromone (Z)-9-tricosene, *J. Agric. Food Chem.,* 28, 343, 1980.

118. **Lensky, Y. and Blum, M. S.,** Chirality in insect chemoreceptors, *Life Sci.,* 14, 2045, 1974.

119. **Kafka, W. A., Ohloff, G., Schneider, D., and Vareshi, E.,** Olfactory discrimination of two enantiomers of 4-methylhexanoic acid by the migratory locust and the honey bee, *J. Comp. Physiol.,* 87, 277, 1973.

120. **Chapman, O. L., Klun, J. A., Mattes, K. C., Sheridan, R. S., and Maini, S.,** Chemoreceptors in Lepidoptera: stereochemical differentiation of dual receptors for an achiral pheromone, *Science,* 201, 926, 1978.

121. **Chapman, O. L., Mattes, K. C., Sheridan, R. S., and Klun, J. A.,** Stereochemical evidence of dual chemoreceptors for an achiral sex pheromone in Lepidoptera, *J. Am. Chem. Soc.,* 100, 4878, 1978.

122. **Riley, R. G., Silverstein, R. M., and Moser, J. C.,** Biological responses of *Atta texana* to its alarm pheromone and the enantiomer of the pheromone, *Science,* 183, 760, 1974.

123. **Riley, R. G., Silverstein, R. M., and Moser, J. C.,** Isolation, identification, synthesis and biological activity of the volatile compounds from the heads of *Atta* ants, *J. Insect Physiol.,* 20, 1629, 1974.

124. **Silverstein, R. M.,** Enantiomeric composition and bioactivity of chiral semiochemicals in insects, in *Chemical Ecology: Odour Communication in Animals,* Ritter, F. J., Ed., Elsevier/North-Holland, Amsterdam, 1979, 133.

125. **Cammaerts, M. C., Attygalle, A. B., Evershed, R. P., and Morgan, E. D.,** The pheromonal activity of chiral 3-octanol for *Myrmica* ants, *Physiol. Entomol.,* 10, 33, 1985.

126. **Elliott, W. J., Hromnask, G., Fried, J., and Lanier, G. N.,** Synthesis of multistriatin enantiomers and their action on *Scolytus multistriatus* (Coleoptera: Scolytidae), *J. Chem. Ecol.,* 5, 279, 1979.

127. **Blight, M. M., Wadhams, L. J., and Wenham, M. J.,** The stereoisomeric composition of the 4-methyl-3-heptanol produced by *Scolytus scolytus* and the preparation and biological activity of the four synthetic stereoisomers, *Insect Biochem.,* 9, 525, 1979.

128. **Blight, M. M., Wadhams, L. J., Wenham, M. J., and King, C. J.,** Field attraction of *Scolytus scolytus* (F) to the enantiomers of 4-methyl-3-heptanol, the major component of the aggregation pheromone, *Forestry,* 52, 83, 1979.

129. **Wood, D. L., Browne, L. E., Ewing, B., Lindahl, K., Badard, W. D., Tilden, P., Mori, K., Pitman, G. B., and Hughes, P. R.,** Western pine beetle: specificity among enantiomers of male and female components of an attractant pheromone, *Science,* 192, 896, 1976.

130. **Payne, T. L., Richardson, J. V., Dickens, J. C., West, J. R., Mori, K., Berisford, C. W., Hedden, R. L., Vite, J. P., and Blum, M. S.,** Southern pine beetle: olfactory receptor and behaviour discrimination of enantiomers of the attractant pheromone frontalin, *J. Chem. Ecol.,* 8, 873, 1982.

131. **Birch, M. C., Light, D. M., Wood, D. L., Browne, L. E., Silverstein, R. M., Bergot, B. J., Ohloff, G., West, J. R., and Young, J. C.,** Pheromonal attraction and allomonal interruption of *Ips pini* in California by the two enantiomers of ipsdienol, *J. Chem. Ecol.,* 6, 703, 1980.

132. **Borden, J. M., Chong, L., McLean, J. A., Slessor, K. N., and Mori, K.,** *Gnathotrichus sulcatus:* synergistic response to enantiomers of the aggregation pheromone sulcatol, *Science,* 192, 894, 1976.

133. **Schneider, D.,** Electrophysiological investigation of insect olfaction, in *Int. Symp. Olfaction and Taste 1st Stockholm 1962,* Zotterman, Y., Ed., Pergamon Press, Oxford, 1963, 85.

134. **Schneider, D.,** Molekulare Grundlagen der chemischen Sinne bei Insekten, *Naturwissenschaften,* 58, 194, 1971.

135. **Klun, J. A. and Brindley, T. A.,** Cis-11-tetradecenyl acetate, a sex stimulant of the European corn borer, *J. Econ. Entomol.,* 63, 779, 1970.

136. **Klun, J. A. and Robinson, J. F.,** Inhibition of European corn borer mating by cis-11-tetradecenyl acetate, a borer sex stimulant, *J. Econ. Entomol.,* 63, 1281, 1970.

137. **Roelofs, W. L. and Arn, H.,** Sex attractant of the red banded leaf roller moth, *Nature (London),* 219, 513, 1968.

138. **Klun, J. A., Chapman, O. L., Mattes, K. C., Wojtkowski, P. W., Beroza, M., and Sonnett, P. E.,** Insect sex pheromones: minor amount of opposite geometrical isomer critical to attraction, *Science,* 181, 661, 1973.

139. **Klun, J. A. and Junk, G. A.,** Iowa European corn borer sex pheromone: isolation and identification of four C_{14} esters, *J. Chem. Ecol.,* 3, 447, 1977.

140. **Roelofs, W. L., Hill, A., and Cardé, R. T.,** Sex pheromone components of the red banded leafroller *Argyrotaenia velutinana* (Lepidoptera: tortricidae), *J. Chem. Ecol.,* 1, 83, 1975.

141. **Cross, J. H., Byler, R. C., Cassidy, R. F., Jr., Silverstein, R. M., Greenblatt, R. E., Burkolder, W. E., Levinson, A. R., and Levinson, H. Z.,** Porapak-Q collection of pheromone components and isolation of (Z)- and (E)-14-methyl-8-hexadecenal sex pheromone components, from the females of four species of *Trogoderma* (Coleoptera: dermestidae), *J. Chem. Ecol.,* 2, 457, 1976.

142. **Bowers, W. S., Nishino, C., Montgomery, M. E., and Nault, L. R.,** Structure-activity relationships of analogs of the aphid alarm pheromone, (E)-β-farnesene, *J. Insect Physiol.,* 23, 697, 1977.

143. **Tai, A., Matsumura, F., and Coppel, H. C.,** Synthetic analogues of the termite trail following pheromone, structure and biological activity, *J. Insect Physiol.,* 17, 181, 1971.

144. **Dumpert, K.,** Alarm Stoffrezeption auf der Antenne von *Lasius fuliginosus* (Latr.) (Hymenoptera, Formicidae), *Z. Vergl. Physiol.,* 76, 403, 1972.

145. **Kafka, W. A.,** Physiochemical aspects of odour reception in insects, *Ann. N.Y. Acad. Sci.,* 237, 115, 1974.

146. **Röller, H., Biemann, K., Bjerke, J. S., Norgard, D. W., and McShan, W. H.,** Sex pheromones of pyralid moths. I. Isolation and identification of the sex attractant of *Galleria mellonella* L. (greater waxmoth), *Acta Entomol. Bohemslov.,* 65, 208, 1968.

147. **Leyner, R. L. and Monroe, R. E.,** Isolation and identification of the scent of the moth *Galleria mellonella,* and a re-evaluation of its sex pheromone, *J. Insect Physiol.,* 19, 2267, 1973.

148. **Payne, T. L. and Finn, W. E.,** Pheromone receptor system in the females of the greater wax moth *Galleria mellonella, J. Insect Physiol.,* 23, 879, 1977.

149. **Boch, R. and Shearer, D. A.,** Chemical releasers of alarm behavior in the honey-bee *Apis mellifera, J. Insect Physiol.,* 17, 2277, 1971.

150. **Moser, J. C., Brownlee, R. C., and Silverstein, R. M.,** Alarm pheromones of the ant *Atta texana, J. Insect Physiol.,* 14, 529, 1968.

151. **Persoons, C. J. and Ritter, F. J.,** Binary sex pheromone mixtures in *Tortricidae,* role of positional and geometrical isomers, *Z. Angew. Entomol.,* 77, 342, 1975.

152. **Norris, D. M., Ferkovich, S. M., Rozental, J. M., Baker, J. E., and Borg, T. K.,** Energy transduction: inhibition of cockroach feeding by naphthoquinone, *Science,* 170, 754, 1970.

153. **Norris, D. M., Baker, J. E., Borg, T. K., Ferkovich, S. M., and Rozental, J. M.,** An energy transduction mechanism in chemo-reception by the bark beetle, *Scolytus multistriatus, Contrib. Boyce Thompson Inst.,* 24, 263, 1970.

154. **Norris, D. M.,** Transduction mechanism in olfaction and gustation, *Nature (London),* 222, 1263, 1969.

155. **Norris, D. M., Ferkovich, S. M., Baker, J. E., Rozental, J. M., and Borg, T. K.,** Energy transduction in quinone inhibition of insect feeding, *J. Insect Physiol.,* 17, 85, 1971.

156. **Borg, T. K. and Norris, D. M.,** Penetration of ^{3}H-catechol, a feeding stimulant into chemoreceptor sensilla of *Scolytus multistriatus, Ann. Entomol. Soc. Am.,* 64, 544, 1971.

157. **Ferkovich, S. M. and Norris, D. M.,** Naphthoquinone inhibitors of *Periplaneta americana* and *Scolytus multistriatus* feeding: ultra-violet difference spectra of reactions of juglone, menadione and 1,4-naphthoquinone with amino acids and the indicated mechanism of feeding inhibition, *Chem. Biol. Interact.,* 4, 23, 1971/1972.

158. **Baker, J. E. and Norris, D. M.,** Neurophysiological and biochemical effects of naphthoquinones on the central nervous system of *Periplaneta, J. Insect Physiol.,* 17, 2383, 1971.

159. **Baker, J. E. and Norris, D. M.,** Effects of feeding-inhibitory quinones on the nervous system of *Periplaneta, Experientia,* 28, 31, 1972.

160. **Ferkovich, S. M. and Norris, D. M.,** Antennal proteins involved in the quinone inhibition of insect feeding, *Experientia,* 28, 978, 1972.

161. **Rozental, J. M. and Norris, D. M.,** Chemosensory mechanism in American cockroach olfaction and gustation, *Nature (London),* 244, 370, 1973.

162. **Rothstein, A.,** Sulfhydryl groups in membrane structure and function, in *Current Topics in Membrane and Transport,* Bronner, F. and Kleinzeller, A., Eds., Academic Press, New York, 1970, 135.

163. **Villet, R. H.,** Involvement of amino and sulphydryl groups in olfactory transduction in silk moths, *Nature (London),* 248, 707, 1974.

164. **Dixon, H. B. F. and Perham, R. N.,** Reversible blocking of amino groups with citraeonic anhydride, *Biochem. J.,* 109, 312, 1968.

165. **Freedman, R. B. and Radda, G. K.,** The reaction of glutamate dehydrogenase with 2,4,6-trinitrobenzene sulphonic acid, *Biochem. J.,* 108, 363, 1968.

166. **Gulun, R., Kosower, E. M., and Kosower, N. S.,** Effect of methyl phenyldiazenecarboxylate (azoester) on the feeding behaviour of blood sucking invertebrates, 224, 181, 1969.

167. **Koyama, N. and Kurihara, K.,** Modification by chemical reagents of proteins in the gustatory and olfactory organs of the fleshfly and cockroach, *J. Insect Physiol.,* 17, 2435, 1971.

168. **Henkin, I. and Bradley, D. F.,** Regulation of taste acuity by thiols and metal ions, *Proc. Natl. Acad. Sci. U.S.A.,* 62, 30, 1969.

169. **Cleland, W. W.,** Dithiothreitol, a new protective reagent for SH groups, *Biochemistry,* 3, 480, 1964.

170. **Frazier, J. L. and Heitz, J. R.,** Inhibition of olfaction in the moth *Heliothis virescens* by the sulfhydryl reagent fluorescein mercuric acetate, *Chem. Senses Flavour,* 1, 271, 1975.

171. **Karush, F., Klinman, N. R., and Marks, R.,** An assay method for disulfide groups by fluorescence quenching, *Anal. Biochem.,* 9, 100, 1964.

172. **Heitz, J. R.,** The reaction of fluorescein mercuric acetate with sorbitol dehydrogenase, *J. Biol. Chem.,* 248, 5790, 1973.

173. **Heitz, J. R. and Anderson, B. M.,** Selective binding of fluorescein mercuric acetate to yeast alcohol dehydrogenase, *Arch. Biochem. Biophys.,* 127, 637, 1968.

174. **Hirsch, J. O. and Margolis, F. L.,** Isolation, separation and analysis of cells from olfactory epithelium, in *Biochemistry of Taste and Olfaction,* Cagan, R. H. and Kae, M. R., Eds., Academic Press, New York, 1981, 311.

175. **Koshland, D. E., Jr.,** A response regulator model in a simple sensory system, *Science,* 196, 1055, 1977.

176. **Jaenicke, L. and Boland, W.,** Signal substances and their reception in the sexual cycle of marine brown algae, *Angew. Chem. Int. Ed.,* 21, 643, 1982.

177. **Taylor, B. L. and Panasenko, S. M.,** Biochemistry of chemosensory behaviour in prokaryotes and unicellular eukaryotes, in *Membranes and Sensory Transduction,* Colombetti, G. and Lenci, F., Eds., Plenum Press, New York, 1984, 71.

178. **Paoni, N. F., Maderis, A. M., and Koshland, D. E., Jr.,** Chemical sensing by bacteria, in *Biochemistry of Taste and Olfaction,* Cagan, R. H. and Kare, M. R., Eds., Academic Press, New York, 1981, 459.

PHEROMONES OF THE LEPIDOPTERA*

Yoshio Tamaki

INTRODUCTION

The term "pheromone" was proposed for a chemical(s) used for communication between individuals of a given species. A chemical or a mixture of chemicals that is secreted by an individual and triggers an immediate response in the sexual behavior of the receiving individual has been called a sex pheromone according to Karlson and Lüscher[1] and Karlson and Butenandt.[2] In the case of the Lepidoptera, intraspecifc communication by a chemical mainly relates to mating behavior of adults. Female adults of many moth species produce and secrete sex pheromones which stimulate and attract male moths. Male adults of some lepidopterous species also produce particular secretions which facilitate copulation. This chapter reviews the chemical characteristics of lepidopterous sex pheromones and related scent substances, their biosynthesis and secretion, and utilization to pest management. Many books dealing with various topics on lepidopterous pheromones are already available.[3-17]

FINDING OF MULTI-COMPONENT SYSTEMS IN FEMALE-PRODUCED SEX PHEROMONES

In 1959, the chemical structure of an active compound which releases sexual behavior in the male silkworm moth, *Bombyx mori*, was determined as (E,Z)-10,12 hexadecadienol (bombykol)[28] and a new term "pheromone" was proposed for this class of biologically active substances.[1,2] Identification of lepidopterous female sex pheromones was reported in 11 other species during the 1960s, following the successful identification of the silkworm moth pheromone.

Although the chemical structure of the pheromones of three of these species was later found to be incorrect, all of the pheromones of these species were reported as single components for each species. The first example of a multi-component sex pheromone in the Lepidoptera was (Z)-9-tetradecenyl acetate and (Z,E)-9,12,-tetradecadienyl acetate in the southern armyworm moth, *Spodoptera eridania*, identified in 1970 by Jacobsen et al.[64] The two components were reported to individually stimulate male sexual behavior and were supposed to have different functions in mating behavior of this moth. The diene may act as a long distance attractant and the monoene as a close-range signal. Though such speculation on the functional differentiation of pheromonal components has not yet been proven for this species, possible differences in the function of different pheromonal components in multi-component pheromonal systems are one of the exciting topics in recent studies on lepidopterous sex pheromones.

The first evidence of synergisms in multiple component lepidopterous sex pheromone systems was reported in the tortricid moths, *Adoxophyes* spp. in 1971.[154,157,160] In the case of the smaller tea tortrix moth, *Adoxophyes* sp. (formerly *A. fasciata*), the major components of the female sex pheromone are comprised of (Z)-9- and (Z)-11-tetradecenyl acetates. Overt behavioral response of male moth of this species was elicited only when the two components were simultaneously given to the sense organ. Only the mixture elicited male behavior and individual components did not. The ratio of the two components was suggested to be an important characteristic of the species-specific chemical signal.[174,177,180] Another interesting finding reported in this year was that the addition of a particular compound to a sex pheromone causes a drastic change in its biological activity. (Z)-11-tetradecenyl acetate was the common

---

* This chapter was submitted for publication in October 1983.

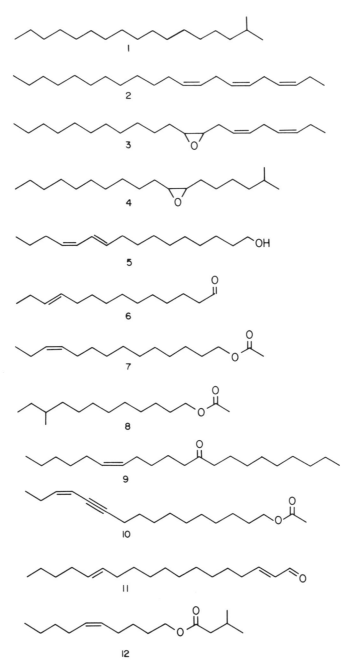

FIGURE 1.   Structures of selected compounds from females (No. 1—14) and males (No. 15—42) of Lepidoptera. (1) 2-Methylheptadecane; (2) (*Z,Z,Z*)-3,6,9-heneicosatriene; (3) *cis*-9,10-epoxy-(*Z,Z*)-3,6-heneicosadiene; (4) *cis*-7,8-epoxy-2-methyloctadecane (disparlure); (5) (*E,Z*)-10,12-hexadecadienol (bombykol); (6) (*E*)-11-tetradecenal; (7) (*Z*)-11-tetradecenyl acetate; (8) 10-methyldodecyl acetate; (9) (*Z*)-15-heneicosen-10-one; (10) (*Z*)-13-hexadecen-11-ynyl acetate; (11) (*E,E*)-2,13-octadecadienal; (12) (*Z*)-5-decenyl 3-methylbutanoate; (13) (*E,Z,Z*)-3,7,10-tridecatrienyl acetate; (14) 11-dodecenyl acetate; (15) 7-oxomethyl-2,3-dihydro-1*H*-pyrrolizin-1-ol; (16) 1-oxomethyl-6,7-dihydro-5*H*-pyrrolizine; (17) 7-methyl-2,3-dihydro-1*H*-pyrrolizin-1-one; (18) 2-phenylethanol; (19) lycopsamine; (20) methyl salicilate; (21) eugenol; (22) 3,4-dimethoxyacetophenone; (23) 1,5,5,9-tetramethyl-10-oxabicyclo[4.4.0]-3-decen-2-one; (24) 2,2,6,8-tetramethyl-7-oxabicyclo[4.4.0]-4-decen-3-one; (25) hydroxytracelanthic acid γ-lactone; (26) benzyl caproate; (27) δ-cadinol; (28) pinocarvone; (29) indole; (30) α-pinene; (31) β-pinene; (32) linalol; (33) nerol; (34) geraniol; (35) β-myrcene; (36) *p*-cymene; (37) limonene; (38) *cis*-jasmone; (39) methyl jasmonate; (40) (*R*)-(−)-mellein; (41) methyl epijasmonate; (42) ethyl cinnamate.

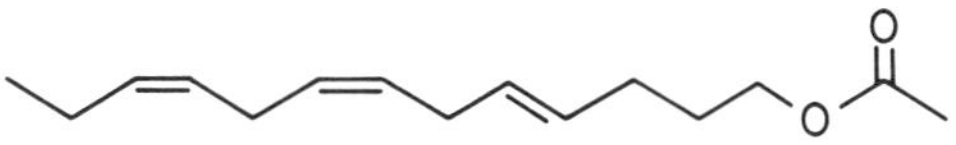

13

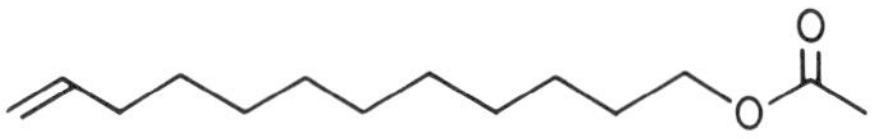

14

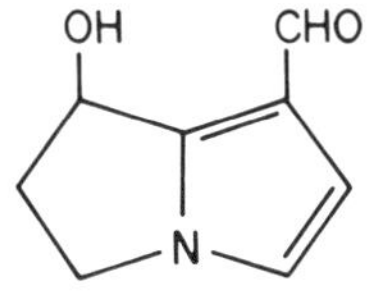

15

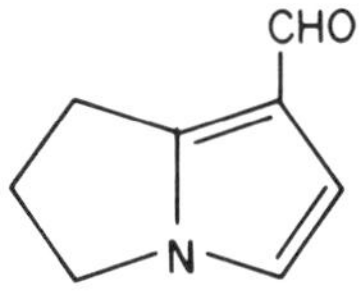

16

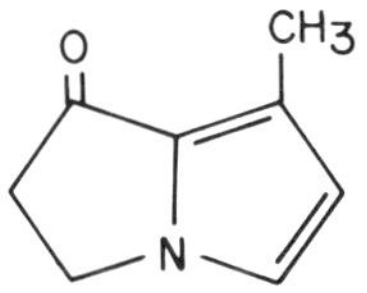

17

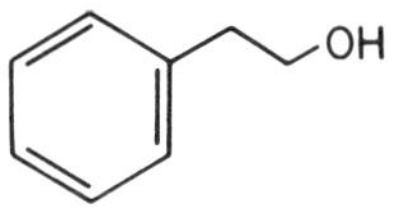

18

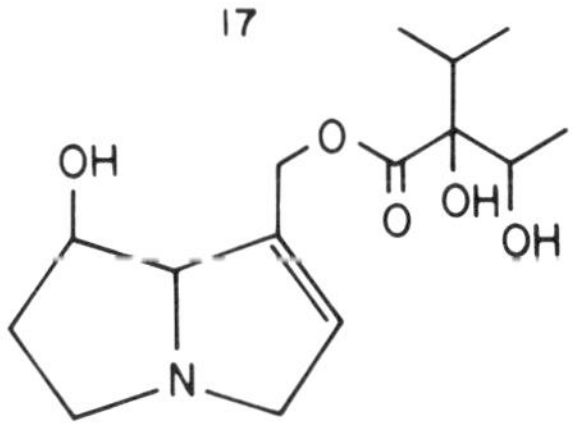

19

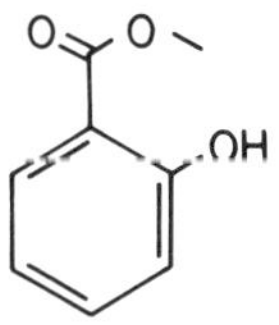

20

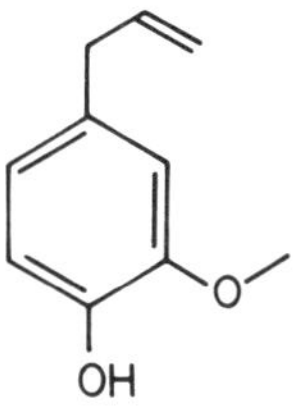

21

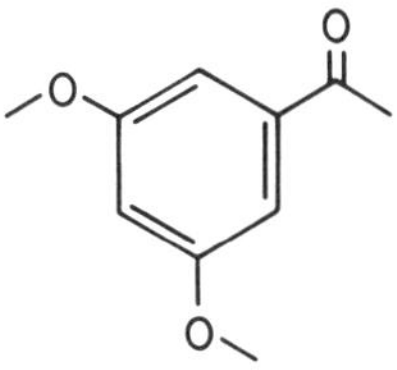

22

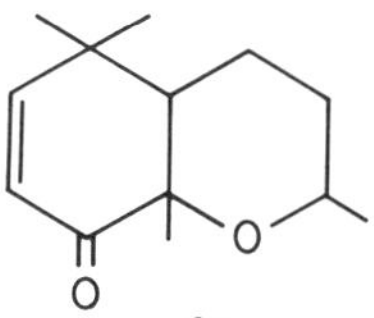

23

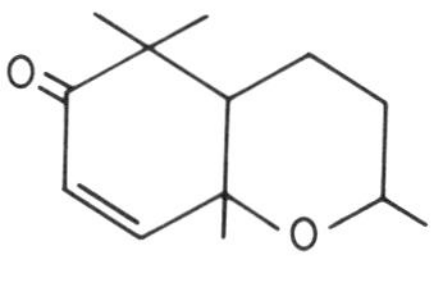

24

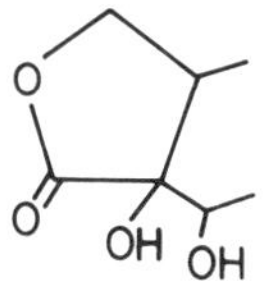

25

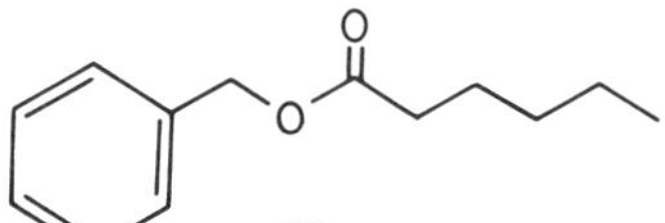

26

FIGURE 1. Continued.

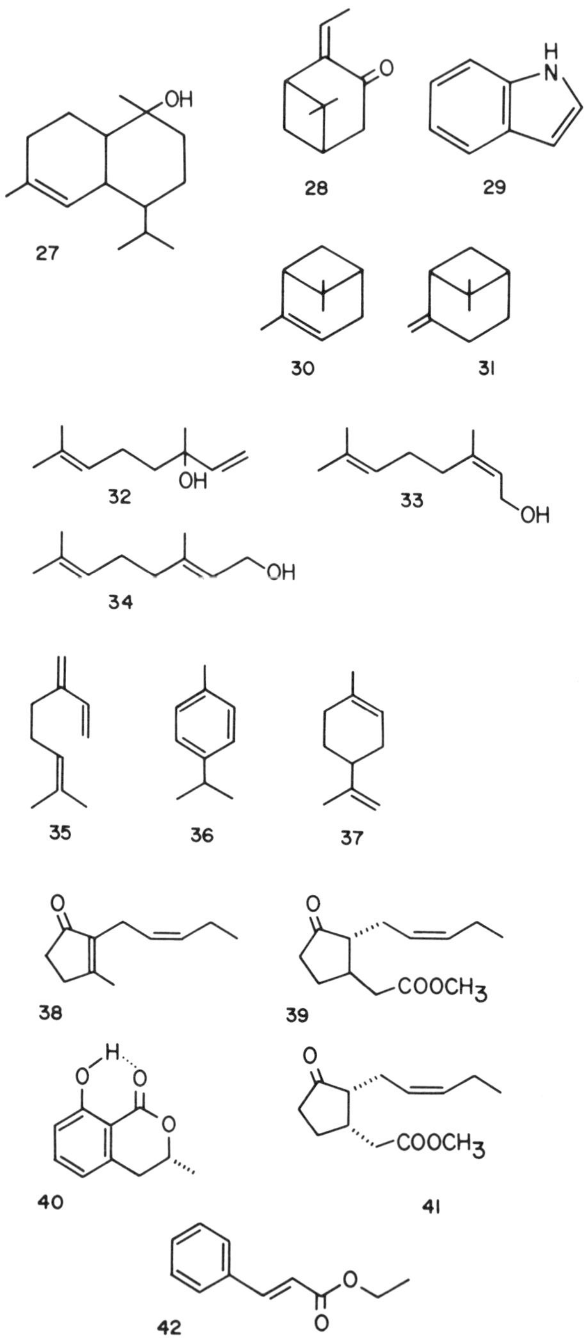

FIGURE 1.   Continued.

sex pheromone of both the redbanded leafroller moth, *Argyrotaenia velutinana*[190] and the oblique-banded leafroller moth, *Choristoneura rosaceana*.[196] Dodecyl acetate was found to act as a synergist for the former species, but was an inhibitor for the latter species.[18] This compound was later realized to be one component of the pheromone of the former species.[191] Importance of admixture and the precise ratio of multiple components in lepidopterous sex pheromones has now been established.

FEMALE-PRODUCED SEX PHEROMONES AND INSECT TAXONOMY

Lepidoptera consists of 5 suborders, 29 superfamilies, 127 families, and more than 130,000 species.[485] For the most part, sex pheromone identifications has been undertaken on economically important insect species, resulting in our information being biased in terms of these forms. This limited information, however, provides us with interesting relationships between the chemical structures of female-produced sex pheromone and the taxonomy of insects. Table 1 lists insect species, the pheromonal components produced by their females, and other related compounds detected from the females until 1983. Table 2 summarizes the pheromonal characteristics of lepidopterous families.

Most of the species belonging to Lepidoptera utilize aliphatic straight-chain compounds such as alcohols, acetates, aldehydes, ketones, and hydrocarbons including epoxyhydrocarbons. Table 1 includes 99 different compounds so far identified as 291 pheromonal components of 160 species. Thus, the mean number of components per species is 1.8, and many compounds are repeatedly utilized in different species. The major lipid class is the alcohol acetates, which comprise 47% of the 99 compounds. This is followed by aldehydes (22%) and alcohols (16%). These three lipid classes comprise 86% of the 99 compounds and 92% of all the 291 pheromonal components. Average molecular weight (mol wt) of the 291 components is 244. The smallest compound is (E)-5-decenol of the peach twig borer, *Anarsia lineatella* (Gelechiidae)[36] with a mol wt of 156, and the largest is (Z)-13-octadecenyl acetate of the sugar cane borer, *Chilo sacchariphagus* (Pyralidae), mol wt 310.[115] The carbon number of the major chain in these compounds ranges from 10 to 21. Those compounds having carbon numbers of 12, 14, and 16, and their acetates comprise 84% of the 282 components. All the compounds in the three lipid classes (alcohol, acetate, and aldehyde) are even number carbon compounds except for the C_{13}-alcohol acetates from the potato tuberworm moth, *Phthorimaea operculella*[41-43] and the smaller tea tortrix moth, *Adoxophyes* sp.[175] and a C_{17}-alcohol acetate from the cabbage armyworm, *Mamestra brassicae*.[383]

The geometry of C-C double bonds shows that 68% are Z oriented bonds and 29% are E. The remaining 3% are terminal double bonds. Only one compound possesses a triple bond. This is (Z)-13-hexadecen-11-yn-1-yl acetate of the processionary moth, *Thaumetopoea pityocampa* (Thaumetopoeidae).[112,113] Double bonds are distributed from positions 1 to 20. Of these, positions 9 and 11 comprise 57% of the total. The major positions from the methyl terminal are 3 and 5, which comprise 60% of the total. In acetates, the double bonds are distributed from positions 2 to 13 and 64% of these bonds are also found in positions 9 and 11. The major positions from the methyl terminal are again 3 and 5, which comprise 71% of the total in this lipid class. Conjugated double bonds are only found in the three major lipid classes, alcohol, acetate, and aldehyde. Nineteen conjugated compounds are utilized as 27 components of various species. Positions of these conjugated systems are 3—5, 5—7, 7—9, 8—10, 9—11, 10—12, and 11—13.

The most intensively investigated lepidopterous families are the Tortricidae, the Noctuidae, and the Pyralidae. These three families comprise 73% of the total species in which female sex pheromones have been identified. The Tortricidae utilize 12 or 14 carbon aldehydes, alcohols, or their acetates as pheromonal components, except for the acetate of a 13 carbon alcohol in *Adoxophyes* sp.[175] This unique acetate, however, is a methyl branched 12 carbon alcohol acetate. Therefore, the female pheromonal components so far identified in the Tortricidae are restricted to compounds with either 12 or 14 carbon atoms in the basic part of their structure. The two major subfamilies, the Tortricidae and the Olethreutinae, each possess their own structural characteristics. More than 90% of 34 species of the Tortricidae utilize 14 carbon alcohols, their acetates, or aldehydes as the major component in their female sex pheromone. The most common compounds are those having 14 carbon atoms with a double bond with Z configuration at the 11th position. These are utilized by about

Table 1
COMPONENTS OF FEMALE-PRODUCED SEX PHEROMONES IDENTIFIED IN LEPIDOPTERA

Insect	Pheromone component, ratio, and activity[a,b]	Ref.
Acrolepiidae		
Acrolepiopsis (Acrolepia) assectella Zeller (leek moth)	Z11-16:Ald; attractive	21, 22
Arctiidae		
Estigmene acrea Drury (Saltmarsh caterpillar)	Z9,Z12-18:Ald/Z9,Z12,Z15-18:Ald/Z3,Z6-cis-9,10-epoxy-21:Hy; 1/6/25 in female, 1/6/27 in effluvia; combinations of the last compound with either the first or the second are attractive	24
Isia isabella	2-Me-17:Hy; slightly attractive	25
Holomelina aurantiaca	2-Me-17:Hy; attractive	25
H. immaculata	2-Me-17:Hy; attractive	25
H. lamae	2-Me-17:Hy; slightly attractive	25
H. nigricans	2-Me-17:Hy; poorly attractive	25
Hyphantria cunea Drury (fall webworm)	Z9,Z12-18:Ald/Z9,Z12,Z15-18:Ald/Z3,Z6-cis-9,10-epoxy-21:Hy; 5/6/13 in females of U.S.S.R., 1/8/21 in females of U.S., 1/6/7 in females of Europe; 1/6/27 in effluvia from U.S. females; the last compound plus either the first or the second is attractive	26, 469
Utetheisa ornatrix (winter moth)	Z3,Z6,Z9-21:Hy; not competitive with female; Z6,Z9-21:Hy and 1,Z3,Z6,Z9-21:Hy identified in females without confirmation of behavioral activity	27, 465
Bombycidae		
Bombyx mori L. (silkworm moth)	E10,Z12-16:OH; sexually stimulative; (E10,Z12-16:Ald, E10,E12-16:OH; aldehyde shows inhibitory action)	28, 70, 71, 439
B. mandarina Moore	E10, Z12-16:OH; no field data	261
Carposinidae		
Carposina niponensis Walsingham (peach fruit moth)	Z13-20-9:Ket/Z12-19-8:Ket; 20/1 in female, the 1st component is more attractive than the 2nd; (Z7-23-11:Ket)	29, 480, 481
Cochylidae		
Eupoecilia (Clysia) ambiguella Hübner	Z9-12:Ac/12:Ac; 1/1 for attraction; (E9-12:Ac,16:Ac,18:Ac, E9-12:OH)	31—33, 466, 467
Cossidae		
Cossus cossus L. (European goat moth)	Z5-12:Ac; addition of 20—25% Z3-10:Ac essential for attraction; (10:Ac,Z5-12:OH,12:Ac,Z5-14:Ac,14:Ac,16Ac)	34, 440
Prionoxystus robiniae Peck (carpenter worm)	Z3,E5-14:Ac; attractive	35, 263
Gelechiidae		
Anarsia lineatella Zeller (peach twig borer)	E5-10:Ac/E5-10:OH; 85/15 in female, this ratio attractive	36
Brachmia macroscopa Harrich-Schäffer (potato leaf folder)	E11-16:Ac; sexually stimulative	37
Pectinophora gossypiella Saunders (pink bollworm)	Z7,Z11-16:Ac/Z7,E11-16:Ac; 1/1 for attraction	38, 39
Phthorimaea operculella Zeller (potato tuberworm moth)	E4,Z7-13:Ac/E4,Z7,Z10-13:Ac; 1/1 in female, 1/4 to 4/1 for attraction	40—43

Table 1 (continued)
COMPONENTS OF FEMALE-PRODUCED SEX PHEROMONES IDENTIFIED IN LEPIDOPTERA

Insect	Pheromone component, ratio, and activity[a,b]	Ref.
Scrobipalpa ocellatella Boyd. (sugar beet moth)	E3-12:Ac; attractive	44
Sitotroga cerealella Olivier (angoumois grain moth)	Z7,E11-16:Ac; attractive	45
Geometridae		
Alsophila pometaria Harris	Z3,Z6,Z9-19:Hy/Z3,Z6,Z9,E11-19:Hy/ Z3,Z6,Z9,Z11-19:Hy; 12/27/7 in female; this mixture attractive	437
Operophtera brumata L. (winter moth)	1,Z3,Z6,Z9-19:Hy; attractive	385, 468
Lasiocampidae		
Dendrolimus punctatus (pine caterpillar)	Z5,E7-12:OH/Z5,E7-12:Ac; no field data	46
D. spectabilis Butler (pine moth)	Z5,E7-12:OH; attractive	47—49
Malacosoma californicus Packard (western tent caterpillar)	E5,Z7-12:Ald; attractive	50
M. disstria Hübner (forest tent caterpillar)	Z5,E7-12:Ald/Z5,E7-12:OH; 1/10 to 1/3 for attraction; (12:Ald)	51
Lymantriidae		
Lymantria dispar L. (gypsy moth)	*cis*-7,8-epoxy-2-Me-18:Hy; attractive	52, 53, 403—405
L. monacha L. (nun moth)	*cis*-7,8-epoxy-2-Me-18:Hy; attractive	54
Orgyia pseudotsugata McDunnough (douglas-fir tussock moth)	Z15-21-10:Ket/Z20,15-21-10:Ket; individually attractive	55, 56
Noctuidae		
(Aconitiinae)		
Naranga aenescens Moor (rice green caterpillar)	Z9-14:Ac/Z9-16:Ac/Z11-16:Ac; 1/1/4 in female, this ratio attractive	57
(Amphipyrinae)		
Busseola fusca Fuller (maize stalk borer)	Z11-14:Ac/E11-14:Ac/Z9-14:Ac; 10/2/2 in female, this ratio attractive	58
Diparopsis castanea Hmps. (red bollworm)	E9,11-12:Ac/Z9,11-12:Ac/11-12:Ac; 80/ 20/25 in female, this mixture attractive; (12:Ac, E9-12:Ac)	59—61
Gortyna xanthenes Ger. (artichoke moth)	Z11-16:Ald/Z9-16:Ald/16:Ald; 96/2/2 for attraction	62
Sesamia inferens Walker (purple stemborer)	Z11-16:Ac; attractive	63
Spodoptera eridania Cramer (southern armyworm)	Z9-14:Ac/Z9,E12-14:Ac; 4/1 in female, sexually stimulative	64
S. exempta Walker (nutgrass armyworm)	Z9-14:Ac/Z9,E12-14:Ac; 20/1 in female, this ratio attractive	65
S. exigua Hübner (beet armyworm)	Z9,E12-14:Ac/Z9-14:OH/Z9-14:Ac; 5/4/ 0, 10/1/0 or 2/0/1 for attraction, Z9,Z12-14:Ac for close-range stimulant; (Z11-14:Ac, 14:Ac, E9-14:OH, Z9,E12-14:OH, Z9,Z12-14:OH, Z9,Z12-14:Ac,Z7-14:Ac)	66—68, 392
S. frugiperda J. E. Smith (fall armyworm)	Z9-12:Ac/Z9-14:Ac; 98/2 or 90/10 for attraction, necessity of Z9-14:Ac doubtful	69, 72, 73, 450

Table 1 (continued)
COMPONENTS OF FEMALE-PRODUCED SEX PHEROMONES IDENTIFIED IN LEPIDOPTERA

Insect	Pheromone component, ratio, and activity[a,b]	Ref.
S. littoralis Boisd. (cotton leafworm)	Z9,E11-14:Ac/Z9,E12-14:Ac; 20/1 in female, the 1st component individually attractive; (14:Ac, Z9-14:Ac, E11-14:Ac)	59, 74—76
S. litura Fabr.	Z9,E11-14:Ac/Z9,E12-14:Ac; 10/1 in female, this ratio attractive; (Z9-14:Ac, E11-14:Ac)	74, 268, 397
(Catocalinae)		
Anticarsia gemmatalis Hübner (velvetbean caterpillar)	Z3,Z6,Z9-20:Hy/Z3,Z6,Z9-21:Hy; 5/3 in female, this ratio attractive	77
Caenurgina erectea (forage looper)	Z3,Z6,Z9-20:Hy/Z3,Z6,Z9-21:Hy; 1/4 in female 9/1 to 1/9 for attraction	438
(Hadeninae)		
Acantholeucania loreyi Duponchel	Z9-14:Ac/Z11-16:Ac; 4/1 in female, addition of ca. 10% of Z7-12:Ac improves the attractiveness	264, 269
Mamestra brassicae L. (cabbage armyworm)	Z11-16:Ac; slightly attractive; Z11-17Ac was proposed as the 2nd component; (Z9-16:Ac, E11-16:Ac, 14:Ac, 16:Ac)	78—82, 383, 522
M. configurata Walker (bertha armyworm)	Z11-16:Ac/Z9-14:Ac; 19/1 for attraction	83, 84
M. oleracea L.	Z11-16:Ac/Z11-16:OH; 1/1 in female; insufficient behavioral data	81, 521
M. (Polia) pisi L.	Z9-14:Ac/Z11-14:Ac; 75/25 in female; insufficient behavioral data	85, 330
M. suasa Den. & Schiff.	Z11-16:Ac; stimulative	86
Pseudaletia (Mythimna) separata Walker	Z11-16:Ac/Z11-16:OH; 8/1 in female, 4/1 or 9/1 attractive	264, 269
P. (Mythimna) unipuncta Haworth (armyworm)	Z11-16:Ac/Z11-16:OH; 500/1 or the 1st component alone attractive; addition of 0.05—0.01% of Z11-16:Ald and Z9-14:Ac improves attractiveness, necessity of Z11-16:OH doubtful; (16:Ac, Z9-16:Ac)	87, 259, 329, 448, 454
Scotogramma trifolii Rottenberg (clover cutworm)	Z11-16:Ac/Z11-16:OH; 1/1 to 9/1 for attraction	88, 396
Xylomyges curialis Grote (citrus cutworm)	Z11-16:OH; attractive	378
(Heliothidinae)		
Heliothis armigera Hübner (old world bollworm)	Z11-16:Ald/Z9-16:Ald; 1000/32 in female, 100/1 to 100/12 for attraction; not competitive with female; (Z11-16:OH, 16:Ald, 16:OH)	89, 90—92
H. punctiger Walengren	Z11-16:Ald/Z11-16:Ac; 60/25 in female; 1/1 plus 3% Z9-14:Ald attractive; (Z11-16:OH, 16:Ald)	436
H. subflexa Gn.	16:Ald/Z9-16:Ald/Z11-16:Ald/Z7-16:Ac/Z9-16:Ac/Z11-16:Ac; 54/198/300/16/43/123 in female; this ratio competitive with female, the two unsaturated aldehydes essential; no data on the necessity of each acetate; (Z9-16:OH, Z11-16:OH, Z9-14:Ald, 14:Ald, Z7-16:Ald, Z7-16:Ac, 16:Ac)	93, 452

Table 1 (continued)
COMPONENTS OF FEMALE-PRODUCED SEX PHEROMONES IDENTIFIED IN LEPIDOPTERA

Insect	Pheromone component, ratio, and activity[a,b]	Ref.
H. virescens F. (tobacco budworm)	Z11-16:Ald/Z9-14:Ald; 77—91/1—3 in female, 16/1 for attraction; addition of Z7-16:Ald, Z9-16:Ald,16:Ald, Z11-16:OH, and 14:Ald increases attractiveness; no data on the necessity of each additional compound	94—97
H. zea Boddie (corn earworm)	Z11-16:Ald/Z9-16:Ald; 90—95/1—2 in female, this ratio attractive; (Z7-16:Ald, 16:Ald)	94, 96, 98
(Noctuinae)		
Agrotis ipsilon Hufnagel (black cutworm)	Z7-12:Ac/Z9-14:Ac; 5/1 from female, 3/1 for attraction	99
A. fucosa (segetum) Butler (Japan)	Z5-10:Ac/Z7-10:Ac; 68/32 in female, 1/9 for attraction	100, 482
A. segetum Schiff. (Europe)	Z5-10:Ac/Z7-12:Ac/Z9-14:Ac; 1/1/1 or 2/100/100 for attraction; 10:Ac seems to be the 4th component; (Z5-10:OH, 12:Ac, E5-12:Ac, Z8-12:Ac, Z9-12:Ac, Z7-12:OH, 14:Ac, Z9-14:OH, 16:Ac)	101—103, 378, 428
Amathes c-nigrum L. (*Xestia adela* Franclemont)	Z7-14:Ac; slightly attractive, addition of 0.1—9% Z5-14:Ac improves the attractiveness	104, 262, 520
Euxoa drewseni	12:Ac/Z5-12:Ac/Z7-14:Ac; 11/100/18 in female, 2/6/1 for attraction; (10:Ac, Z5-12:OH, 14:Ac, Z5-14:Ac, Z7-14:OH, 16:Ac, Z7-16:Ac, Z9-16:Ac)	446
E. ochrogaster Guenée (redbacked cutworm)	Z5-12:Ac/Z7-12:Ac/Z9-12:Ac; 764/31/trace in female, this mixture plus Z5-10:Ac (200/2/1/1) is attractive; (10:Ac,12:Ac,E5-12:Ac)	105, 106
Scotia (Agrotis) exclamationis	Z5-14:Ac/Z9-14:Ac; 9/1 for attraction	107, 518
Spaelotis clandestina Harris (w-marked cutworm)	Z7-14:Ald; attractive	435
(Plusiinae)		
Autographa bilova Stephens	Z7-12:Ac; attractive	108, 109
A. californica Speyer (alfalfa looper)	Z7-12:Ac; attractive, 10% Z7-12:OH or 20% Z7-12:For acts as a synergist	109, 304, 306
A. gamma L.	Z7-12:Ac; attractive	519
Chrysodeixis (Plusia) chalcites Esper (tomato looper)	Z7-12:Ac/Z9-14:Ac; 5/1 in female, this ratio attractive; (12:Ac, 14:Ac, 16:Ac, Z9-12:Ac)	110
C. eriosoma Doubleday (silver y moth)	Z7-12:Ac/Z9-12:Ac; 97/3 in female, this ratio attractive; (12:Ac, Z9-14:Ac, 16:Ac)	443
Pseudoplusia includens Walker (soybean looper)	Z7-12:Ac; attractive	108, 265
Rachiplusia ou Guenée	Z7-12:Ac; attractive	108
Trichoplusia ni Hübner (cabbage looper)	Z7-12:Ac/12:Ac; 452/48 in female, 9/1 for attraction	266, 267
(Westermanniinae)		
Earias insulana Boisd. (spiny bollworm)	E10,E12-16:Ald; attractive	111
Pterophoridae		
Platyptilia carduidactyla Riley (artichoke plume moth)	Z11-16:Ald; attractive	150

Table 1 (continued)
COMPONENTS OF FEMALE-PRODUCED SEX PHEROMONES IDENTIFIED IN LEPIDOPTERA

Insect	Pheromone component, ratio, and activity[a,b]	Ref.
Pyralidae		
(Crambinae)		
Chilo partellus Swinhoe (spotted stalk borer)	Z11-16:Ald; attractive; (Z11-16:OH)	114
C. sacchariphagus Bojer (sugar cane borer)	Z13-18:Ac/Z13-18:OH; 7/1 in female, this ratio attractive	115
C. suppressalis Walker (rice stem borer)	Z11-16:Ald/Z13-18:Ald/Z9-16:Ald; 48/6/5 in female, this ratio attractive; (16:Ald,Z11-16:OH,18:Ald)	116, 117, 483
C. zacconius Blesz. (African rice borer moth)	Z11-16:OH/16:OH; this mixture not attractive, addition of Z13-18:OH is necessary for attraction	441
Chrysoteuchia topiaria Zeller (cranberry girdler)	Z11-16:Ald; attractive, Z9-16:Ald acts as a synergist; (Z11-16:OH)	118—120
(Phycitinae)		
Anagasta kuehniella (Mediterranean flour moth)	Z9,E12-14:Ac; attractive	121, 122
Amyelois transitella Walker (navel orangeworm)	Z11,Z13-16:Ald; attractive	123
Cadra cautella Walker (almond moth)	Z9,E12-14:Ac/Z9-14:Ac sexually stimulative	124—127
C. figulilella Gregson (raisin moth)	Z9,E12-14:Ac stimulative	128
Dioryctria disclusa Heinrich (webbing coneworm moth)	Z9-14:Ac; attractive	453
Ephestia elutella Hübner (tobacco moth)	Z9,E12-14:Ac; attractive	127, 129, 140
Homoeosoma electellum Hulst (sunflower moth)	Z9,E12-14:OH/Z9-14:OH; 1/10 to 10/1 for attraction; necessity of Z9-14:OH doubtful; (14:OH)	141, 456
Hypsipyla robusta Moore	Z9,E12-14:Ac; attractive; (Z9-14:Ac, Z11-16:Ac)	470
Plodia interpunctella Hübner (Indian meal moth)	Z9,E12-14:Ac/Z9,E12-14:OH; 4/6 for attraction	122, 124, 125, 127, 142, 143
Vitula edmandsae serratilineella Ragonot (driedfruit moth)	Z9,E12-14:OH/Z9,E12-14:Ald; 100/5 in female wash; addition of 0.5% Z9-14:OH and 0.1% Z9-14:Ald may be beneficial for the attraction; (12:Ald, 12:Ac, 14:Ald, Z9-14:Ald, 12:OH, 16:Ald, Z9-16:Ald, 14:OH, Z9-14:OH, 16:Ac, 18:Ald, 16:OH, Z9-16:OH, 18:Ac)	478
(Pyraustinae)		
Dichocrosis punctiferalis Guenée (yellow peach moth)	E10-16:Ald; addition of 10% Z-isomer improves attractiveness	144
Hellula undalis Fabr. (cabbage webworm)	E11,E13-16:Ald; attractive	384
Ostrinia furnacalis Guenée (Oriental corn borer)	Z12-14:Ac/E12-14:Ac; 3/2 (Japan) and 47/53 (China) in female; additional component(s) suggested in Japanese population; (14:Ac in Chinese population)	145—147

Table 1 (continued)
COMPONENTS OF FEMALE-PRODUCED SEX PHEROMONES IDENTIFIED IN LEPIDOPTERA

Insect	Pheromone component, ratio, and activity[a,b]	Ref.
Ostrinia nubilalis Hübner (European corn borer)	Z11-14:Ac/E11-14:Ac; 4/96 (New York) or 96/4 (Iowa) in female, these ratios attractive for each population	148, 149
Sceliodes cordalis Doubleday	E11-16:OH/E11-16:Ac; 1/1 for attraction	433
Saturniidae		
Antheraea polyphemus Cramer (polyphemus moth)	E6,Z11-16:Ac/E6,Z11-16:Ald; 9/1 in female, this ratio attractive	151
Nudaurelia cytherea cytherea F. (pine emperor moth)	Z5-10:MeBut; sexually stimulative; insufficient behavioral data	152, 153
Sesiidae		
Aegeria tibialis Harris	Z3,Z13-18:Ac/Z3,Z13-18:OH; tentative identification; 4/1 for attraction	132
Sanninoidea exitiosa Say (peachtree borer)	Z3,Z13-18:Ac; other component(s) possible	23
Synanthedon pictipes Grote & Robinson (lesser peachtree borer)	E3,Z13-18:Ac; attractive	23
Vitacea polistiformis Harris	E2,Z13-18:Ac	445
Sphingidae		
Manduca sexta Joh. (sphinx moth)	E10,Z12-16:Ald; attractive, other components suggested	154
Thaumetopoeidae		
Thaumetopoea pityocampa Den. & Schiff. (processionary moth)	Z13-11yn-16:Ac; attractive; (12:Ac)	112, 113
Tineidae		
Tineola bisselliela (webbing clothes moth)	E2-18:Ald/E2,E13-18:Ald; no biological data	257
Tortricidae (Olethreutinae)		
Cryptophlebia batrachopa Meyrick	Z8-12:Ac/E8-12:Ac; 200/1	434
C. leucotreta Meyrick (false codling moth)	E8-12:Ac/Z8-12:Ac/12:Ac; 69/23/8 in female and in effluvia, 10/10/1 for attraction	155—157, 471
Cydia nigricana F. (pea moth)	E8,E10-12:Ac; attractive; E10-12:Ac also attractive	371, 388
C. pomonella L. (codling moth)	E8,E10-12:OH; attractive; additional component(s) suggested	161, 162, 381
Grapholitha molesta Busck (Oriental fruit moth)	Z8-12:Ac/E8-12:Ac/Z8-12:OH/12:OH; 100/7/30/6 in female, this mixture attractive; the last component is not essential; (Z8-12:OH in effluvia)	158, 159
Hedya nubiferana Haworth (green budworm)	12:Ac/Z8-12:Ac/E8,E10-12:Ac; 16/30/44 in female, 100/32/48 for attraction	160
Lobesia botrana Den. & Schiff. (grape vine moth)	E7,Z9-12:Ac; attractive	163
Paralobesia viteana (grape berry moth)	Z9-12:Ac; attractive	164
Rhyacionia buoliana Den. & Schiff. (European pine shoot moth)	E9-12:Ac; attractive	165
R. frustrana Comstock (nantucket pine tip moth)	E9-12:Ac/E9,11-12:Ac; 96/4 in female, 95/5 to 97.5/2.5 for attraction; (12:OH, E9-12:OH, 12:Ac)	166
R. rigidana Fernald (pitch pine tip moth)	E8,E10-12:Ac; attractive	167

Table 1 (continued)
COMPONENTS OF FEMALE-PRODUCED SEX PHEROMONES IDENTIFIED IN LEPIDOPTERA

Insect	Pheromone component, ratio, and activity[a,b]	Ref.
R. subtropica Miller (subtropical pine tip moth)	E9-12:Ac; attractive; (*E9*-12:OH)	168
Spilonota ocellana Den. & Schiff. (eye-spotted budmoth)	Z8-14:Ac; attractive	169, 170
Tetramoera (Argyroploce) shistaceana Snellen (sugarcane borer)	Z9-12:Ac; not competitive with female	369, 398
Zeiraphera diniana Guenée (larch budmoth)	E11-14:Ac; attractive	171, 372, 353, 374
(Sparganothinae)		
Cenopis (Sparganothis) acerivorana MacKay (maple leaf roller)	Z11-14:Ald/E11-14:Ald/Z11-14:Ac; 31/16/16 in female, 65/35/10 for attraction; (14:Ald, 14:Ac, *E*11-14:Ac)	462
Sparganothis sp. (woodbine leafroller)	E11-14:OH/Z11-14:OH/E11-14:Ac; 127/93/17 in female, this ratio attractive; (Z11-14:Ac, 14:Ac, 14:OH)	172
S. directana Walker (chokecherry leafroller)	E9,11-12:Ac/E11-14:Ac/Z11-14:Ac; 35/19/28 in female, 50/12/38 for attraction; (Z9,11-12:Ac, E9-12:Ac, Z9-12:Ac)	173
(Tortricinae)		
Acleris minuta Robinson	E11,13-14:Ald	386
Adoxophyes sp. (smaller tea tortrix)	Z9-14:Ac/Z11-14:Ac/E11-14:Ac/10-Me-12:Ac; 63/31/4/2 in female, 30—70/70—30/1—40/20—200 for attraction; (12:Ac, 12:1:Ac, 13:1:Ac, 14:Ac, 12-Me-14:Ac, 15:1:Ac, 16:1:Ac)	174—176
A. orana Fischer von Röslerstamm (summerfruit tortrix)	Z9-14:Ac/Z11-14:Ac; 9/1 in female, this ratio attractive; addition of E11-14:Ac, Z9-14:OH, or Z11-14:OH improves the attractiveness; (12:Ac, Z9-12:Ac, 14:Ac, E9-14:Ac, E11-14:Ac, Z9-14:OH, Z11-14:OH, Z11-16:Ac)	177—179
A. orana fasciata Walsingham	Z9-14:Ac/Z11-14:Ac; 9/1 in male, this ratio attractive	180
Amorbia cuneana Walsingham	E10,E12-14:Ac/E10,Z12-14:Ac; attractive; (14:Ac, E10-14:Ac)	181
Archippus breviplicanus Walsingham (Asiatic leafroller)	E11-14:Ac/Z11-14:Ac; 53/23 in female, 7/3 for attraction; (14:Ac)	182
Archips argyrospilus Walker (fruittree leafroller)	Z11-14:Ac/E11-14:Ac/Z9-14:Ac/12:Ac; 60/40/4/200 is optimum for attraction	183, 184
A. cerasivoranus Fitch (uglynest caterpillar)	E11-14:Ac/Z11-14:Ac; 83/17 in female, this ratio attractive	185
A. podana Scopoli (fruittree tortrix)	Z11-14:Ac/E11-14:Ac; 1/1 for attraction	186
A. rosanus L. (European leafroller)	Z11-14:OH/Z11-14:Ac; 1/9 in female, 1/9 to 9/1 for attraction	187, 347
A. semiferanus Walker (oak leafroller)	E11-14:Ac/Z11-14:Ac; 67/33 in female, 7/3 for attraction	188
A. xylosteana L. (apple leafroller)	Z11-14:Ac/E11-14:Ac; 90/10 in female; 92/8 for attraction	344, 345, 472
Argyrotaenia citrana Fernald (orange tortrix)	Z11-14:Ald/Z11-14:Ac; 1/100 in female and 15/1 in effluvia, 5/1 to 1/10 for attraction	189

Table 1 (continued)
COMPONENTS OF FEMALE-PRODUCED SEX PHEROMONES IDENTIFIED IN LEPIDOPTERA

Insect	Pheromone component, ratio, and activity[a,b]	Ref.
A. velutinana Walker (redbanded leafroller)	Z11-14:Ac/E11-14:Ac/12:Ac; 36/4/60 in female, this ratio attractive	190, 191
Choristoneura fumiferana Clemens (spruce budworm)	E11-14:Ald/Z11-14:Ald; 94/6 in female, 95/5 in effluvia, 95/5 to 98/2 for attraction; (E11-14:Ac, Z11-14:Ac, E11-14:OH, Z11-14:OH, 14:Ald, 14:Ac, 14:OH)	192, 193, 258
C. murinana Hübner (European fir budworm)	Z9-12:Ac; attractive; addition of 10% Z11-14:Ac enhances attractiveness	353, 395, 442
C. occidentalis Freeman (western spruce budworm)	E11-14:Ald/Z11-14:Ald; 92/8 for attraction	382
C. parallela Robinson	E11-14:OH/Z11-14:OH/E11-14:Ac/ 14:OH; 87.5-81.0/0.5-0.1/5.6-12.5/6.6-8.9 in female; this mixture attractive; necessity of each component is not conclusive	455
C. rosaceana Harris (obliquebanded leafroller)	Z11-14:Ac/E11-14:Ac/Z11-14:OH; 95/5/ 10 for attraction; (E11-14:OH)	196, 197
Clepsis spectrana Treitschke	Z9-14:Ac/Z11-14:Ac; 1/9 for attraction	198, 199
Cnephasia pumicana Zeller (cereal tortrix)	E9-12:Ac/Z9-12:Ac/12:Ac; 5/3/2 in female, 6/4/10 for attraction	200
Epichoristodes acerbella Walker	Z9-14:Ac/Z11-14:Ac/Z11-14:OH; 8/64/ 26 in female, this ratio attractive; (Z9-14:OH)	201
Epiphyas postvittana Walker (light-brown apple moth)	E11-14:Ac/E9,E11-14:Ac; 20/1 in female, 10/1 to 50/1 for attraction	479
Homona coffearia Neitner (tea tortrix)	12:Ac/12:OH/E9-12:Ac; 49/15/36 in female, 1/3/1 for attraction	202
H. magnanima Diakonoff (tea tortrix)	Z11-14:Ac/Z9-12:Ac/11-12:Ac; 30/3/1 in female, this ratio attractive	203, 484
Pandemis cerasana Hübner	E11-14:Ac/Z11-14:Ac; 64/21 in female, 3/1 for attraction; (E11-14:OH, Z11-14:OH, 14:Ac)	204, 205
P. heparana Den. & Schiff.	Z11-14:Ac/Z9-14:Ac/Z11-14:OH; 90/5/5 or 90/4/6 in female, this ratio attractive; necessity of Z11-14:OH is not conclusive; (12:Ac, 14:Ac)	206, 207
P. limitata Robinson (threelined leafroller)	Z9-14:Ac/Z11-14:Ac; 9/91 in female, 6/ 94 for attraction	208
P. pyrusana Kearfott	Z11-14:Ac/Z9-14:Ac; 94/6 in female, this ratio attractive	209
Platynota flavedana Clemens (variegated leafroller)	E11-14:OH/Z11-14:OH; 9/1 in female, 84/16 for attraction; (E11-14:Ac, Z11-14:Ac, 14:OH, 14:Ac)	210
P. idaeusalis Walker (tufted apple budmoth)	E11-14:OH/E11-14:Ac; 2/1 in female, this ratio attractive	211
P. stultana Walsingham (omnivorous leafroller)	E11-14:Ac/Z11-14:Ac; 88/12 in female, 94/6 for attraction; (E11-14:OH, Z11-14:OH)	212
Sperchia intractana Walker	E11-14:Ac/Z11-14:Ac; 2/1 for attraction	213
Tortrix viridana L.	Z11-14:Ac; attractive; 9/1 mixture of E11-13:Ac/Z11-13:Ac also attractive	387, 506

Table 1 (continued)
COMPONENTS OF FEMALE-PRODUCED SEX PHEROMONES IDENTIFIED IN LEPIDOPTERA

Insect	Pheromone component, ratio, and activity[a,b]	Ref.
Yponomeutidae		
Plutella xylostella L. (diamondback moth)	Z11-16:Ald/Z11-16:Ac; 1/1 to 2/3 in female, this ratio attractive; Z11-16:OH and/or Z9-14:Ac act as synergists	204, 205, 389, 399, 402, 447
Prays citri Mill. (citrus flower moth)	Z7-14:Ald; attractive	216
Prays oleae Bern. (olive moth)	Z7-14:Ald; attractive	217, 218

[a]   Compounds in parentheses are those detected in female or its effluvia but their biological functions, if any, were not necessarily verified.

[b]   Abbreviation of compounds: Z, E, cis, and trans indicate geometry of double bonds; the 1st numbers following these letters show the position of double bonds and the 2nd indicate the number of carbon atoms in the major moiety of the compound. Ac, Bu, For, Ald, OH, Ket, Hy, and Me indicate acetate, butanoate, formate, aldehyde, alcohol, ketone, hydrocarbon and methyl, respectively. In the case of ketone, Z13-20-9:Ket means (Z)-13-eicosen-9-one; and for acetylenic compound, Z13-11yn-16:Ac shows (Z)-13-hexadecen-11-yn-1-yl acetate. 16:1:Ac means a hexadecenyl acetate.

80% species of this subfamily. Exceptions are the tea tortrix moth in Sri Lanka, *Homona coffearia*, which utilizes a 12 carbon alcohol and its acetates[182] and *Cnephasia pumicana* and *Choristoneura muriana*, which utilize acetates of 12 carbon alcohols.[200,353,395] In contrast, about 87% of 15 species in the Olethreutinae utilize 12 carbon compounds or their acetates. The eye-spotted budmoth, *Spilonota ocellana*,[169,170] and the larch budmoth, *Zeiraphera diniana*,[171,372,373] are the exceptions in this subfamily. They utilize acetates of 14-carbon alcohols.

Compared to the pheromones of the Tortricidae, the compounds in 45 species of the Noctuidae range from 10 to 21 in the number of carbon atoms in their basic structural skeleton. However, 91% of the components have carbon numbers of 12, 14, or 16. Alcohol acetates, free alcohols, and aldehydes are their principal components. In addition, two triene hydrocarbons with carbon numbers of 20 and 21 are interestingly found in two species of the subfamily Catocalinae. Unique structural characteristics are found in particular subfamilies or genera of this family. Most species of the subfamilies Aconitiinae and Amphipyrinae utilize (Z)-9-tetradecenyl or (Z)-11-hexadecenyl compounds, though five species out of six belonging to the genus *Spodoptera* (subfamily Amphipyrinae) utilize (Z,E)-9-12-tetradecadienyl acetate as their important pheromonal component. (Z)-11-hexadecenyl compounds are also utilized by almost all of the species in the Hadeninae. Eight species out of ten hadenine moths utilize (Z)-11-hexadecenyl acetate as the major pheromonal component of the female sex pheromones. (Z)-11-hexadecenal is the major pheromonal component of five species of the Heliothidinae. (Z)-11-hexadecenyl and (Z)-9-tetradecenyl compounds seem to be basic pheromone components in the Noctuidae except for the subfamilies Plusiinae, Noctuidae, and Catocalinae. (Z)-7-dodecenyl acetate is the principal pheromonal component of all the Plusiinae species so far investigated. The subfamily Noctuinae utilizes monosaturated alcohol acetates of 10, 12, and 14 carbon numbers in their alcohol moieties. Thus, pheromonal diversity in the Noctuidae seems to be larger than in the Tortricidae.

In the family Pyralidae, (Z)-11-hexadecenal and (Z)-13-octadecenal are the major components of female sex pheromones in the subfamily Crambinae. (Z,E)-9-12-tetradecadienol or its acetate is commonly utilized by six species in the subfamily Phycitinae.

Table 2
STRUCTURAL CHARACTERISTICS OF PHEROMONAL COMPONENTS IN THE LEPIDOPTEROUS FAMILIES

	Acrolepidae	Arctiidae	Bombycidae	Carposinidae	Cochylidae	Cossidae	Gelechiidae	Geometridae	Lasiocampidae	Lymantriidae	Noctuidae	Pterophoridae	Pyralidae	Saturniidae	Sesiidae	Sphingidae	Thaumetopoeidae	Tineidae	Tortricidae	Yponomeutidae	Total or mean
No. of species:	1	8	2	1	1	2	6	2	4	3	45	1	20	2	4	1	1	1	52	3	160
No. of components:	1	12	2	2	2	2	9	4	6	4	88	1	31	3	5	1	1	2	111	4	291
No. of compounds in each family																					
Total	1	5	1	2	2	2	8	4	4	3	30	1	21	3	4	1	1	2	26	3	99[a]
Acetates	—	—	—	—	2	2	7	—	1	—	20	—	8	1	3	—	1	—	17	1	47[a]
Alcohols	1	—	1	—	—	—	1	—	—	—	2	1	6	—	1	—	—	—	6	—	16[a]
Aldehydes	—	2	—	—	—	—	—	—	2	—	6	—	7	1	—	1	—	2	3	2	22[a]
Ketones	—	—	—	2	—	—	—	—	—	2	—	—	—	—	—	—	—	—	—	—	4[a]
Hydrocarbons[b]	—	3	—	—	—	—	—	4	—	1	2	—	—	—	—	—	—	—	—	—	9[a]
Others[b]	—	—	—	—	—	—	—	—	—	—	—	—	—	1	—	—	—	—	—	—	1[a]
Mean mol wt:	238	269	238	287	227	239	242	261	188	295	246	282	245	260	300	236	278	265	236	235	244
Carbon no. of components[c] (%)																					
10											3										2
12					100	50	22		100		23			33					32		22
13							11														1
14						50	22				34		55						69	50	43
16	100		100				44				33	100	35	66		100	100			50	19
17		42									1										2
18		33											10		100			100			6
19				50				100		50											2
20				50							2										1
21		25								50	2										2
No. of C-C double bonds (%)																					
0		42			50					50	3		3						9		8
1	100			100	50	50	44		100	25	79	100	58	50				50	81	100	67
2		50	100			50	44			25	13		39	50	100	100		50	10		20

Table 2 (continued)
STRUCTURAL CHARACTERISTICS OF PHEROMONAL COMPONENTS IN THE LEPIDOPTEROUS FAMILIES

	Acrolepiidae	Arctiidae	Bombycidae	Carposinidae	Cochylidae	Cossidae	Gelechiidae	Geometridae	Lasiocampidae	Lymantriidae	Noctuidae	Pterophoridae	Pyralidae	Saturniidae	Sesiidae	Sphingidae	Thaumetopoeidae	Tineidae	Tortricidae	Yponomeutidae	Total or mean
3	—	8	—	—	—	—	11	25	100	—	5	—	—	—	—	—	100[d]	—	—	—	4
4	—	—	—	—	—	—	—	75	—	—	—	—	—	—	—	—	—	—	—	—	1
Geometry of double bonds (%)																					
Z	100	100	50	100	100	66	47	87	50	66	86	100	60	60	80	50	100	—	52	100	68
E	—	—	50	—	—	33	53	7	50	—	11	—	40	40	20	50	—	100	45	—	29
Terminal	—	—	—	—	—	—	—	7	—	33	3	—	—	—	—	—	—	—	4	—	3
Position of double bonds (%)																					
1	—	—	—	—	—	—	—	7	—	—	—	—	—	—	—	—	—	—	—	—	1
2	—	—	—	—	—	—	—	—	—	—	—	—	—	—	10	—	—	66	—	—	1
3	—	18	—	—	—	33	7	27	—	—	4	—	—	—	40	—	—	—	—	—	5
4	—	—	—	—	—	—	13	—	—	—	—	—	—	—	—	—	—	—	—	—	1
5	—	18	—	—	—	66	13	—	50	—	5	—	—	—	—	—	—	—	—	—	4
6	—	—	—	—	—	—	—	27	—	—	—	—	—	20	—	—	—	—	—	—	4
7	—	—	—	—	—	—	33	—	50	—	14	—	—	40	—	—	—	—	1	50	9
8	—	—	—	—	—	—	—	—	—	—	—	—	—	—	—	—	—	—	12	—	4
9	—	29	—	—	100	—	—	27	—	—	36	—	33	—	—	—	—	—	21	—	24
10	100	—	50	—	—	—	—	—	—	—	1	—	2	—	—	50	—	—	5	—	3
11	—	24	—	—	—	—	27	13	—	—	30	100	24	40	—	—	—	—	59	50	33
12	—	—	50	50	—	—	—	—	—	—	8	—	29	—	—	50	—	—	2	—	8
13	—	12	—	50	—	—	—	—	—	66	—	—	12	—	50	—	100	33	1	—	4
15	—	—	—	—	—	—	—	—	—	33	—	—	—	—	—	—	—	—	—	—	1
20	—	—	—	—	—	—	—	—	—	—	—	—	—	—	—	—	—	—	—	—	1

a   Values are total no. of pheromonal compounds in Lepidoptera.
b   (Z)-5-decenyl methylbutanoate.
c   Carbon no. of basic skeleton of each compound.
d   (Z)-13-hexadecen-11-yn-1-yl acetate.

The four species belonging to the family Lasiocampidae all utilize 5,7-dodecadienols or 5,7-dodecadienals. Thus, the mean molecular weight of the pheromonal components of this family is the smallest in the Lepidoptera. In contrast, the mean molecular weight in the Lymantriidae and the Sesiidae are the largest. Hydrocarbons, epoxyhydrocarbons, or ketones are the major components of the sex pheromones in the Arctiidae, the Carposinidae, the Geometridae, and the Lymantridae. 2,13- and 3,13-octadecadienyl compounds are the principal components in the Sesiidae. Many species belonging to this family are attracted to the particular geometrical isomer or combinations of isomers of 3,13-octadecadienol and its acetates.

Although much more data are required for detailed discussion on the relation between taxonomy and sex pheromones, available information suggests the following: (1) species belonging to the same genus or the same subfamily usually share a common pheromone component or have components of closely related structure; (2) particular classes of compounds are repeatedly utilized in several different families as illustrated by the hydrocarbon compounds in the Arctiidae, the Geometridae, the Lymantriidae, and the Noctuidae (Catocalinae); (3) pheromonal diversity within families is not necessarily uniform as is indicated by the comparison between the Tortricidae and the Noctuidae.

SPECIFICITY OF FEMALE-PRODUCED SEX PHEROMONES

Sex pheromone specificity is an important basis of sexual isolation between sympatric species especially when other premating isolation mechanisms, such as habitat and seasonal and diel differentiation in mating are not working.[19,20] The specificity of lepidopterous sex pheromone is based on the chemical structure of particular components and specific combinations of multiple components. Fifty-eight species reportedly utilize single-component pheromones (Table 1). However, attractiveness of synthetic compounds for some of these single-component species is not necessarily competitive with virgin females. Even in two-component pheromone systems, synthetic mixtures are often less active than live females. Presence of additional components is suggested in these insect species, as has been indicated in many others (see References 23, 25, 27, 120, 144, 154, 262, 269, 373, 381, and 395). Reinvestigations of reported single-component pheromone systems may reveal the presence of additional components. Sex pheromone specificity principally depends on the blend of multiple components, and the specificity of the blend is provided by both the quality and quantity of each pheromonal component. In addition, particular pheromonal components possibly inhibit the orientation behavior of other sympatric species and may cause strong specificity in particular cases.

Multicomponent pheromone systems seem to be more likely to occur from the point of view of speciation.[19] If we suppose a hypothetical ancestral species possessed a single 11-tetradecenyl moiety as its basic pheromonal component, the development of multicomponent pheromone systems in the present Tortricidae could be explained by gradual changes in pheromonal signals by qualitative and quantitative alteration in additional components through speciation. Compounds possessing the 11-tetradecenyl moiety are utilized by about 81% of the tortricine species and these compounds comprise about 64% of the total pheromonal components in this subfamily (Table 1). However, the geometry of the double bond at the 11th position and the functionality of this basic compound show definite variation among species. (Z)-9-tetradecenyl acetate is utilized in some tortricine species as a pheromonal component in addition to the 11-tetradecenyl compounds. For example, the fruit tree leafroller, *Archips argyrospilus,* utilizes a small amount of (Z)-9-tetradecenyl acetate, about 4% of the amount of 11-tetradecenyl compounds ((Z)- and (E)-11-tetradecenyl acetates).[183,184]*Adoxophyes* spp. also utilize the 9-tetradecenyl acetate, but in amounts larger than the 11-tetradecenyl components.[174-180] Saturated compounds such as dodecyl alcohol, dodecyl acetate, and 10-methyldodecyl acetate, are also utilized by some tortricine species as additional pheromonal components. Thus, the specificity of sex pheromones in the Tortricidae is based on a combination of one of a series of compounds having 11-tetradecenyl moiety and additional related compounds.

Various compounds related to the pheromonal components have been identified in many species (Table 1). In the smaller tea tortrix, *Adoxophyes* sp., at least seven related compounds were detected in female extracts in addition to the four pheromonal components. From the summerfruit tortrix, *A. orana,* eight related compounds were identified in addition to the two pheromonal components.[179] Similarly, at least 7 related compounds were detected in female abdominal extracts of *Spodoptera exigua,*[66-68] 9 compounds in *Agrotis segetum* in Europe,[101-103,378,428] and 14 compounds in *Vitula edmandsae serratilineela.*[478] No biological activity of these related compounds has been found at the present. Even if these chemicals are not biologically active, it is possible that these structurally related compounds would become pheromonal components during speciation. Furthermore, these related compounds may have a role as interspecific signals to strengthen the pheromonal specificity.

STRUCTURE-ACTIVITY RELATIONSHIPS IN FEMALE-PRODUCED SEX PHEROMONES

The species specificity of a single component pheromone depends on the chemical structure of the compound and this suggests high specificity in structure-activity relationships. For example, most of the 55 epoxyhydrocarbons, structural analogs of disparlure, 7,8-epoxy-2-methyloctadecane, the sex pheromone of the gypsy moth, *Lymantria dispar,* were biologically inactive. Compounds most closely related to disparlure attracted male moths, but even the best of these was 40 to 100 times less active than racemic disparlure.[403] Changes in chain length of one or two carbon atoms or in the position of the methyl branch to the next carbon are the only allowable modifications in disparlure which still exhibit appreciable behavioral activity for the gypsy moth. In addition, disparlure is a chiral compound. Only (7R,8S)(+)-disparlure is active and racemic disparlure shows very low attractiveness in the field[404] and sometimes the racemic mixture is entirely inactive.[405] Therefore, structural specificity of disparlure, a single component sex pheromone of the gypsy moth, seems to be very high. However, there is an interesting as yet unexplained decrease in the activity of the racemic form from 1972 to 1977.[403,405] Another example of a single-component sex pheromone is bombykol, (E,Z)-10,12-hexadecadienol, of the silkworm moth, *Bombyx mori.*[28] Although bombykal, (E,Z)-10,12-hexadecadienal, was originally reported as a pheromonal component,[70,71] the inhibitory activity of bombykal suggests that this compound is not really a pheromonal component. Similar inhibitory activity by related compounds in female moths has also been reported in several other species. Biological activity of bombykol and its 11 derivatives was compared by measuring the minimal amount of odor sources required to evoke nerve impulses by single cell recordings.[406] The data indicated that the derivative with highest activity was (Z,E)-10,12-hexadecadienol, a geometrical isomer of bombykol. But even this activity was 100 times lower than bombykol.

In multicomponent pheromone systems, the structural specificity of each pheromonal component seems to be lower than for single-component pheromones. Changes in chain length and position of the double bond in (Z)-7-dodecenyl acetate, a pheromonal component of the cabbage looper moth, *Trichoplusia ni,* caused more than a tenfold decrease in attractiveness. But (Z)-7-undecenyl acetate exhibited about 90% of the activity of the pheromonal compound under the field conditions.[407] This species utilizes a two-component pheromone system. The sex pheromones of *Adoxophyes orana* and *Clepsis spectrana* are binary mixtures of (Z)-9- and (Z)-11-tetradecenyl acetate in different ratios. For both of these species, (Z)-11-tetradecenyl acetate could be replaced by (Z)-11-tridecenyl acetate, although the substituted mixtures were less attractive in the field than the natural pheromonal system.[408] A similar phenomenon was also found in the redbanded leafroller, *Argyrotaenia velutinana*, which has a three-component pheromone system, a 92/8/150 mixture of (Z)-11- and (E)-11- tetradecenyl acetates, and dodecyl acetate. A binary mixture of (Z)-11- and (E)-11-tridecenyl acetate at the ratios of 91/9 or 85/5 produced a trap catch equivalent to or higher than the one obtained with the three-component pheromone blend.[409] In this species both (Z)-11- tetradecenyl acetate and dodecyl acetate can be replaced by (Z)-11-tridecenyl acetate and (E)-11-tetradecenyl acetate replaced by (E)-11-tridecenyl acetate. The smaller tea tortrix moth, *Adoxophyes* sp., uses a four-component pheromonal system, a 63/31/4/2 ratio of (Z)- 9-, (Z)-11-, and (E)-11-tetradecenyl acetates and 10-methyldodecyl acetate. Of the four components, (Z)-9-tetradecenyl acetate and 10-methyldodecyl acetate can be replaced by a mixture of (Z,E)-9-12-tetradecadienyl acetate and 11-methyl-(Z)-9,12-tridecadienyl ace- tate.[410] In summary, these data on lepidopterous insects indicate that of particular pheromonal components the structural specificity in multicomponent pheromone systems is not neces- sarily high. This suggests a flexibility in multicomponent pheromone systems and easier change in the composition of the sex pheromone through the course of speciation.

SEX ATTRACTANTS FOR MALE LEPIDOPTERA

During field tests of proposed sex pheromone compounds and field screening of pheromone analogs, male moths of a variety of species were attracted to a number of synthetic compounds and their mixtures (Table 3). It is not yet certain whether these compounds are the sex pheromones of the attracted species, but they appear to act as sex attractants for the males of these species. Most of these compounds are major candidates of the female sex pheromones in these species. The relationships between insect taxonomy and chemical structure, which were discussed above, also hold for these apparent sex attractants (Table 3).

It should be noted that the exact potency of attractiveness of these compounds is not known for most of the species because comparisons of attractiveness between the compounds and virgin females have not been undertaken except for the several economically important species.

MALE-PRODUCED SEX PHEROMONES AND MALE SCENTS

Although a variety of compounds have been identified as components of male-specific scents or as aphrodisiac pheromones, the biological activity of these compounds has been confirmed for only a few insect species. A variety of possible functions has been proposed for these male specific substances: (1) as an aphrodisiac to cause inhibition of locomotion in the female, promoting successful mating by blocking her normal escape-flight response to a strange object; (2) as a species-specific signal to prevent cross breeding; (3) as a primer pheromone to initiate oogenesis; (4) as a repellent pheromone to deter other males; and (5) as a defensive substance to repel predators.[411]

Observed behavioral responses to these so-called aphrodisiac pheromones in experiments may simply be the eventual copulation that would occur irrespective of the presence of the chemicals. This results in great difficulties in obtaining experimental proof of behavior mediated by these aphrodisiacs. Although a variety of chemicals has been isolated as can- didate aphrodisiacs (until 1983), biological activities have not been verified for most of these chemicals (Table 4).

Males of the queen butterfly, *Danaus gilippus berenice,* utilize an alkaloidal ketone, 2,3- dihydro-7-methyl-1*H*-pyrrolizin-1-one as a pheromonal component. The biological activity of this compound was verified in a series of elegant experiments.[223-225] The female queen butterfly alights on vegetation when her antennae are brushed during flight with the male

Table 3
SYNTHETIC SEX ATTRACTANTS FOR MALE LEPIDOPTERA FOUND BY FIELD TESTS

Insect	Attractant and ratio[a]	Ref.
Acrolepiidae		
Acrolepia alliella Semenov and Kuznesov	Z11-16:Ac/Z11-16:Ald/Z11-16:OH = 45/	130, 131
(= *Acrolepiopsis sapporensis* Matsumura)	45/10 or 1/3/1	
Acrolepiopsis issikiella Moriuti	Z11-16:Ald	131
A. suzukiella Matsumura	Z11-16:Ald/Z11-16:Ac = 9/1	131
Amatidae		
Syntomis phegia L.	cis-7,8-epoxy-2-Me-18:Hy	275
Arctiidae		
Holomelina laeta	2-Me-17:Hy	25
H. rubicundaria	2-Me-17:Hy	25
Argyresthiidae		
Argyresthia beta Friese and Moriuti	Z11-14:OH	131
A. chamaecypariae Moriuti	E11-16:Ald	131
A. nemorivaga Moriuti	Z11-16:Ald	131
Paraargyresthia japonica Moriuti	E11-16:Ac	131
Cochylidae		
Aethes rubigana Treitschke	Z11-14:Ac/E11-14:Ac = 1/1	131
Commophila umbrabsana Kearfott	E11-14:Ac	276
Eupoecilia angustana Hübner	12:Ac/Z11-14:Ac/E11-14:Ac	277
E. kobeana Razowski	Z11-14:OH	278
Hysterosia cartwrightana Kearfott	E11-14:Ac	279
Phalonidia minimana Caradja	Z11-14:Ac	278
Phtheochroides clandestina Razowski	Z11-14:Ac/Z9-14:Ac = 1/1	131
Coleophoridae		
Coleophora laricella Hübner	Z5-10:OH	444
Cosmopterigidae		
Cosmopterix fulminella Stringer	Z7-12:Ac	278
C. victor Stringer	Z10-13:Ac or Z10-12:Ac	278
Labdia citracma Meyrick	E11-14:Ald	131
L. semicoccinea Stainton	E11-14:Ald	131
Limnaecia phragmitella Stainton	Z11-14:Ac/E11-14:Ac = 1/10	279
Stagmatophora niphosticta Meyrick	E11-14:Ac/Z11-14:Ac = 7/3	131
Cossidae		
Acossus centerensis Lintner	E3,E5-14:Ac	263
Epiplemidae		
Epiplema foenella L.	Z8-12:Ac	278
E. moza Butler	E11-16:Ac	302
E. plaqifera Butler	E9-14:OH	278
E. scudderiana	Z8-12:OH	284
Ethmidae		
Ethmia monticola Walsingham	E11-14:Ac	276
Eustrotiinae		
Micardia argentata Butler	E7-12:Ac	278
Gelechiidae		
Brachmia arotraea Meyrick	E7-14:Ac	278
Bryotropha senectella Zeller	Z9-14:Ac	280
B. similis Stainton	Z9-14:Ac/Z11-14:Ac	281
B. terrella Den. & Schiff.	E9-14:Ac	169
Chionodes fuscomaculella	Z5-12:Ac	281
C. psiloptera Barnes & Busck	Z7,Z11-16:Ac	283
Dactylethra tegulifera Meyrick	E7-12:Ac	131
Dichomeris ligulella Hübner	E11-14:Ac	284
Gaesa sparsella Christoph	Z8-11:Ac	278
Lixodessa ochrofasciella Toll	Z7-12:Ac	280

Table 3 (continued)
SYNTHETIC SEX ATTRACTANTS FOR MALE LEPIDOPTERA FOUND BY FIELD TESTS

Insect	Attractant and ratio[a]	Ref.
Metzneria sp.	E3-12:Ac	44
Pectinophora endema Common	Z7,Z11-16:Ac/Z7,E11-16:Ac = 2/1	285
P. scutigera Holdaway	Z7,Z11-16:Ac/Z7,E11-16:Ac = 100/0 to 80/20	285, 464
Protobathra leucostola Meyrick	Z7-12:OH	278
Scrobipalpa atriplicella	Z5-12:Ac	281
Syncopacma cinctella Clerck	Z7-12:Ac or Z7-11:Ac	280
Geometridae		
Operophtera bruceata	1,Z3,Z6,Z9-19:Hy	385
O. occidentalis	1,Z3,Z6,Z9-19:Hy	385
Perizoma grandis Hulst	E11-14:Ald	276
Scoplula personata Prout	Z9,E11-14:Ac	286
Sterrha biselata Hufnagel	E9,Z12-14:Ac	287
S. imbecilla Inoue	Z7-12:Ac	278
Glyphipterigidae		
Glyphipterix eulucotoma Diakonoff and Arita	Z11-16:Ac/Z11-16:Ald = 1/1	131
G. nigromarginta Issiki	Z11-16:Ac/Z11-16:Ald = 1/1	131
Gracillariidae		
Caloptilia theivora Walsingham	E11-16:Ald	131
Gracilaria syringella Fab.	E11-14:Ald/Z11-14:Ald = 9/1	131
Lithocolletis blancardella Fab.	E10-12:Ac	288, 289
L. corylifoliella Hübner	E4,Z7-13:Ac	288
L. orientalis Kumata	Z10-14:Ac	278
L. pygmaea Kumata	Z8-14:OH	278
L. watanabei Kumata	Z10-13:Ac	278
Heliodinidae		
Stathmopoda theoris Meyrick	Z7-16:Ac or Z7,Z11-16:Ac/Z7,E11-16:Ac	286
Limacodidae		
Microleon longipalpis Butler	Z7-12:Ald	131
Lymantriidae		
Dasychira grisefacta ella Bryk	Z6-21-11:Ket	290
D. plagiata Walker	Z6-21-11:Ket	291
D. vagans grisea Barnes & McDunnough	Z6-21-11:Ket	290
L. dispar japonica Motschulsky	cis-7,8-epoxy-2-Me-18:Hy	292
L. fumida Butler	cis-7,8-epoxy-2-Me-18:Hy	292
L. obfuscata Walker	cis-7,8-epoxy-2-Me-18:Hy	293
Orgyia antiqua L.	Z6-21-11:Ket	294
O. cana Edwards	Z6-21-11:Ket	290
O. leucostigma J. E. Smith	Z6-21-11:Ket	291
Noctuidae		
Abagrotis placida Grote	Z7-14:Ac/Z11-14:Ac = 10000/1	295
Acronicta grisea Walker	Z7-12:Ac/Z9-12:Ac = 1/1	296, 297
A. separata Grote	Z7-12:Ald/Z7-12:Ac = 1/9	298
Actebia fennica Tauscher	Z11-14:Ac/Z7-12:Ac = 1/1	297
Agroperina cogitata Smith	Z11-16:Ac/Z9-14:Ac = 20/1	299
A. dubitans Walker	Z11-16:Ac/Z9-14:Ac = 20/1	84
Agrotis collaris G. & R.	Z7-14:Ac/Z5-14:Ac = 20/1	297
A. obliqua Smith	Z5-14:Ac	299
A. orthogonia Morrison	Z5-10:Ac/Z7-12:Ac = 1/2 to 1/5	300
A. venerabilis Walker	Z5-10:Ac/Z7-12:Ac/Z5-12:Ac = 100/10/1	295
A. vetusta Walker	Z11-16:Ac/Z9-14:Ac = 50/1	299
A. volubilis Harvey	Z5-14:Ald/Z5-14:Ac = 1/1 or Z5-14:Ac/Z7-14:Ac = 100/1	298, 299

Table 3 (continued)
SYNTHETIC SEX ATTRACTANTS FOR MALE LEPIDOPTERA FOUND BY
FIELD TESTS

Insect	Attractant and ratio[a]	Ref.
Aletia oxygala Grote	Z11-16:Ald/Z11-16:Ac = 7/3	301
Amathes c-nigrum L., small type (*Xestia dolosa* Franclemont)	E7-14:Ac	281
A. collaris Grote & Robinson	Z7-14:Ac/Z9-14:Ac = 9/1	262
Amphipoea interoceanica Smith	Z9-14:Ac	284
A. velata Walker	Z11-16:Ac	284
Amphipyra monolitha Guenée	Z11-16:Ac	302
Anagrapha falcifera Kirby	Z7-12:Ac/Z7-12:OH = 10/1 or Z7-12:Ac/ Z7-12:For = 5/1	303, 304
Andropolia contacta Walker	Z5-12:Ac/Z7-12:Ac = 1/1	296
Anhimella contrahens Walker	Z11-16:Ald/Z11-16:OH = 10/1	299
Apamea indela Smith	Z11-16:Ac/Z11-16:OH/Z11-16:Ald = 200/ 50/1	295
A. interoceanica Smith	Z9-14:Ac	281
A. velata Walker	Z11-16:Ac	284
Argyrogramma verruca Fabr.	Z7-12:Ac/E7-12:Ac = 94/6	305
Athetis dissimilis Hampson	Z9-14:OH	278
Autographa ampla Walker	Z7-12:Ac	281
A. confusa Stephens	Z7-12:Ac/Z9-14:Ac = 1/1	131
A. egena Guenée	Z5-12:Ac	307
A. falcifera	Z7-12:Ac	281
A. flagellum Walker	Z7-12:Ac/Z7-14:Ac	297
A. gamma L.	Z7-12:Ac	308
A. jessica Butler	Z7-12:Ac/Z7-12:OH = 9/1	131
A. precationis Guenée	Z7-12:Ac	284
Axylia putris L.	Z9-14:Ac	131
Blepharosis costalis Butler	Z9-14:Ald/Z9-14:Ac = 9/1	131
Caradrina morpheus Hufnagel	E9-12:Ac	169
Ceramica picta Harris	Z11-14:Ac/Z11-16:Ac/Z9-14:Ac = 200/2/1 or Z11-14:Ald/Z11-14:Ac	298, 299
Chersotis juncta Grote	Z11-16:Ac/Z11-16:Ald	299
Chrysaspidia contexta Grote	Z7-12:Ac	281
C. putnami Grote	Z5-12:Ac; or Z5-12:Ac/Z7-12:Ac = 100/1	295, 296
C. venusta Walker	Z5-12:Ac/Z7-12:Ac = 1/1	296
Conistra grisescens Draudt	Z7-12:Ac	278
C. unimacula Sugi	Z11-16:Ac	278
Crymodes devastator Brace	Z11-16:Ald/Z11-16:Ac = 7/3 to 1/1; Z11-16:Ald/Z11-16:Ac/Z7-12:Ac = 1/1/1; or Z11-16:Ald/Z11-16:Ac/ Z11-16:OH/Z9-14:Ac = 70/30/1/0.01	298, 301, 309, 447
Cryptocala acadiensis Bethune	Z11-16:Ac/Z9-14:Ac/Z9-16:Ald	299
Cucillia florea Guenée	Z9-14:Ac	299
C. intermedia Speyer	Z9-14:Ac	281
C. omissa Dod.	Z9-14:Ac	299
C. posteva Guenée	Z9-14:Ac/Z9-14:Ald = 100/1	299
C. speyeri Lintner	Z9-14:Ac	299
C. umbratica L.	Z9-14:Ac	310
Dadica lineosa Moore	Z11-14:Ac/Z9-14:Ac = 1/1	131
Diarsia deparca Butler	Z9-14:Ac	278
Earias biplaga Walker	Z11-16:Ald	90
Enargia infumata Grote	Z11-16:Ac/Z9-14:Ac = 500/1	295
Euagrotis tepperi Smith	Z11-16:Ac/Z9-14:Ac/Z7-12:Ac = 10/1/1	299
Eupsilia quadrilinea Leeth	Z11-16:OH or E11-16:OH	278
E. tripunctata Butler	Z11-16:Ac/Z11-16:Ald/Z11-16:OH = 9/9/2	131

Table 3 (continued)
SYNTHETIC SEX ATTRACTANTS FOR MALE LEPIDOPTERA FOUND BY FIELD TESTS

Insect	Attractant and ratio[a]	Ref.
Eurois astricta Morrison	Z11-16:Ald/Z11-16:Ac/Z9-14:Ac = 1/8/2	298
E. occulta L.	Z11,E14-16:Ac/Z9-14:Ac = 25/75; or Z11-16:Ac/Z9-14:Ac = 25/75 or Z9-14:Ac/Z11-16:Ac/Z11-16:Ald = 100/30/1	311, 312
Euxoa acornis Smith	Z11-16:Ac/Z9-14:Ac/Z9-14:Ald = 5/4/1	298
E. albipennis Grote	Z5-10:Ac/Z7-12:Ac/Z7-12:OH = 10/1/1	313
E. altera Smith	Z11-16:Ac/Z9-14:Ac	299
E. auxiliaris Grote	Z5-14:Ac/Z7-14:Ac/Z9-14:Ac = 100/1/10	314, 315
E. basalis Grote	Z7-12:Ac/Z7-12:OH/Z5-12:Ac = 5/5/1; or Z7-12:Ac/Z7-12:OH/Z5-10:Ac = 5/5/1	299
E. campestris Grote	Z5-10:Ac; or Z5-10:Ac/Z5-12:Ald = 10/1	316, 317
E. cicatricosa G. & R.	Z7-12:Ac	299
E. dargo Strecker	Z9-14:Ac/Z11-16:Ac = 9/1	299
E. declarata Walker	Z5-10:Ac; or Z5-10:Ac/Z5-12:Ac = 10/1	316, 317
E. divergens Walker	Z9-12:Ac/Z7-12:Ac = 200/1	296
E. fessellata	Z7-16:Ac	281
E. flavicollis Smith	Z9-14:Ac/Z7-12:Ac = 500/1	295
E. lutulenta Smith	Z9-14:Ac/Z11-16:Ac	312
E. manitobana McDunnough	Z9-14:Ac/Z7-14:Ac = 4/1	299
E. messoria Harris	Z11-16:Ac/Z7-16:Ac = 20/1; or Z11,E14-16:Ac/Z7-16:Ac = 25/1	311, 319
E. mimallonis Grote	Z5-10:Ac	299
E. obeliscoides Guenée	Z11-16:Ald/Z9-14:Ac = 1/1; or Z9-14:Ac/Z7-12:Ac = 500/1	295, 298
E. olivia Morrison	Z5-10:Ac	316
E. perolivalis Smith	Z11-16:Ac/Z9-14:Ac/Z11-16:OH = 4/5/1	320
E. pestula Grote	Z5-10:Ac/Z7-12:Ac/Z7-12:Ald = 100/1/1	299
E. plagigera Morrison	Z9-14:Ac/Z9-14:OH = 5/1	299
E. pleuritica Grote	Z7-12:Ac/Z7-12:OH/Z5-12:Ac = 5/1/1 or 5/5/1	321
E. ridingsiana Grote	Z9-14:Ac/Z11-16:Ac/Z7-12:Ac = 500/20/1	299
E. rockburnei Hardwick	Z5-10:Ac; or Z5-10:Ac/Z5-12:Ald = 10/1	316, 317
E. scandens Riley	Z7-12:OH	299
E. servita Smith	Z7-14:Ac/Z7-14:Ald	299
E. tessellata Harris	Z5-14:Ac/Z7-16:Ac = 200/1; or Z5-16:Ac/Z7-16:Ac = 1/1	322
E. tristicula Morrison	12:Ac/Z9-12:Ac	323
Evergestis forficalis L.	E11-14:Ac	324
Faronta diffusa Walker	Z11-16:Ald/Z11-16:Ac = 1/9	298
Feltia ducens Walker	Z11-16:Ac/Z9-14:Ac/Z7-12:Ac = 2000/1/1 or Z11-16:Ac/Z9-14:Ald = 1/1	298, 299
Fishia derelicta Hampson	Z9-14:Ald/Z9-14:Ac/Z7-12:Ac = 1/8/4	298
F. hanhami Grote	Z5-14:Ac/Z9-14:Ac = 8/1	297
Heliothis ononis D. & S.	Z11-16:Ald	299
H. phloxiphaga G. & R.	Z11-16:Ald	109
Helotropha reniformis Grote	Z11-16:Ac/Z7-16:Ac = 20/1	319
Hermonassa cecilia Butler	Z9-14:Ac; or Z9-14:Ac/Z9-14:Ald = 9/1	131, 302
Homohadena infixa Walker	Z7-14:Ac/Z9-14:Ac = 100/1 to 1000/1	262, 295
Hypocoena rufostrigata Packard	Z11-16:Ald/Z11-16:OH/Z9-14:Ald = 100/1/1	299
Ipimorpha pleonectusa Grote	Z11-16:Ald/Z11-16:OH = 100/1	295, 299
Isopolia strigidisca Moore	Z11-14:Ac	131
Lacinipolia lorea Guenée	Z7-14:Ac	284
L. renigera Stephens	Z9-14:Ac	299

Table 3 (continued)
SYNTHETIC SEX ATTRACTANTS FOR MALE LEPIDOPTERA FOUND BY FIELD TESTS

Insect	Attractant and ratio[a]	Ref.
L. vicina Grote	Z11-16:Ac/Z9-14:Ac/Z7-14:Ac = 5/5/1	299
Lamprothripa lactaria Graeser	Z9,E12-14:Pro	278
Leucania commoides Guenée	Z11-16:Ac/Z9-14:Ac/Z11-16:OH = 4/4/1	320
L. linda Franclemont	Z9-14:Ac/Z11-14:Ac = 19/1	325
L. multilinea Walker	Z11-16:Ac/Z9-14:Ac/Z11-16:OH = 4/4/1	320
L. phragmatidicola Guenée	Z9-14:Ac	284
Lithacodia albidula Guenée	Z5-12:Ac/Z7-14:Ac	299
Lithomoia solidaginis Hübner	Z11-16:Ald	298
L. thaxteri Grote	Z9-14:Ac	297
Lithophane unimoda Lintner	Z9-14:Ac	299
Lucania linda Franclemont	Z9-14:Ac/Z11-14:Ac = 95/5	208
Melanchra (Ceramica) picta Harris	Z11-14:Ac/Z11-14:Ald	284, 298
Morrisonia confusa Hübner	Z11-16:Ac	281
Nedra ramulosa Guenée	Z11-14:OH	281
Nephelodes emmedonia Cramer	Z11-16:Ald	298
Ochropleura plecta L.	E11-16:Ac/Z11-16:Ac = 1/1	324
O. triangularis Moore	Z9-14:Ac	278
Olgia bridghami G. & R.	Z11-16:Ac/Z11-16:OH = 500/1	295, 299
O. mactata Guenée	Z11-16:Ac/Z11-16:OH = 500/1	295, 299
Oncocnemis lepipuloides McDunnough	Z7-12:Ac/Z9-14:Ac	299
O. piffardi Walker	Z9-14:Ac/Z7-12:Ac = 500/1	295
Orthodes crenulata Butler	Z11-16:Ac	281
Orthogonia sera Felder	Z7-12:Ac	302
Orthosia angustipennis Matsumura	Z9,E12-14:Ac	278
O. carnipennis Butler	Z9-14:Ac/Z11-16:Ac = 1/1	131
O. ella Butler	Z11-16:Ac	278
O. evanida Butler	Z9-14:Ald	131
O. gothica askoldensis Straudinger	Z9-14:Ald/Z9-14:Ac = 9/1	131
O. hibisci Guenée	Z9-14:Ac; or Z9,E12-14:Ac; or Z9-14:Ald/ Z11-14:Ald = 100/1	295, 326
O. limbata Butler	Z9-14:Ac/Z9-14:Ald = 9/1	131
O. lizetta Butler	Z11-16:Ac/Z11-16:Ald = 1/1	131
O. munda Schiff.	Z11-16:OH/Z11-16:Ald/Z11-16:Ac = 9/9/2	131
O. odiosa Butler	Z7-12:Ac	278
O. paromoae Hampson	Z7-14:Ac	278
Panolis flammea Schiff	Z9-14:Ac/Z11-14:Ac	327
Paradiarsia littoralis Packard	Z7-12:Ac/Z5-12:Ac = 100/1	295
Parastichtis discivaria Walker	Z11-16:Ac/Z11-16:Ald	299
Peridroma saucia Hübner	Z9-14:Ac/Z11-16:Ac = 1/1 to 1/2	328
Phyllophila obliferata cretacea Butler	Z9-14:Ac or Z9,E12-14:Ac	286
Platysenta videns Gn.	Z11-16:Ald/16:Ald/Z9-14:Ald/14:Ald = 48/ 12/16/24	449
Plusia aereoides Grote	Z7-12:Ac	281
Polia assimilis Morrison	Z11-14:Ac/Z9-14:Ac = 100/1	295
P. atlantica Grote	Z11-16:Ac/Z11-16:Ald/Z11-14:Ald = 200/ 1/1; or Z11-16:Ac/Z7-12:Ac = 9/1; or Z11-16:Ac/Z11-16:Ald = 500/1	295, 296, 329
P. discalis Grote	Z9-14:Ac/Z11-14:Ac	299
P. grandis Boisd.	E9-14:Ac	284
P. ingravis Smith	Z9-14:Ac/Z11-16:Ac = 100/1	295
P. lilacina Harvey	Z7-14:Ac/Z9-14:Ac = 10/1	299
P. nevadae Grote	Z11-16:Ac/Z11-16:Ald	299
P. purpurissata Grote	Z9-14:Ac/Z7-14:Ac = 100/1	299
P. segregata Smith	Z7-14:Ac/Z11-14:Ac = 10/1	299
P. tacoma Strecker	Z9-14:Ac/Z9-14:Ald = 100/1 or 9/1	295, 298, 299

Table 3 (continued)
SYNTHETIC SEX ATTRACTANTS FOR MALE LEPIDOPTERA FOUND BY FIELD TESTS

Insect	Attractant and ratio[a]	Ref.
Protagrotis niveivenosa Grote	Z11-16:Ald/Z9-14:Ac	299
P. obscura Barnes & McDunnough	Z11-16:Ald	331
Pseudorthodeos vecors Guenée	Z11-16:Ac	281
Pyreferra citrombra Franclemont	Z7-12:Ac/Z9-14:Ac	284
Rachiplusia ou Guenée	Z7-12:Ac	281
Rancora albida Guenée	Z9-14:Ac	299
Raphia frater Grote	Z7-12:OH	332
Rhynchaglaea fuscipennis Sugi	Z7-12:Ac or Z9-14:Ald	131, 278
R. scitula Butler	Z9,E12-14:OH	278
Rusidrina depravata Butler	Z9,E12-14:Ac	302
Sarcopdia illoba Butler	Z11-16:Ac	278
Schinia bina Guenée	Z11-16:Ald/Z11-16:Ac = 3/7	298
S. meadi Grote	Z11-16:Ald	299
Scotogramma farnhami Grote	Z11-16:Ald/Z9-14:Ald/Z9-16:Ald = 100/1/1	295
Sesamia cretica Lederer	Z9-14:OH/Z9-14:Ac	333
Sideridis rosea Harvey	Z9-14:Ac/Z11-14:Ac = 10/1; or Z9-14:Ac/ Z11-16:Ac = 2/1	328
Simyra henrici Grote	Z7-14:Ald	299, 435
Sineugraphe disgnosta Boursin	Z7-12:OH	131
Spaelotis clandestina Harris	7Z7-14·Ald	297
Spodoptera cilium Guenée	Z9-14:Ac/Z9,E12-14:Ac	121
S. dolichos Fab.	Z9,E12-14:Ac	282
S. triturata Boisd.	Z9-14:Ac/E9-14:Ac/Z9,E12-14:Ac	335—337
Sunira bicolorago Guenée	Z11-16:Ac/Z9-16:Ac	299
Sutyna profunda Smith	Z11-16:Ac/Z9-16:Ac	299
Syngrapha epigaea Grote	Z7-12:Ac/14:Ac = 20/1	299
S. rectangula Kirby	Z7-12:Ac	299
Telorta acuminata Butler	Z11-16:Ald	131
T. divergens Butler	Z11-16:Ac	302
T. edentata Leech	Z11-16:Ac or E11-16:Ac	278
Teratoglaea pacifica Sugi	Z9-14:Ac or Z9,E12-14:Ac	278
Trichoplusia oxygramma Geyer	Z7-12:Ac/E7-12:Ac = 96/4	305
Valeriodes viridimacula Graeser	Z11-16:Ac	278
Xanthia lutea Ström	Z9,E12-14:Ald	299
Xylena formosa Butler	Z11-16:Ald	278
Xylomyges curialis Walker	Z7-12:Ac	297
X. dolosa Grote	Z7-12:Ac	297
Zanclognatha lunaris Scopoli	cis-7,8-epoxy-2-Me-18:Hy	338
Notodontidae		
Oligocentria (Dicentria) semirufescens Walker	Z7-12:OH	284
Shizura semirufescens Walker	Z7-12:OH	284
Oecophoridae		
Agonopterix encentra Meyrick	Z7-12:OH/Z7-12:Ac = 9/1	131
Xenomicta cupreifera Butler	Z11-14:Ac	278
Pterophoridae		
Nippoptilia issikii Yano	Z7-12:Ac	286, 302
N. vitis Sasaki	Z7-12:Ac	278
Oidaematophorus cretidactylus Fitch (or *O. rileyi* Fern.)	E11-14:Ac	276
O. guttatus Walsingham	E11-14:Ald	276
O. homodactylus Walker	E11-14:Ald	276
O. mathewianus Zeller	E11-14:Ac	276
Platyptilia jezoensis Matsumura	Z11-16:Ac/Z11-16:Ald = 9/1	131

Table 3 (continued)
SYNTHETIC SEX ATTRACTANTS FOR MALE LEPIDOPTERA FOUND BY FIELD TESTS

Insect	Attractant and ratio[a]	Ref.
Pterophorus jezonicus Matsumura	Z9-12:Ald	131
P. tenuidactylus Fitch	Z7-12:Ac	284
Trichoptilus wahlbergi Zeller	Z7-12:Ald/Z7-12:OH = 9/1	131
Pyralidae		
Acrobasis rufilimbalis Wileman	Z9-15:Ac	278
Diasemia litterata Scopoli	E11-16:Ac	278
Dioryctria clarioralis	Z9-14:Ac	453
Ephestiodes infimella Ragonot	Z9,E12-14:Ac	339
Eurythmia hospitella Zeller	Z9,E12-14:Ac	296
Herculia glaucinalis L.	Z7-12:Ac	286
Loxostege chortalis Grote	E9-12:Ac; or E9-12:Ac/Z9-12:Ac = 1000/1 to 2000/1	284, 329
L. neobliteralis Capps	E9-14:Ac	281
L. sticticalis L.	E11-14:Ac	279
Mutuuraia mysippusalis Walker	Z7-10:Ac/Z7-12:Ac =1/1	296
Ostrinia obumbratalis Lederer	Z11-14:Ac/E11-14:Ac = 1/1	340
Pagyda arbiter Butler	Z11-14:Ald/Z11-14:Ac = 9/1	131
Paliga minnehaha Pryer	E11-14:Ac	131
Pareromene exosectella Christopoh	Z11-16:Ac	278
Phlyctaenia terrealis Treitschke	Z7-12:Ac	281
Procerus venosatus Walker	Z13-18:Ac/Z13-18:OH/Z11-16:Ac = 4/2/4	401
Pyrausta aurata Scopoli	E11-14:Ac	341
P. ochosalis	E11-14:Ac	284
P. purpuralis L.	Z11-14:Ac/E11-14:Ac	342
Sitochroa chortalis Grote	E9-12:Ac	296
Tehama bonifatella Hulst	Z11-16:Ald/Z13-18:Ald = 1/1 to 7/3	377
Sesiidae		
Albuna fraxini Hy. Edwards	E3,Z13-18:Ac/Z3,Z13-18:OH	133
A. pyrimidalis Walker	E3,Z13-18:OH	134
Alcathoe carolinensis Engelhart	Z3,E13-18:Ac/E3,Z13-18:Ac	135
Carmenta bassiformis Walker	Z3,E13-18:Ac/Z3,Z13-18:OH	135
C. odda Duckworth & Eichlin	Z3,E13-18:Ac	135
C. suffusata Engelhart	Z3,Z13-18:OH	135
C. teta	Z3,Z13-18:Ac	136
C. texana Hy. Edwards	Z3,E13-18:Ac	135
C. welchelonm Duckworth & Eichlin	Z3,E13-18:Ac	137
Paranthrene asilipennis Boisd.	Z3,E13-18:Ac	135
P. dollii Neumoegen	Z3,Z13-18:OH/E3,Z13-18:OH/E3,Z13-18:Ac	133, 135
P. robiniae Hy. Edwards	Z3,E13-18:Ac/E3,Z13-18:Ac	138
P. simulans Grote	Z3,Z13-18:Ac	136
P. tabaniformis Rottemburg	Z3,E13-18:Ac/E3,Z13-18:OH	139
Podosesia aureocincta Purrington & Nielsen	Z3,Z13-18:OH/Z3,E13-18:Ac/E3,Z13-18:Ac	133, 135, 139
P. syringae Harris	Z3,Z13-18:Ac	136
Sannia uroceriformis Walker	Z3,Z13-18:OH/E3,Z13-18:OH/E3,Z13-18:Ac	133, 135
Sylovora acerni	Z3,Z13-18:Ac	136
Synanthedon acerni Clemens	E3,Z13-18:OH/E3,Z13-18:Ac	139
S. acerrubri Engelhardt	E2,Z13-18:Ac	445
S. alleri Engelhardt	E3,Z13-18:Ac	135
S. arkansasensis Duckworth & Eichlin	E3,Z13-18:Ac	135
S. bibionipennis Boisd.	E3,Z13-18:OH/E3,Z13-18:Ac	270
S. decipiens Hy. Edwards	E3,Z13-18:Ac	135

Table 3 (continued)
SYNTHETIC SEX ATTRACTANTS FOR MALE LEPIDOPTERA FOUND BY FIELD TESTS

Insect	Attractant and ratio[a]	Ref.
S. fatifera Hodges	Z3,Z13-18:Ac	136
S. fulvipes Harris	Z3,E13-18:Ac	139
S. geleformis Walker	Z3,Z13-18:Ac	136
S. (Conopia) hector Butler	Z3,Z13-18:Ac/E3,Z13-18:Ac = 1/1	271
S. proxima	Z3,Z13-18:Ac	136
S. (Conopia) quercus Butler	Z3,E13-18:Ac/E3,E13-18:Ac = 1/1	272
S. kathyae Duckworth & Eichlin	Z3,E13-18:Ac	273
S. myopaeformis Borkhausen	Z3,E13-18:Ac	274
S. rileyana	E3,Z13-18:OH/E3,Z13-18:Ac	133
S. rubrofascia Hy. Edwards	Z3,Z13-18:Ac/E3,Z13-18:Ac	133
S. sapygaeformis Walker	Z3,Z13-18:Ac	136
S. scitula Harris	Z3,Z13-18:Ac	136
S. sequoiae Hy. Edwards	Z3,Z13-18:OH	270
S. (Conopia) tenuis Butler	Z3,Z13-18:Ac	272
Vespamime sequoiae	Z3,Z13-18:Ac/E3,Z13-18:Ac	136
Vitacea scepsiformis Hy. Edwards	Z3,E13-18:Ac	135
Zenodoxus tineiformis Esper	E3,Z13-18:Ac	133
Thyatiridae		
Epipsestis perornata Inoue	Z9-14:Ald	131
Thyrididae		
Thyris maculata Harris	Z11-14:Ac	284
T. usitata Butler	Z11-14:OH	278
Tineidae		
Ereunetis (Decadarchis) simulans Butler	Z3,E13-18:Ac	343
Nemapogon apicisignalellus	Z9-14:Ac	284
Tortricidae		
Acleris emargana Fabr.	E11-14:Ac	332
A. enitescens Meyrick	E11-14:Ac/E11-14:Ald = 1/1	131
A. filipjevi Obraztsov	E11-14:Ald/E11-14:Ac = 9/1	131
A. paradiseana Walsingham	E11-14:Ac	344
A. perfundana Kuznetsov	Z11-14:Ald/E11-14:Ald = 1/1	131
A. rhombana Den. & Schiff.	E11-14:Ac/Z11-14:Ac = 9/1 to 8/10	324, 345
A. scabrana Den. & Schiff.	E11-14:Ald	346
Aphania auricristana Walsingham	Z9-14:Ac; or Z9-14:Ac/Z11-14:Ac = 9/1 or 7/3	286, 302
A. infida Heinrich	Z8-12:Ac	284
Aphelia paleana Hübner	12:Ac/Z11-14:Ac/E11-14:Ac	277
Apotomis lineana Den. & Schiff.	Z8-12:Ac/Z10-14:Ac	169
A. turbidana Hübner	Z10-14:Ac	390
Archippus asiaticus Walsingham	Z11-14:Ac/E11-14:Ac = 9/1 to 5/5	131, 344
A. ingentanus Christoph	Z11-14:Ac; or Z11-14:Ac/Z11-14:OH = 9/1	278, 344
A. peratratus Yasuda	Z11-14:Ac/E11-14:Ac = 9/1	131
A. piceanus similis Butler	Z11-14:Ac; or Z11-14:Ac/E11-14:Ac = 8/2	131, 278, 344
A. tsuganus Powell	E11-14:Ac	276
Archips eapsigernus Kennel	Z11-14:Ac/E11-14:Ac = 1/1	131
A. fuscocupreanus Walsingham	Z11-14:Ac/E11-14:Ac = 8/2	131
A. griseus	Z11-14:Ald/E11-14:Ald = 75/25	462
A. (Hoshinoa) longicellanus Wals.	Z11-14:Ac	278, 286, 302
A. mortuanus Kearfott	Z9-14:Ac/Z11-14:Ac/E11-14:Ac/12:Ac = 1/90/10/200	184
A. semistructus Meyrick	Z11-14:Ac	278
A. viola Falkovitsh	E11-14:Ac/Z11-14:Ac = 7/3	131

Table 3 (continued)
SYNTHETIC SEX ATTRACTANTS FOR MALE LEPIDOPTERA FOUND BY
FIELD TESTS

Insect	Attractant and ratio[a]	Ref.
Argyroploce aurofasciana Haworth	E10-12:Ac	169
A. lacunana Den. & Schiff.	Z8-12:Ac	348
Argyrotaenia angustilineata Walsingham	Z11-14:Ac/Z9,E12-14:Ac = 1/1; or Z11-14:Ac/Z9-12:Ac = 1/1	131, 344
A. dorsalana Dyar	Z11-14:Ac; or Z11-14:Ac/Z11-14:OH = 1/1	276, 296
A. pulchellana Haworth	Z11-14:Ac/E11-14:Ac = 9/1	277, 345
A. quadrifasciana Fernald	Z10-12:Ac	347
Barbara colfaxiana Kearfott	Z9-12:OH	349
Cacoecimorpha pronubana Hübner	Z9-14:Ac; Z9-14:Ac/Z11-14:Ac = 4—40/100	207, 345, 350
Choristoneura biennis Freeman	E11-14:Ald	351
C. conflictana Walker	Z11-14:Ald	352
C. fractivittana Clemens	Z11-14:OH/Z11-14:Ac	281
C. lambertiana Busck	E11-14:Ac or E11-14:Ald	276
C. retiniana Walsingham	E11-14:Ac	351
C. sorbiana Hübner	Z11-14:OH	345
C. viridis Freeman	E11-14:Ac	276, 351
Clepsis furcana Walsingham	E11-14:Ac	276
C. melaleucana Walker	Z11-14:OH/Z11-14:Ac = 1/1	284
Cnephasia communana Herrich-Schäffer	Z8-12:Ac/E8-12:Ac	355
C. stephensiana Doubleday	12:OH/Z8-12:Ac/E8-12:Ac	277, 356, 357
Croesia askoldana Christoph	Z11-14:Ac/Z11-14:Ald = 5/5 to 1/9	344
C. bergmannia L.	E11-14:Ac	324
C. conchyloides Walsingham	Z11-14:Ald; or Z11-14:Ac/Z11-14:Ald = 1/9 to 0/10	131, 344
C. dealbata Yasuda	E11-14:Ald	131
C. holmiana L.	E11-14:Ac	149, 324
C. semipurpurana Kearfott	Z11-14:Ald/E11-14:Ald = 15/85	358, 462, 463
C. tigricolor Walsingham	E11-14:Ald/Z11-14:Ald = 9/1	131
Cryptophlebia ombrodelta Lowoer	Z8-12:Ac	302
Cydia (Pammene) gallicana Guenée	Z8-12:Ac	348
C. (Grapholitha) janthinana Duponchel	Z8-12:Ac/E8-12:Ac	356
C. medicaginis Kuzn.	E8,E10-12:Ac	160
C. nigricana F.	E10-12:Ac or E8,E10-12:Ac	359
C. tenebrosana Duponchel	Z8-12:Ac/E8-12:Ac	277
Ecdytolopha insiticiana Zeller	E8-12:Ac	284
Enarmonia formosana Scopoli	Z8-12:Ac/E8-12:Ac; or Z9-12:Ac/E9-12:Ac	277, 360
Endothenia benausopis Meyrik	Z10-14:Ac	278
E. ustulana Haworth	Z10-12:Ac	169
Epiblema desertana Zeller	Z8-12:Ac	284
E. foenella L.	Z8-12:Ac/E8-12:Ac	169, 277, 356
E. scudderiana Clemens	Z8-12:OH/Z8-12:Ac	284, 361
E. scutulana Den. & Schiff.	12:Ac/Z8-12:Ac/E8-12:Ac	277, 361
E. atistriga Clark	Z7-12:Ac	284
E. leucantha Meyrick	Z9-12:OH	131
E. zandana Kearfott	Z7-12:Ac	284
Epiphyas postvittana Walker	E11-14:Ac/E9,E11-14:Ac	362
Episimus argutanus Clemens	Z9-12:Ac	284
Eucosma campoliliana Den. & Schiff.	E8,E10-12:Ac	389
E. nigromaculana	E8,E10-12:OH	361
E. sonomana Kearfott	Z9-12:Ac/E9-12:Ac	363
Exartema appendiceum Zeller	E9-14:Ac	296

Table 3 (continued)
SYNTHETIC SEX ATTRACTANTS FOR MALE LEPIDOPTERA FOUND BY FIELD TESTS

Insect	Attractant and ratio[a]	Ref.
E. fusticanum McDunnough	Z7-12:OH	296
Grapholitha endrosias Meyrick	E7-12:Ac	131
G. funebrana Teitschke	Z8-12:Ac	391, 393, 430
G. inopinata Heinrich	Z8-12:Ac	278
G. janthinana Duponchell	Z8-12:Ac/E8-12:Ac = 3/1	356
G. (Prunivorana) lobarzewskii Rag.	Z8-12:Ac/E8-12:Ac = 1/9	430
G. packardii Zeller	E8-12:Ac	281
G. prunivora Walsingham	Z8-12:Ac/E8-12:Ac = 50/1	281
G. bolliana Slingerland	Z8-12:Ac/E8-12:Ac/12:OH = 100/7/1000	364
Hedya atropunctana Zeller	Z10-12:Ac	324
H. chinosema Zeller	E8-12:OH/E8-12:Ac = 1/2	284
Hedya ochroleucana Fröl.	Z8,E10-12:Ac	160
Homonopsis foederatana Kennel	Z11-14:Ac/E11-14:Ac = 1/1	131, 344
H. illotana Kennel	Z11-14:Ac/Z11-14:OH = 9/1	344
Hoshinoa adumbratana Walsingham	Z11-14:Ald	131
H. longicellanus Walsingham	Z11-14:Ac	278, 286, 302
Kawabea ignavana Christoph	Z7-12:Ac	131
K. razawskii Kawabe	Z5-12:Ac	278
Laspeyresia aurana F.	Z11-12:Ac	324
L. exquisitana Rebel	E8,E10-12:OH	473
L. youngana Kft.	E7-12:OH/Z7-12:OH = 9/1	349
Lobesia aeolopa Meyrick	Z7-10:Ac	302
Matsumuraeses ussuriensis Caradja	E8-12:Ac	131
Melissopus latiferranus Walsingham	E8,E10-12:Ac	365
Olethreutes humeralis Walsingham	Z8-12:OH/Z8-12:Ac = 1/1	131
Pammene albuginana Guenée	Z8-12:Ac	341
P. amygdalana Duponchel	E8-12:Ac/Z8-12:Ac = 9/1	430
P. argyrana Hübner	Z8-12:Ac/E8-12:Ac	348, 356, 361
P. aurantiana Staudinger	Z8-12:Ac	348
P. fasciana L.	Z8-12:Ac	348, 356, 361
P. inquilina Fletcher	Z8-12:Ac	348, 361
P. insulana Guenée	Z8-12:Ac	348
P. querceti Gozmany	Z8-12:Ac	348
P. nemorosa V. Kuznetsov	Z8-12:Ac	302
P. regiana Zeller	Z8-12:Ac	348
P. rhediella Clerck	Z8,E10-12:OH	366
P. splendidulana Guenée	Z8-12:Ac	348
P. spiniana Duponchel	Z8-12:Ac	348
P. suspectana Zeller	12:OH/Z8-12:Ac/E8-12:Ac	277
Pandemis borealis	Z11-14:Ac	276
P. canadana Kearfott	Z11-14:Ac/Z9-14:Ac = 9/1	296
P. chlorograpta Meyrick	Z11-14:Ac/E11-14:Ac = 9/1	344
P. cinnamomeana Treitschke	Z11-14:Ac/E11-14:Ac = 8/2	344
P. dumetana Teitschke	Z8-12:Ac	348
Parapandemis (Archepandemis) borealis Freeman	Z11-14:Ac	276
Petrova luculentana Heinrich	Z7-12:OH/Z7-12:Ac/E9-12:Ac	367
P. metallica Busk	Z7-12:OH/Z7-12:Ac	368
Pseudexentera maracana Kearfott	Z8-12:Ac	284
Ptycholoma circumclusana Christoph	Z11-14:OH	278

Table 3 (continued)
SYNTHETIC SEX ATTRACTANTS FOR MALE LEPIDOPTERA FOUND BY FIELD TESTS

Insect	Attractant and ratio[a]	Ref.
P. lecheanum	Z11-14:OH/Z11-14:Ac = 8/2	345
P. lecheana circumclusana Christoph	Z11-14:Ac/Z11-14:OH = 1/1 to 1/9	344
Rhopobota naevana Hübner	Z9-12:Ac	278
Rhyacionia busckana Heinr.	E8,E10-12:Ac	400
R. fumosana Powell	E8,E10-12:Ac	400
R. jenningsi	E9-12:Ac, or E9-12:Ac/Z9-12:Ac = 1/1	400
R. monophylliana	E9-12:Ac	400
R. multilineata	E9-12:Ac	400
R. neomexicana Dyar	Z7-12:Ac/E7-12:Ac/E9-12:Ac, E9-12:Ac/ Z9-12:Ac = 1/1, or E9-12:Ac	354, 400
R. salmonicolor Powell	Z9-12:Ac	400
R. zozana Kearfott	E9-12:Ac, or E9-12:Ac/Z9-12:Ac = 1/1	363, 400
Sparganothis acerivorana MacKay	Z11-14:Ald/E11-14:Ald	358
S. albicaudana Busck	E11-14:Ac	284
S. niveana groteana Wals.	Z11-14:OH	284
S. pilleriana Den. & Schiff.	Z11-14:Ald	131
S. sulfureana Clemens	E11-14:Ac	281
S. tunicana Walsingham	E11-14:Ac	276
Spatalistis bifasciana Hübner	Z11-14:Ac/Z9-14:Ac = 9/1 or 1/1	131, 344
Strophedra nitidana Fabr.	E8-12:OH or E8-12:Ac	278
Terricula violetana Kawabe	Z11-14:Ac/Z9-14:Ac = 1/1; or Z11-14:Ac/ Z9,E12-14:Ac = 1/1	131, 344
Thiodia alterana	E10-12:Ac	347
Tortricodes alternella Den. & Schiff.	Z8-12:Ac	389
Tortrix sinapina Butler	Z11-14:Ac/Z11-14:OH = 10/0 to 9/1	344
T. viridana sinapina Butler	Z11-14:Ac	131
Zeiraphera diniana Guenée (cembran pine form)	E9-12:Ac/E11-14:Ac	256, 373, 374
Xyloryctidae		
Rhyzosthenes falciformis Meyrick	Z7-12:Ac/Z7-12:Ald = 9/1	131
Yponomeutidae		
Ethmia monticola Walsingham	E11-14:Ac	276
Euhyponomeutoides trachydeltus Meyrick	Z11-14:Ac	131
Harpipteryx "xylostella" auct.	Z7-12:OH/Z7-12:Ac	284
Niphonympha anas Stringer	E11-14:Ac/E11-14:Ald	131
Yponomeuta evonymella L.	Z12-15:Ac/E12-15:Ac = 3/1	375
Y. padella (rorella) L.	Z11-14:Ac/E11-14:Ac; Z12-14:Ac/E12-14:Ac; or Z12-15:Ac/E12-14:Ac	376
Y. plumbella Den. & Schiff.	Z12-15:Ac/E12-14:Ac = 1/3	375
Y. yanagawanus Matsumura	Z11-14:Ac	278
Y. longus Moriuti	Z7-12:OH	131
Xyrosaris lichneuta Meyrick	Z11-14:Ac	131
Zygaenidae		
Doratopteryx plumigera Butler	Z11-14:Ac/E11-14:Ac	307
Staphylinochrous whytei Butler	Z11-14:OH/Z11-14:Ac/E11-14:Ac	307
Zygaena ephialtes L.	12:Ac/Z11-14:Ac/E11-14:Ac	277, 474
Z. hippocrepidis Hb.	Z11-14:Ac	474, 475
Z. transalpina Esper	E11-14:OH/Z11-14:Ac	372

[a] Abbreviation of compounds is the same as in Table 1.

Literature search stopped by September 1983.

Table 4
COMPONENTS OF MALE-PRODUCED SEX PHEROMONES AND MALE SCENTS

Insect and source	Compound detected, activity, and function	Ref.
Agaristidae		
Phalaenoides glycinae Lew., abdominal scent brush	2-Phenylethanol and its acetate	236
Arctiidae		
Creatonotos gangis L., coremata	7-Oxomethyl-2,3-dihydro-1*H*-pyrrolizin-1-ol	394
C. transiens Walker, coremata	7-Oxomethyl-2,3-dihydro-1*H*-pyrrolizin-1-ol	394
Utetheisa lotrix Cram., coremata	7-Oxomethyl-2,3-dihydro-1*H*-pyrrolizin-1-ol/1-oxomethyl-6,7-dihydro-5*H*-pyrrolizine	477
Utetheisa pulchelloides Hamps., coremata	7-Oxomethyl-2,3-dihydro-1*H*-pyrrolizin-1-ol/1-oxomethyl-6,7-dihydro-5*H*-pyrrolizine	477
Danaidae		
Amauris albimaculata, hairpencil	7-Methyl-2,3-dihydro-1*H*-pyrrolizin-1-one	245
A. echeria, hairpencil	7-Methyl-2,3-dihydro-1*H*-pyrrolizin-1-one	245
Amauris niavius, hairpencil	7-Methyl-2,3-dihydro-1*H*-pyrrolizin-1-one/3,4-dimethoxyacetophenone	245
A. ochlea, hairpencil	7-Methyl-2,3-dihydro-1*H*-pyrrolizin-1-one/octanal/nonanal/3-hexenoic acid/methyl salicylate/eugenol/*cis*-jasmone	245, 246
A. tartarea, hairpencil	3,4-Dimethoxyacetophenone	245
Danaus affinis affinis Fabr., hairpencil	7-Methyl-2,3,dihydro-1*H*-pyrrolizin-1-one/7-oxomethyl-2,3-dihydro-1*H*-pyrrolizine	247
Danaus chrysippus L., hairpencil	7-Methyl-2,3-dihydro-1*H*-pyrrolizin-1-one/3,7-dimethyl-(*E*)-2-octene-1,8-diol	248
D. chrysippus dorippus, hairpencil	7-Methyl-2,3-dihydro-1*H*-pyrrolizin-1-one	245
D. chrysippus petilea Stoll, hairpencil	7-Methyl-2,3-dihydro-1*H*-pyrrolizin-1-one	460
D. gilippus berenice Cramer, hairpencil	7-Methyl-2,3-dihydro-1*H*-pyrrolizin-1-one/3,7-dimethyl-(*E,E*)-2,6-decadien-1,10-diol; the first elicits female acceptance behavior	223—225
D. hamatus hamatus, hairpencil	7-Methyl-2,3-dihydro-1*H*-pyrrolizin-1-one/7-methyl-2,3-dihydro-1*H*-pyrrolizin-1-ol/lycopsamine	247, 461
D. limniace petiverana, hairpencil	7-Methyl-2,3-dihydro-1*H*-pyrrolizin-1-one	245
D. plexippus L., hairpencil	Benzyl caproate/1,5,5,9-tetramethyl-10-oxabicyclo[4.4.0]-3-decen-2-one or 2,2,6,8-tetramethyl-7-oxabicyclo[4.4.0]-4-decen-3-one/10-hydroxy-3,7-dimethyl-(*E,E*)-2,6-decadienoic acid/3,7-dimethyl-(*E,E*)-2,6-decadien-1,10-dioic acid	249, 457, 458

Table 4 (continued)
COMPONENTS OF MALE-PRODUCED SEX PHEROMONES AND MALE SCENTS

Insect and source	Compound detected, activity, and function	Ref.
Euploea sylvesta sylvesta Fabr., hairpencil	7-Oxomethyl-2,3-dihydro-1*H*-pyrrolizin-1-ol	247
E. tulliola tulliola Fabr., hairpencil	7-Oxomethyl-2,3-dihydro-1*H*-pyrrolizin-1-ol/lycopsamine	247, 461
Lycorea ceres ceres Cramer, hairpencil	7-Methyl-2,3-dihydro-1*H*-pyrrolizin-1-one/hexadecyl acetate/(Z)-11-octadecenyl acetate	250, 251
Geometridae		
Bapta temerata Schiff., abdominal scent brush	*n*-Butyric acid	237
Ithomiidae		
Godyris kedema Hewitson, wing costal fringe	γ-Lactone of hydroxytrachelanthic acid	459
Hymenitis dercetis Herrich-Schaeffer, wing costal fringe	γ-Lactone of hydroxytrachelanthic acid	459
H. andromica Hewitson, wing costal fringe	γ-Lactone of hydroxytrachelanthic acid	459
Ithomia iphianassa Doubleday and Hewitson, wing costal fringe	γ-Lactone of hydroxytrachelanthic acid	459
Pteronymia beebei Fox & Fox, wing costal fringe	γ-Lactone of hydroxytrachelanthic acid	459
P. nubivage, wing costal fringe	γ-Lactone of hydroxytrachelanthic acid	459
P. veia, wing costal fringe	γ-Lactone of hydroxytrachelanthic acid	459
Lycaenidae		
Lycaeides argyrognomon Bgstr., wing scale	Nonanal/hexadecanal/hexadecyl acetate/δ-cadinol	252
Noctuidae		
Agrocola helvola, abdominal scent brush	2-Phenylethanol	238
Apamea monoglypha Huf., abdominal scent brush	Pinocarvone	237

Table 4 (continued)
COMPONENTS OF MALE-PRODUCED SEX PHEROMONES AND MALE SCENTS

Insect and source	Compound detected, activity, and function	Ref.
Cucullia umbra-tica, abdominal scent brush	2-Methylpropanoic acid/2-methylbutanoic acid	476
Leucania coni-gera Schiff., abdominal scent brush	Benzaldehyde/*iso*-butyric acid	237, 239
L. impura Hübner, abdominal scent brush	Benzaldehyde/*iso*-butyric acid	237, 239
L. pallens L., abdominal scent brush	Benzaldehyde/*iso*-butyric acid	237
Mamestra brassi-cae L., abdominal scent brush	2-Phenylethanol/benzylalcohol	238
M. configurata Walker, abdominal scent brush	2-Phenylethanol	250
M. persicariae L., abdominal scent brush	2-Phenylethanol/benzylalcohol/phenylacetaldehyde/benzaldehyde	237
Peridroma sau-cia Hübner, abdominal scent brush	2-Phenylethanol	241
Phlogophora me-ticulosa, abdominal scent brush	6-Methyl-5-hepten-2-one/6-methyl-5-hepten-2-ol/2-methylbutanoic acid	237, 239
P. nebulosa Huf., abdomi-nal scent brush	Benzaldehyde/2-phenylethanol/benzylalcohol	237
Polia tincta, ab-dominal scent brush	2-Phenylethanol/phenylacetaldehyde	238
Pseudaletia se-parata Walker, abdominal scent brush	Benzylalcohol/benzaldehyde/benzoic acid/*n*-butanol benzaldehyde; in-hibit female movement (or inhibit other male approach)	219, 242, 431, 432
P. unipuncta Haworth, abdominal scent brush	Benzaldehyde/benzylalcohol; inhibit other male approach	220, 221, 242, 243
Trichoplusia ni Hübner, genital scent brush	2-Phenylethanol; wing flutter and elevation of abdominal tip of female	222

Table 4 (continued)
COMPONENTS OF MALE-PRODUCED SEX PHEROMONES AND MALE SCENTS

Insect and source	Compound detected, activity, and function	Ref.
Papilionidae		
Atrophaneura alcinous alcinous, wing scale	Benzaldehyde/phenylacetaldehyde/2-phenylpropanal/*h*-heptanal/6-methyl-hept-5-en-2-one/linalool	380
Pieridae		
Colias eurytheme, wing	13-Methylheptacosane; female acceptance behavior	226
C. philodice, wing	Hexyl myristate/hexyl palmitate/hexyl stearate; the 2nd component is the major, elicited female acceptance behavior	226
Pieris melete Ménétriès, wing scale	Decane/undecane/dodecane/tridecane/neral/ geranial/nerol/geraniol/α-pinene/β-pinene/ myrcene/*p*-cymene/limonene	253, 254
P. napi L., wing scale	Neral/geranial	255
P. napi japonica Shirôzu, wing scale	Undecane/α-pinene/myrcene/β-pinene/*p*-cymene/limonene/neral/geranial/linalool	253
P. rapae crucivora, wing scale	3-Methylbutanal/phenylacetaldehyde/2-phenylpropenal/methyl salicylate/indol/methyl palmitate/methyl *n*-heptadecanoate/methyl linolenate/methyl linolate/methyl stearate; indole, methyl salicylate, and unidentified 3 compounds are male specific	379
Pyralidae		
Achroia grisella Fabr., aerial collection	Undecanal/Z-11-octadecenal 100/1 from male, attractive for female	227
Aphomia gularis Zeller, aerial collection	2-Phenylethanol/*cis*-2,6-nonadien-4-olide; short-range attraction for female, synthetic racemate of the 2nd has no activity	228
Eladana saccharina Walker, wing gland and hairpencil	3,7-Dimethyl-6-octen-4-olide/vanillin/*p*-hydroxybenzaldehyde; the 1st component in wing gland attracts female	229—231
Galleria mellonella L., aerial collection	Undecanal/nonanal; 2/5 from male, sexual stimulation and short-range attraction for female	232, 233
Tortricidae		
Grapholitha molesta Busck., hairpencil	Ethyl *trans*-cinnamate/methyl 2-epijasmonate/methyl jasmonate/*R*-(−)-mellein; mixtures of 1st and 2nd or 1st, 3rd, and 4th act as a short-range attractant	234, 235

Literature search stopped by September 1983.

hair pencils, the source of the ketone. The active ketone functions by inhibiting locomotion in the female, promoting successful mating by suppressing her escape-flight. In the case of a pierid butterfly, *Colias eurytheme,* 13-methylheptacosane derived from wing scales elicit female acceptance behavior. Males of another related species, *C. philodice,* utilize a three-component mixture of fatty acid esters as a signal to conspecific females (Table 4). In these pierid butterflies the male scents apparently act as species-specific aphrodisiacs. Similar functions for aphrodisiac pheromones have also been reported for *Pseudaletia separata* and *Trichoplusia ni.* Benzaldehyde was identified from the abdominal brush organs of male *P. separata* and may function as an arrestant though the results are inconclusive.[219] Females of the cabbage looper moth, *T. ni*, have been reported to show wing fluttering and elevation of the abdominal tip in response to 2-phenylethanol, which was reported to occur in male genital scent brush.[222] However, according to other reports, the presence of 2-phenylethanol in the cabbage looper moth is doubtful.[412]

Apparent short-range sexual stimulation of females by a a male scent has been reported in four pyralid moths and one torticid moth. Males of the lesser wax moth, *Achroia grisella* release undecanal and (Z)-11-octadecenal that, in combination with sound, elicit short-range orientation behavior in conspecific females.[227] The greater wax moth, *Galleria mellonella*, utilizes a mixture of undecanal and nonanal which elicit a sexual stimulation and a short-range attraction for females.[232,233] 2-Phenylethanol and *cis*-2,6-nonadien-4-olide identified from the aerial collection from males of another pyralid moth, *Aphomia gularis*, attract females within a short range.[228] Similarly, 3,7-dimethyl-6-octen-4-olide from male wing glands of *Eladana saccharina* attracts conspecific females.[229-231] A particular combination of scent substances such as ethyl *trans*-cinnamate, methyl 2-epijasmonate, methyl jasmonate, and R-(−)-mellein from male hairpencils of the oriental fruit moth, *Grapholitha molesta* (Tortricidae), also attracts sex pheromone-releasing females from a distance of several centimeters.[234,235]

In the case of *Pseudaletia unipuncta*, benzaldehyde and benzylalcohol from male abdominal scent brushes inhibited the approach of other males to calling females.[220,221] A similar effect of benzaldehyde was also noticed in *P. separata*.[431] These data suggest that the first male that comes to a calling female may deter other competitive males from approaching the pheromone-releasing female. We await the confirmation of similar functions for male scents of other lepidopterous species, especially those possessing similar scent chemicals. In the cabbage armyworm, *Mamestra brassicae*, the male scent was observed to have only a slight effect, if any, on male responsiveness.[451]

These results indicate that male scents of the Lepidoptera function (1) as aphrodisiac pheromones to inhibit locomotion or elicit acceptance behavior in the female, (2) as attractant pheromones to elicit short-range orientation of females to males, and (3) as repellent pheromones to deter other competing males.

BIOSYNTHESIS AND SECRETION

To understand the specificity of sex pheromones, biosynthetic pathways and regulatory mechanisms are important targets for investigation. In addition, to understand the mechanisms of pheromonal communication in the Lepidoptera, various endogenous and exogenous factors that affect the production and release of pheromones have been investigated. This section summarizes the main endogenous biosynthetic pathways leading to the production and release of sex pheromones.

The utilization of plant components as lepidopterous pheromone sources has been demonstrated only for the aphrodisiac pheromone, 2,3-dihydro-7-methyl-1H-pyrrolizin-1-one, of the male queen butterfly, *Danaus gilippus berenice*. Behavioral observations and chemical analyses on butterflies fed on a diet with or without plant materials apparently suggest that the pyrrolizidinone pheromone is derived from pyrrolizidine alkaloids of particular plants.[223,225] This is supported by the evidence of no incorporation of radioactivity to the pheromone by using Na (2-^{14}C)acetate, Na (2-^{14}C)mevalonate, (5-^{14}C)ornithine, and Na (2-^{14}C)α-aminolevulinic acid.[414] Most of the female-produced sex pheromones are believed to be synthesized by females themselves.

Although *de novo* synthesis of pheromonal components has been suggested for several species,[414-416] the utilization of the major fatty acyl moieties of the body lipids for biosynthesis of pheromonal components seems to make economical use of body lipids. In the case of nonanal and undecanal, the male pheromonal components of *Galleria mellonella*, the *de novo* synthetic pathways do not appear to be of primary importance. Nonanal was found to incorporate a much larger amount of radiolabeled carbon from $(10\text{-}^{14}C)$oleic acid than from $(1,2\text{-}^{14}C)$acetic acid and Na $(3\text{-}^{14}C)$propionate.[416] Desaturation, chain-shortening, and chain elongation of fatty acyl compounds could produce various pheromonal components. The sex pheromone of the silkworm moth, (E,Z)-10,12-hexadecadienol, was produced in vivo from a $(1\text{-}^{14}C)$palmitate.[417] This may proceed through (Z)-11-hexadecenoate which was detected in the female abdominal gland.[418] On the other hand, hexadecanyl phosphate was dehydrogenated in vitro, at the 10th and 12th carbons by an oxygenase, which occurs in the abdominal tips of female silkworms during the last pupal stage.[417]

A study of the biosynthesis of the pheromone components of the redbanded leafroller moth, *Argyrotaenia velutinana,* indicated that a 16-carbon acyl compound was shortened to a 14-carbon acyl compound, dehydrogenated at position 11 to produce 11-tetradecenoate, and then reduced and acetylated to 11-tetradecenyl acetate.[419] The ratio of Z to E in the tetradecenoyl moieties in triglycerides and phospholipids was different from the Z/E ratio observed from the pheromonal components.[419] This suggests that the regulation of the ratio of Z and E pheromonal components occurs just prior to the reduction of the tetradecenoyl moieties and subsequent acetylation.

The glandular epithelium of the pheromone gland seems to be an important site of biosynthesis of pheromonal compounds.[414] Female sex pheromone glands of almost all moth species are situated in the intersegmental membrane between the eighth and ninth abdominal segments. These glands are eversible folds or eversible sacs, dorsally or ventrally located in the intersegmental membrane. In some species, the gland has a protrusible ring on the membrane around the abdomen. These glands are normally folded between the two abdominal segments. At the time of calling, females evert or protrude the glands and expose the gland surface to the air to release the pheromonal compounds which have been secreted onto the glandular surface. The detailed mechanisms of secretion of the pheromonal compounds onto the glandular surface are not yet known. Histological and histochemical observations on various lepidopterous pheromone glands suggest a direct association between biosynthesis and secretion. It has also been suggested that different glandular tissues are specialized to secrete particular compounds or the mixtures of pheromonal compounds of multicomponent sex pheromone.[420]

Biosynthesis and secretion of pheromonal compounds are under the control of various physiological and environmental factors. Of these, female age and reproductive status, time of day, food, and host factors are of prime importance.[421] Pheromone biosynthesis usually starts in the pharate stage, whereupon the titer increases for a definite time up to a peak level, and then declines with advancing age of the females. In species with an extended adult life, sex pheromone production decreases for a definite period after mating and then increases again. Thus, biosynthesis and release of sex pheromone coincide with the reproductive status of each species. The calling behavior, a pheromone-releasing posture, of female insects is usually restricted to a definite time of the day. Periodicity of calling is endogenously controlled by a circadian rhythm and also controlled exogenously by photoperiod and ambient temperature. The amount of pheromone in virgin females shows a peak just prior to the onset of calling. Once calling has commenced, the quantity drops rapidly and reaches a minimum value during the calling period, probably reflecting a fast emission of pheromone from the gland surface.[422] However, some species do not show any significant diel pattern of sex pheromone titer.[423-425]

Larval diet or host plant could indirectly affect the production of sex pheromone through the nutritional status of growing insects. Only a few investigations have established this in the Lepidoptera. In the case of the smaller tea tortrix moth, *Adoxophyses* sp., reared on a meridic diet composed of soybean powder and tea leaf powder as major ingredients, sex pheromone titers of females reared on a tea-leaf-free diet were approximately 6000 times less active than the extracts of females reared on the standard diet.[426] The tea-leaf-free diet is apparently not suitable for the production of sex pheromone in this species. However, growth rate, percentage of pupae obtained, weight of pupae of both sexes, and percentage of emergence are not significantly different from those obtained on the standard diet.[427] This suggests that quality of food may significantly affect the biosynthesis of sex pheromones even when it does not affect growth and development.

UTILIZATION IN PEST MANAGEMENT

The strong biological activity of insect pheromones, especially of sex pheromones, has attracted the attention of many scientists. They have recognized the potential of utilizing these chemicals to monitor and to manipulate pest populations. A great deal of effort has been concentrated to develop technologies for the practical application of sex pheromones to pest control as has been shown in recent reviews.[486-494] Applications for lepidopterous sex pheromones are usually categorized as follows: (1) monitoring and survey for early detection of introduced foreign insects for forecasting pest outbreaks and for estimating population density; (2) mass trapping for population suppression and detailed monitoring; and (3) communication disruption to inhibit mate-finding and to suppress the population.

Application to Monitoring and Survey

Recent developments in transportation have increased the possibility of accidental introduction of important foreign pest insects. Traps baited with synthetic sex pheromones can be used to monitor for these pests. Expansion of the distribution of particular insects can also be monitored by pheromone traps. Routine trapping systems for particular target species are apparently valuable for the early detection of their presence, which allows fast control. International cooperation and exchange of information is desirable for the establishment of routine monitoring systems for these important pest insects.

Forecasting the time of pest occurrence and estimating their population densities are important components in integrated pest management. Seasonal trapping of male moths with pheromone traps has been routinely conducted on various insects, and many pheromone formulations are now commercially available. Analysis of accumulated seasonal trapping data for a particular target species provides the information necessary for forecasting the time of peak occurrence. Estimates of the developmental threshold and thermal constants for each species can be used to estimate the rate of development under the particular temperature conditions.[495-497]

Pheromone trap records can be used to forecast the degree of pest outbreaks and to decide whether to apply insecticides. For this purpose, the relationship between the trap catch of the adult insects during a given period and the rate of injury caused by the larvae in the subsequent generation must be elucidated. Unfortunately, trap catches do not necessarily reflect the population density of adult stages in any strict sense. Numerous factors influencing not only the traps but also the responding insects, often adversely affect the relationship between trap catches and adult population densities. Although these factors have been the target of detailed investigations, often simple correlations between trap catches and population densities or pest damage are analyzed.[498,499] For the correct estimation of the control threshold density of a target pest species, however, construction of a system model which describes the relation between crop yield and the number of adult insects caught by pheromone traps might be recommended.[500] Parameters used in this type of model are the number of insects caught per trap, the number of eggs laid per plant, the rate of development from hatch to the instar that causes serious damage to the plant, the larval survival rate, and the injury threshold density of the larvae.

Important topics for further investigation to advance the use of sex pheromone traps to monitor pest population levels are as follows: (1) seasonal stability of attractant lures; (2) trap structure and trap location for stable trapping efficiency; (3) immigration of adult insects; (4) optimum trap density and the distribution of traps in and around the target area; (5) correlation between trap catches and oviposition in the subsequent generation; and (6) construction of a system model relating trap catches and control threshold density.

Mass Trapping

Mass trapping is one of the potential applications of sex pheromones to pest control. Since most of the lepidopterous attractant pheromones attract males, competition between traps and wild females for males is one of the most important factors affecting the effectiveness of this method. Theoretical calculations suggest that a trap:female ratio of 5:1 would be needed to obtain 95% mating suppression for lepidopterous insects.[501,502] Thus, the effect of mass trapping depends on the population density of adults in the field. In addition, the size of the experimental plot would seriously affect the effectiveness of mass trapping because influx of gravid females from outside the plot will adversely affect the results of mass trapping. The flight distances of gravid females should be determined for each target species, however this is not known for most of the species so far investigated. Optimal trap density for effective mass trapping might be obtained from flight distances of male insects. On the other hand, this can be determined from the effective trapping radius of a single trap. This effective trapping radius could be determined by a simulation model based on the recovery of marked males released at various distances from the trap.[503]

Although many attempts to mass trap lepidopterous insects have been conducted, the largest tests to date are those on *Spodoptera littoralis* in Israel[504,505] and *S. litura* in Japan.[507] In the 1977 test with *S. littoralis* on 1,800 ha with a trap density of 0.9 trap per hectare, capture of 4,171,100 males reduced insecticide treatment by 26% and reduced the number of egg clusters by 40%. Similar tests conducted with *S. litura* from 1977 to 1978 on 740 ha with a trap density of 1 trap per hectare suppressed damage on taro to less than 10% in more than 96% of the treated area, while about 45% of the check area suffered between 31 to 51% damage. In May of 1977, the Ministry of Agriculture, Forestry and Fisheries of Japan (MAFF) registered the attractant formulation, known commercially as Pherodin SL (Takeda Chem. Ind.), as a control agent for this insect.

The mass trapping method is usually applied on a large area to avoid influx of gravid females. However, it can be effective in very small areas in particular cases. One example is experiments on *Adoxophyes* sp. conducted in Japan. Trap-out of males with 35 traps in a 0.08-ha tea field suppressed leaf damage to 50 to 60% of the check field which had conventional insecticide treatments.[508] Similar trials on a 0.06-ha plot suppressed damage to 11 to 25% that of the check.[509] Analysis of 11 trials on this species gave a regression between the number of males per trap (X) and the percentage reduction of larval density (Y) as $Y = 95.42 - 0.42X$, $r^2 = 0.83$.[492] Although the efficacy of mass trapping would vary with species, this method can be used as a control measure under certain circumstances especially when the population is low. In addition, this technique is suitable for detailed monitoring of populations of target species, to make it easier to find locations where additional control measures are needed. Important subjects to be studied for improved application of mass trapping are as follows: (1) pheromone formulation for optimum attractiveness; (2) trap structure for the optimum male catch and for easy handling; (3) trap location for the optimum male catch; (4) minimum field area for successful mass trapping; (5) trap density and distribution pattern of traps in and around the plot; (6) determination of the threshold density of adults for the success of this method.

Communication Disruption

Another application of sex pheromones for the control of pest insects is the disruption of sexual communication between males and females during mating. In general, a synthetic sex pheromone is evaporated and evenly distributed in the air in and around the habitat of the target insects. This idea was proposed[510] just after the first identification of an insect sex pheromone, bombykol, of the silkworm moth.[28] The feasibility of this method was established for the first time on the cabbage looper, *Trichoplusia ni,* by Shorey et al.[511] This was followed by extensive research on the pink bollworm, *Pectinophora gossypiella,* and led to the first registration of a communication disruptant, known commercially as Gossyplure H.F. (Conrel Co.), by the U.S. EPA as a control agent for this species in February of 1978. Numerous experiments on communication disruption have been conducted on various lepidopteran species. Roelofs[486] listed examples of disruption tests on 21 lepidopteran species and Rothschild[512] tabulated examples of 39 species. A typical example of this kind of experiment is the pink bollworm.[513,514] A disruption experiment in 1977 in California, using 16 g/ha-season Gossyplure in 214 ha, resulted in 55% lower insect control costs, compared to conventional practice.[513]

The mechanism of communication disruption by sex pheromone is still largely speculative. By reviewing much research, Bartell[491] categorized possible mechanisms as follows: (1) neurophysiological effects on the peripheral and central nervous system; (2) false trail-following resulting in "confusion" of males; (3) inability of males to distinguish individual odor trails from the odor background; (4) imbalance in the pattern of sensory input; and (5) modification of pheromone response by related chemicals. Fundamental studies on insect behavior and the mechanism of pheromonal communication are highly desirable for developing effective disruptants and application technologies.

Pheromone mimics are promising candidates for disruptants, especially when the pheromone is unstable. In addition, pheromone mimics and particular pheromone components could be used as a common disruptant for multiple target species. For pheromonal control of a pest insect complex on a particular agricultural crop, the high species specificity of sex pheromones is an inconvenient characteristic. However, many taxonomically related species use a common chemical as the key component of their pheromones. This key component or mixtures with other additives could be used as a common disruptant for several species. This has been evaluated on the two tortricid moth species attacking tea plant.[515,517] (Z)-11-tetradecenyl acetate is a common pheromonal component for *Adoxophyes* sp. and *Homona magnanima,* the two major leaf feeders of tea plants.[174-176,203] This compound is a good disruptant for both species and was recently registered in Japan as a control agent for both species.

In addition to the selection of chemicals as disruptants, there are many factors affecting the efficiency of communication disruption. Most of these are the same as those affecting mass trapping. Population density and the size of experimental plots are important factors for the successful application of this method.

Subjects for further investigation for increased application of communication disruption are as follows: (1) selection of disruptant chemicals or mixtures of chemicals based on behavioral analysis of target species; (2) formulation of disruptants; (3) application methods of the disruptant; (4) minimum area of treatment; and (5) determination of threshold density for successful use.

CONCLUSION

Most lepidopterous sex pheromones consist of multiple components of aliphatic compounds with relatively simple structure. Although a definite relationship between taxonomy and pheromone chemicals seems to exist, accumulation of much more data is needed for detailed discussion. Pheromone specificity principally depends on the blend of multiple components. The regulatory mechanisms during biosynthesis of these components are an interesting target for further investigation of pheromonal specificity. To better understand pheromone signaling it is necessary to investigate in detail various endogenous and exogenous factors affecting the production and release of pheromones. Despite continuing research and successful practical application of a few sex pheromones, a variety of problems remain to be solved before establishing pheromones as an important control technique in pest management. Behavioral responses of male moths to pheromonal chemicals under natural conditions are one of the most important problems to be analyzed. It is not necessary to consider pheromones as substitutes for insecticides. Pheromonal control should be utilized as a preventive measure against the outbreak of particular key pests. Final evaluation of pheromonal application should be conducted when integrated with other control measures.

ACKNOWLEDGMENTS

The author thanks to D. A. Andow, H. Arn, J. R. Clearwater, C. Descoins, D. R. Hall, K. Hirai, K. Honda, B. G. Kovalev, Y. Kuwahara, S. Takahashi, S. Wakamura, E. W. Underhill, and P. Zagatti for their helpful suggestions.

REFERENCES

1. **Karlson, P. and Lüscher, M.,** ''Pheromones'': a new term for a class of biologically active substances, *Nature (London),* 183, 55, 1959.
2. **Karlson, P. and Butedandt, A.,** Pheromones (ectohormones) in insects, *Ann. Rev. Entomol.,* 4, 49, 1959.
3. **Jacobson, M.,** *Insect Sex Attractants,* John Wiley & Sons, New York, 1965, 154.
4. **Ishii, S.,** *Insect Physiologically Active Substances,* Nankodo, Tokyo, 1969, 196.
5. **Jacobson, M.,** *Insect Pheromones,* Academic Press, New York, 1972, 382.
6. **Birch, M. C., Ed.,** *Pheromones,* American Elsevier, New York, 1974, 495.
7. **Suzuki, K.,** *Pheromones: Semiochemicals in Animals,* Sankyo Shuppan, Tokyo, 1974, 246.
8. **Pimentel, D., Ed.,** *Insect, Science, & Society,* Academic Press, New York, 1975, 284.
9. **Shorey, H. H.,** *Animal Communication by Pheromone,* Academic Press, New York, 1976, 167.
10. **Yushima, T.,** *Insect Pheromones,* Tokyo University Press, Tokyo, 1976, 166.
11. **Shorey, H. H. and McKelvey, J. J., Jr., Ed.,** *Chemical Control of Insect Behavior: Theory and Application,* John Wiley & Sons, New York, 1977, 414,
12. **Ishii, S., Hirano, C., Tamaki, Y., and Takahashi, S.,** *Chemistry of Insect Behavior,* Baifukan, Tokyo, 1978, 242.
13. **Ritter, F. J., Ed.,** *Chemical Ecology: Odour Communication in Animals,* Elsevier, Amsterdam, 1979, 427.
14. **Roelofs, W. L., Ed.,** *Establishing Efficacy of Sex Attractants and Disruptants for Insect Control,* Entomol. Soc. Am., College Park, Md., 1979, 97.
15. **Kloza, M., Ed.,** *Regulation of Insect Development and Behaviour,* Parts 1 and 2, Tech. Univ. Wroclaw, Karpacz, 1981, 1146.
16. **Mitchell, E. R., Ed.,** *Management of Insect Pests with Semiochemicals,* Plenum Press, New York, 1981, 514.
17. **Nakamura, K. and Tamaki, Y.,** *Sex Pheromones and Insect Pest Management: Experiments and Feasibility,* Kokin-Shoin, Tokyo, 1983, 202.
18. **Roelofs, W. L. and Comeau, A.,** Sex pheromone perception: synergists and inhibitor for the red-banded leafroller attractant, *J. Insect Physiol.,* 17, 435, 1971.
19. **Tamaki, Y.,** Insect sex pheromone and species speciation, *Seibutsu-Kagaku,* 24, 119, 1972.
20. **Roelofs, W. L. and Cardé, R. T.,** Sex pheromones in the reproductive isolation of lepidopterous species, in *Pheromones,* Birch, M. C., Ed., American Elsevier, New York, 1974, 96.
21. **Renou, M., Descoins, C., Priesner, E., Gallois, M., and Lettere, M.,** Etude de la phéromone sexuelle de la teigne du poireau *Acrolepiopsis assectella, Entomol. Exp. Appl.,* 29, 198, 1981.
22. **Rahm, R. and Renou, M.,** Vers l'utilisation d'une phéromone de synthèse, L'hexadécène 11Z,A1 1 dans la lutte contre la teigne du poireau, *Acrolepiopsis* (ex. *Acrolepia*) *assectella* Z., Lepidoptera, Plutellidae, *C. R. Acad. Sci. Agric. France,* p.759, 1979.

23. **Tumlinson, J. H., Yonce, C. E., Doolittle, R. E., Heath, R. R., Gentry, C. R., and Mitchell, E. R.,** Sex pheromones and reproductive isolation of the lesser peachtree borer and the peachtree borer, *Science,* 185, 614, 1974.

24. **Hill, A. S. and Roelofs, W. L.,** Sex pheromone of the saltmarsh caterpillar moth, *Estigmene acrea, J. Chem. Ecol.,* 7, 655, 1981.

25. **Roelofs, W. L. and Cardé, R. T.,** Hydrocarbon sex pheromone in tiger moths (Arctiidae), *Science,* 171, 684, 1971.

26. **Hill, A. S., Kovalev, B. G., Nikolaeva, L. N., and Roelofs, W. L.,** Sex pheromone of the fall webworm moth, *Hyphantria cunea, J. Chem. Ecol.,* 8, 383, 1982.

27. **Conner, W. E., Eisner, T., Vander Meer, R. K., Guerrero, A., Chiringelli, D., and Meinwald, J.,** Sex attractant of an arctiid moth (*Utethesia ornatrix*): a pulsed chemical signal, *Behav. Ecol. Sociobiol.,* 7, 55, 1980.

28. **Butenandt, A., Beckman, R., Stamm, D., and Hecker, E.,** Uber den Sexuallockstoff des Seidenspinners *Bombyx mori.* Reidarstellung und Konstitution, *Z. Naturforsch.,* B14, 283, 1959.

29. **Tamaki, Y., Honma, K., and Kawasaki, K.,** Sex pheromone of the peach fruit moth, *Carposina niponensis* Walsingham (Lepidoptera: Carposinidae): isolation, identification and synthesis, *Appl. Entomol. Zool.,* 12, 60, 1977.

30. **Honma, K., Kawasaki, K., and Tamaki, Y.,** Sex-pheromone activities of synthetic 7-alken-11-ones for male peach fruit moths, *Carposina niponensis* Walsingham (Lepidoptera: Carposinidae), *Jpn. J. Appl. Entomol. Zool.,* 22, 87, 1978.

31. **Arn, H., Rauscher, S., Buser, H.-R., and Roelofs, W. L.,** Sex pheromone of *Eupoecilia ambiguella: cis*-9-dodecenyl acetate as a major component, *Z. Naturforsch.,* 31c, 499, 1976.

32. **Rauscher, S., Arn, H., and Guerin, P.,** Effects of dodecyl acetate and Z-10-tridecenyl acetate on attraction of *Eupoecilia ambiguella* males to the main sex pheromone component, Z-9-dodecenyl acetate, *J. Chem. Ecol.,* 10, 253, 1984.

33. **Arn, H.,** Some recent problems in analysis and formulation of Lepidoptera sex phéromones, *Les Phéromones Sexuelles,* Les Colloques de l'INRA, Paris, 1981, 27.

34. **Pasqualini, E., Bortolotti, A., Maini, S., Baronio, P., and Campadelli, G.,** *Cossus cossus* L. (Lep., Cossidae) male catches in Emilia-Romagna (Italy) with synthetic pheromones, Les Colloques de l'INRA, Paris, 1982, 399.

35. **Doolittle, R. E., Roelofs, W. L., Solomon, J. D., Cardé, R. T., and Beroza, M.,** (Z,E)-3,5-tetradecadien-1-ol acetate, sex attractant for the carpenter worm moth, *Prionoxystus robiniae* (Peck) (Lepidoptera: Cossidae), *J. Chem. Ecol.,* 2, 399, 1976.

36. **Roelofs, W., Kochansky, J., Anthon, E., Rice, R., and Cardé, R.,** Sex pheromone of the peach twig borer moth (*Anarsia lineatella*), *Environ. Entomol.,* 4, 580, 1975.

37. **Hirano, C., Muramoto, H., and Horiike, M.,** Sex pheromone produced by female potato leaffolder moth, *Naturwissenschaften,* 63, 439, 1976.

38. **Hummel, H. E., Gaston, L. K., Shorey, H. H., Kaae, R. S., Byrne, K. J., and Silverstein, R. M.,** Clarification of the chemical status of the pink bollworm sex pheromone, *Science,* 181, 873, 1973.

39. **Bierl, B. A., Beroza, M., Staten, R. T., Sonnet, P. E., and Adler, V. E.,** The pink bollworm sex attractant, *J. Econ. Entomol.,* 67, 211, 1974.

40. **Roelofs, W. L., Kochansky, J. P., Cardé, R. T., and Kennedy, G. G.,** Sex pheromone of the potato tuberworm moth, *Phthorimaea operculella, Life Sci.,* 17, 699, 1975.

41. **Yamaoka, R., Fukami, H., and Ishii, S.,** Isolation and identification of the female sex pheromone of the potato tuberworm moth, *Phthorimaea operculella* (Zeller), *Agric. Biol. Chem.,* 40, 1971, 1976.

42. **Persoons, C. J., Voerman, S., Verwiel, P. E. J., Nooijen, P. J. F., Nooijen, W. J., Ritter, F. J., and Minks, A. K.,** Sex pheromone of the potato tuberworm moth, *Phthorimaea operculella:* isolation and identification, *Meded. Fac. Landbouwwet. Rijksuniv. Gent,* 41, 945, 1976.

43. **Persoons, C. J., Voerman, S., Verwiel, P. E. J., Ritter, F. J., Nooijen, W. J., and Minks, A. K.,** Sex pheromone of the potato tuberworm moth, *Phthorimaea operculella:* isolation, identification, and field evaluation, *Entomol. Exp. Appl.,* 20, 289, 1976.

44. **Renou, P. M., Descoins, C., Lallemand, J. Y., Priesner, E., Lettere, M., and Gallois, M.,** L'acétoxy-1 dodécène 3E, composant principal de la phéromone sexuelle de la teigne de la betterave: *Scrobipalpa ocellatella* Boyd. (Lépidoptère, Gelechiidae), *Z. Angew. Entomol.,* 90, 275, 1980.

45. **Vick, K. W., Su, H. C. F., Woser, L. L., Mahany, P. G., and Drummond, P. C.,** (Z,E)-7,11-hexadecadien-1-ol acetate: the sex pheromone of the angoumois grain moth, *Sitotoroga cerealella, Experientia,* 30, 17, 1974.

46. Laboratory of Insect Pheromone, Isolation, identification and synthesis of EAG-active components of the pine caterpillar moth sex pheromone, *Kagaku-tsuho (China),* p.1004, 1979.

47. **Vu, M. H., Ando, T., Yoshida, S., Takahashi, N., Tatsuki, S., Katagiri, K., Yamane, A., and Ikeda, T.,** 5,7-Dodecadien-1-ol: candidate for the sex pheromone of the pine moth, *Agric. Biol. Chem.,* 43, 1615, 1979.

48. **Vu, M. H., Ando, T., Takahashi, N., Tatsuki, S., Yamane, A., Ikeda, T., and Yamasaki, S.,** Identification of the female pheromone of the pine moth, *Agric. Biol. Chem.,* 44, 231, 1980.

49. **Ando, T., Vu, M. H., Yoshida, S., Takahashi, N., Tatsuki, S., Katagiri, K., Yamane, A., Ikeda, T., and Yamasaki, S.,** (5Z,7E)-5,7-dodecadien-1-ol: female sex pheromone of the pine moth *Dendrolimus spectabilis* Butler, *Agric. Biol. Chem.,* 46, 709, 1982.

50. **Underhill, E. W., Chisholm, M. D., and Steck, W.,** (E)-5,(Z)-7-dodecadienal, a sex pheromone component of the western tent caterpillar, *Malacosoma californicus* (Lepidoptera: Lasiocampidae), *Can. Entomol.,* 112, 629, 1980.

51. **Chisholm, M. D., Underhill, E. W., Steck, W., Slessor, K. N., and Grant, G. G.,** (Z)-5,(E)-7-dodecadienal and (Z)-5,(E)-7-dodecadien-1-ol, sex pheromone components of the forest tent caterpillar, *Malacosoma disstria, Environ. Entomol.,* 9, 278, 1980.

52. **Bier, B. A., Beroza, M., and Collier, C. W.,** Potent sex attractant of the gypsy moth; its isolation, identification, and synthesis, *Science,* 170, 87, 1970.

53. **Bierl, B. A., Beroza, M., and Collier, C. W.,** Isolation, identification, and synthesis of the gypsy moth sex attractant, *J. Econ. Entomol.,* 65, 659, 1972.

54. **Bierl, B. A., Beroza, M., Adler, V. E., Kasang, G., Schröter, H., and Schneider, D.,** The presence of disparlure, the sex pheromone of the gypsy moth, in the female nun moth, *Z. Naturforsch.,* 30c, 672, 1975.

55. **Smith, R. G., Daterman, G. E., and Daves, G. D., Jr.,** Douglas-fir tussock moth: sex pheromone identification and synthesis, *Science,* 188, 63, 1975.

56. **Smith, L. M., Smith, R. G., Loehr, T. M., Daves, G. D., Jr., Daterman, G. E., and Wohleb, R. H.,** Douglas-fir tussock moth pheromone: identification of a diene analogue of the principal attractant and synthesis of stereochemically defined 1,6-, 2,6-, and 3,6-heneicosadien-11-ones, *J. Org. Chem.,* 43, 2361, 1978.

57. **Ando, T., Kishino, K., Tatsuki, S., Nakajima, H., Yoshida, S., and Takahashi, N.,** Identification of the female sex pheromone of the rice green caterpillar, *Agric. Biol. Chem.,* 41, 1819, 1977.

58. **Hall, D. R., Beevor, P. S., Cork, A., Lester, R., Nesbitt, B. F., Nyirenda, G. K. C., Nota Phiri, D. D., Blair, B. W., and Tannock, J.,** The female sex pheromone of the maize stalk-borer *Busseola fusca* (Fuller) (Lepidoptera: Noctuidae): identification and initial field trials, *Zimbabwe J. Agric. Res.,* 19, 111, 1981.

59. **Nesbitt, B. F., Beevor, P. S., Cole, R. A., Lester, R., and Poppi, R. G.,** Sex pheromones of two noctuid moths, *Nature (London) New Biol.,* 244, 208, 1973.

60. **Nesbitt, B. F., Beevor, P. S., Cole, R. A., Lester, R., and Poppi, R. G.,** The isolation and identification of the female sex pheromones of the red bollworm moths, *Diparopsis castanea, J. Insect Physiol.,* 21, 1091, 1975.

61. **Marks, R. J.,** Field studies with the synthetic sex pheromone and inhibitor of the red bollworm *Diparopsis castanea* Hmps. (Lepidoptera, Noctuidae) in Malawi, *Bull. Entomol. Res.,* 66, 243, 1976.

62. **Rotundo, G., Piccardi, P., and Tremblay, E.,** Preliminary report on the artichoke moth (*Gortyna xanthenes* Ger.) sex pheromone and its possible use in integrated control, Les Colloques de l'INRA, Paris, 1982, 177.

63. **Nesbitt, B. F., Beevor, P. S., Hall, D. R., Lester, R., and Dyck, V. A.,** Identification of the female sex pheromone of the purple stem borer moth, *Sesamia inferens, Insect Biochem.,* 6, 105, 1976.

64. **Jacobson, M., Redfern, R. E., Jones, W. A., and Aldridge, M. H.,** Sex pheromones of the southern armyworm moth: isolation, identification, and synthesis, *Science,* 170, 542, 1970.

65. **Beevor, P. S., Hall, D. R., Lester, R., Poppi, R. G., Read, J. S., and Nesbitt, B. F.,** Sex pheromones of the armyworm moth, *Spodoptera exempta* (Wlk.), *Experientia,* 31, 22, 1975.

66. **Brady, U. E. and Ganyard, M. C., Jr.,** Identification of the sex pheromone of the female beet armyworm, *Spodoptera exigua, Ann. Entomol. Soc. Am.,* 65, 898, 1972.

67. **Tumlinson, J. H., Mitchell, E. R., and Sonnet, P. E.,** Sex pheromone components of the beet armyworm, *Spodoptera exigua, J. Environ. Sci. Health,* A16, 189, 1981.

68. **Persoons, C. J., Graan, C., Nooijen, W. J., Ritter, F. J., Voerman, S., and Baker, T. C.,** Sex pheromone of the beet armyworm, *Spodoptera exigua:* isolation, identification and preliminary field evaluation, *Entomol. Exp. Appl.,* 30, 98, 1981.

69. **Sekul, A. A. and Sparks, A. N.,** Sex pheromone of the fall armyworm moth: isolation, identification, and synthesis, *J. Econ. Entomol.,* 60, 1270, 1967.

70. **Kasang, G., Kaissling, K. E., Vostrowsky, O., and Bestmann, H. J.,** Bombykal, a second pheromone component of the silkworm moth, *Bombyx mori* L., *Angew. Chem. Int. Ed.,* 17, 60, 1978.

71. **Kaissling, K. E., Kasang, G., Bestmann, H. J., Stransky, W., and Vostrowsky, O.,** A new pheromone of the silkworm moth, *Bombyx mori.* Sensory pathway and behavioural effect, *Naturwissenschaften,* 65, 382, 1978.

72. **Sekul, A. A. and Sparks, A. N.,** Sex attractant of the fall armyworm moth, *USDA Tech. Bull.,* 1542, 1, 1976.

73. **Jones, R. L. and Sparks, A. N.,** (Z)-9-tetradecen-1-ol acetate. A secondary sex pheromone of the fall armyworm, *Spodoptera frugiperda* (J. E. Smith), *J. Econ. Entomol.*, 5, 721, 1979.

74. **Tamaki, Y., Noguchi, H., and Yushima, T.,** Sex pheromone of *Spodoptera litura* (F.) (Lepidoptera: Noctuidae): isolation, identification, and synthesis, *Appl. Entomol. Zool.*, 8, 200, 1973.

75. **Tamaki, Y. and Yushima, T.,** Sex pheromone of the cotton leafworm, *Spodoptera littoralis, J. Insect Physiol.*, 20, 1005, 1974.

76. **Campion, D. G., Bettany, B. W., Nesbitt, B. F., Beevor, P. S., Lester, R., and Poppi, R. G.,** Field studies of the female sex pheromone of the cotton leafworm *Spodoptera littoralis* (Boisd.) in Cyprus, *Bull. Entomol. Res.*, 64, 89, 1974.

77. **Heath, R. R., Tumlinson, J. H., Leppla, N. C., McLaughlin, J. R., Dueben, B., Dundulis, E., and Guy, R. H.,** Identification of a sex pheromone produced by female velvetbean caterpillar moth, *J. Chem. Ecol.*, 9, 645, 1983.

78. **Hirai, Y., Kimura, H., Kawasaki, K., and Tamaki, Y.,** (Z)-11-hexadecenyl acetate: a sex pheromone component of the cabbage armyworm moth, *Mamestra brassicae* Linné, *Appl. Entomol. Zool.*, 13, 136, 1978.

79. **Bestmann, H. J., Vostrowsky, O., Koschantzky, K. H., Platz, H., Szmanska, A., and Knauf, W.,** Pheromone. XVII. (Z)-11-Hexadecenylacetat, ein Sexual-lockstoff des Pheromonsystems der Kohleule *Mamestra brassicae, Tetrahedron Lett.*, 6, 605, 1978.

80. **Knauf, W., Bestmann, H. J., Vostrowsky, O., und Platz, H.,** Die biologische Prüfung des Sexualpheromones der Kohleule (*Mamestra brassicae*) im Labor und unter Freidlandbedingungen, *Mitt. Dtsch. Ges. Allg. Angew. Entomol.*, 1, 170, 1978.

81. **Descoins, C., Priesner, E., Gallois, M., Arn, H., and Martin, G.,** Sur la sécrétion phéromonale des femelles vierges de *Mamestra brassicae* L. et de *Mamestra oleracea* L. (Lépidoptères, Noctuidae, Hadeninae), *C. R. Acad. Sci. Paris Ser. D*, 286, 77, 1978.

82. **Struble, D. L., Arn, H., Buser, H. R., Städler, E., and Feuler, J.,** Identification of 4 sex pheromone components isolated from calling females of *Mamestra brassicae, Z. Naturforsch.*, 35c, 45, 1980.

83. **Chisholm, M. D., Steck, W. F., Arthur, A. P., and Underhill, E. W.,** Evidence for *cis*-11-hexadecen-1-ol acetate as a major component of the sex pheromone of the bertha armyworm, *Mamestra configurata* (Lepidoptera: Noctuidae), *Can. Entomol.*, 107, 361, 1975.

84. **Underhill, E. W., Steck, W. F., and Chisholm, M. D.,** A sex pheromone mixture for the bertha armyworm moth, *Mamestra configurata*: (Z)-9-tetradecen-1-ol acetate and (Z)-11-hexadecen-1-ol acetate, *Can. Entomol.*, 109, 1335, 1977.

85. **Renou, M., Lalanne-Casson, B., Frérot, B., Gallois, M., and Descoins, C.,** Composition de la sécrétion phéromonale émise par les femelles vierges de *Mamestra (Polia) pisi* (L.)(Lépidoptère, Noctuidae, Hadeninae), *C. R. Acad. Sci. Paris*, 292(III), 1117, 1981.

86. **Vrkoc, J., Kovalev, B. G., and Starets, V. A.,** Isolation and identification of the main component of the female sexual pheromone of *Mamestra suasa* (Lepidoptera, Noctuidae), *Acta Entomol. Bohemoslov.*, 78, 353, 1981.

87. **Hill, A. S. and Roelofs, W. L.,** A female-produced sex pheromone component and attractant for males in the armyworm moth, *Pseudaletia unipuncta, Environ. Entomol.*, 9, 408, 1980.

88. **Underhill, E. W., Steck, W. F., and Chisholm, M. D.,** Sex pheromone of the clover cutworm moth, *Scotogramma trifolli*: isolation, identification, and field studies, *Environ. Entomol.*, 5, 307, 1976.

89. **Piccardi, P., Capizzi, A., Cassani, G., Spinelli, P., Arsura, E., and Massado, P.,** A sex pheromone component of the old world bollworm *Heliothis armigera, J. Insect Physiol.*, 23, 1443, 1977.

90. **Nesbitt, B. F., Beevor, P. S., Hall, D. R., and Lester, R.,** Female sex pheromone components of the cotton bollworm, *Heliothis armigera, J. Insect Physiol.*, 25, 535, 1979.

91. **Nesbitt, B. F., Beevor, P. S., Hall, D. R., and Lester, R.,** (Z)-9-hexadecenal: a minor component of the female sex pheromone of *Heliothis armigiera* (Hübner) (Lepidoptera, Noctuidae), *Entomol. Exp. Appl.*, 27, 306, 1980.

92. **Kehat, M., Gothilf, S., Dunkelblum, E., and Greenberg, S.,** Field evaluation of female sex pheromone components of the cotton bollworm, *Heliothis armigera, Entomol. Exp. Appl.*, 27, 188, 1980.

93. **Teal, P. E. A., Heath, R. R., Tumlinson, J. H., and McLaughlin, J. R.,** Identification of a sex pheromone of *Heliothis subflexa* (Gn.) (Lepidoptera: Noctuidae) and field trapping studies using different blends of components, *J. Chem. Ecol.*, 7, 1011, 1981.

94. **Roelofs, W. L., Hill, A. S., Cardé, R. T., and Baker, T. C.,** Two sex pheromone components of the tobacco budworm moth, *Heliothis virescens, Life Sci.*, 14, 1555, 1974.

95. **Tumlinson, J. H., Hendricks, D. E., Mitchell, E. R., Doolittle, R. E., and Brennan, M. M.,** Isolation, identification, and synthesis of the sex pheromone of the tobacco budworm, *J. Chem. Ecol.*, 1, 203, 1975.

96. **Klun, J. A., Plimmer, J. R., Bierl-Leonhardt, B. A., Sparks, A. N., and Chapman, O. L.,** Trace chemicals: the essence of sexual communication system in *Heliothis* species, *Science*, 204, 1328, 1979.

97. **Klun, J. A., Bierl-Leonhardt, B. A., Plimmer, J. R., Sparks, A. N., Primiani, M., Chapman, O. L., Lepone, G., and Lee, G. H.,** Sex pheromone chemistry of the female tobacco budworm moth, *Heliothis virescens, J. Chem. Ecol.,* 6, 177, 1980.

98. **Klun, J. A., Plimmer, J. R., Bierl-Leonhardt, B. A., Sparks, A. N., Primiani, M., Chapman, O. L., Lee, G. H., and Lepone, G.,** Sex pheromone chemistry of female corn earworm moth, *Heliothis zea, J. Chem. Ecol.,* 6, 165, 1980.

99. **Hill, A. S., Rings, R. W., Swier, S. R., and Roelofs, W. L.,** Sex pheromone of the black cutworm moth, *Agrotis ipsilon, J. Chem. Ecol.,* 5, 439, 1979.

100. **Wakamura, S.,** Sex attractant pheromone of the common cutworm moth, *Agrotis fucosa* Butler (Lepidoptera: Noctuidae): isolation and identification, *Appl. Entomol. Zool.,* 13, 290, 1978.

101. **Bestmann, H. J., Vostrowsky, O., Koschatzky, K.-H., Platz, H., Brosche, T., Kantardjiew, I., Rheinwald, M., and Knauf, W.,** (Z)-5-decenyl acetate, a sex attractant for the male turnip moth, *Agrotis segetum* (Lepidoptera), *Angew. Chem.,* 90, 815, 1978.

102. **Arn, H., Städler, E., Rauscher, S., Buser, H. R., Mustaparta, H., Esbjerg, P., Philipsen, H., Zethner, O., Struble, D. L., and Bues, R.,** Multicomponent sex pheromone in *Agrotis segetum:* preliminary analysis and field evaluation, *Z. Naturforsch.,* 35c, 986, 1980.

103. **Toth, M., Jakab, J., and Novák, L.,** Identification of two components from the sex pheromone system of the white-line dart moth, *Scotia segetum* (Schiff.) (Lep., Noctuidae), *Z. Angew. Entomol.,* 90, 505, 1980.

104. **Hirai, Y.,** *Cis-*7-tetradecen-1-ol acetate: a sex pheromone component of the stored cutworm moth, *Amathes c-nigrum* L., *Proc. Symp. Insect Pheromones and Their Applications,* Japan Plant Protection Assoc., Tokyo, 1976, 165.

105. **Struble, D. L., Buser, H. R., Arn, H., and Swailes, G. E.,** Identification of sex pheromone components of redbacked cutworm, *Euxoa ochrogaster,* and modification of sex attractant blend for adult males, *J. Chem. Ecol.,* 6, 573, 1980.

106. **Struble, D. L.,** A four-component pheromone blend for optimum attraction of redbacked cutworm males, *Euxoa ochrogaster* (Guenée), *J. Chem. Ecol.,* 7, 615, 1981.

107. **Bestmann, H. J., Brosche, T., Koschatzky, K. H., Michaelis, K., Platz, H., Vostrowsky, O., and Knauf, W.,** Pheromone. XXX. Identifizierung eines neuartigen Pheromonkomplexes aus der Graseule *Scotia exclamationis* (Lepidoptera), *Tetrahedron Lett.,* 21, 747, 1980.

108. **Berger, R. S. and Canerday, T. D.,** Specificity of the cabbage looper sex attractant, *J. Econ. Entomol.,* 61, 452, 1968.

109. **Kaae, R. S., Shorey, H. H., McFarland, S. U., and Gaston, L. K.,** Sex pheromones of Lepidoptera. XXXVII. Role of sex pheromones and other factors in reproductive isolation among ten species of Noctuidae, *Ann. Entomol. Soc. Am.,* 66, 444, 1973.

110. **Dunkelblum, E., Gothilf, S., and Kehat, M.,** Sex pheromone of the tomato looper, *Plusia chalcites* (Esp.), *J. Chem. Ecol.,* 7, 1081, 1981.

111. **Hall, D. R., Beevor, P. S., Lester, R., and Nesbitt, B. F.,** (*E,E*)-10,12-hexadecadienal: a component of the female sex pheromone of the spiny bollworm, *Earias insulana* (Boisd.) (Lepidoptera: Noctuidae), *Experientia,* 36, 152, 1980.

112. **Guerrero, A., Camps, F., Coll, J., Riba, M., Einhorn, J., Descoins, C., and Lallemand, J. Y.,** Identification of a potential sex pheromone of the processionary moth, *Thaumetopoea pityocampa* (Lepidoptera: Notodontidae), *Tetrahedron Lett.,* 22, 2013, 1981.

113. **Cuevas, P., Montoya, R., Belles, S., Camps, F., Coll, J., Guerrero, A., and Riba, M.,** Initial field trials with the synthetic sex pheromone of the processionary moth, *Thaumetopoea pityocampa* (Denis and Schiff.), *J. Chem. Ecol.,* 9, 85, 1983.

114. **Nesbitt, B. F., Beevor, P. S., Hall, D. R., Lester, R., Davies, J. C., and Seshu Reddy, K. V.,** Components of the sex pheromone of the female spotted stalk borer, *Chilo partellus* (Swinhoe) (Lepidoptera: Pyralidae): identification and preliminary field trials, *J. Chem. Ecol.,* 5, 153, 1979.

115. **Nesbitt, B. F., Beevor, P. S., Hall, D. R., Lester, R., and Williams, J. R.,** Components of the sex pheromone of the female sugar cane borer, *Chilo sacchariphagus* (Bojer) (Lepidoptera: Pyralidae): identification and field trials, *J. Chem. Ecol.,* 6, 385, 1980.

116. **Nesbitt, B. F., Beevor, P. S., Hall, D. R., Lester, R., and Dyck, V. A.,** Identification of the female sex pheromones of the moth, *Chilo suppressalis, J. Insect Physiol.,* 21 1883, 1975.

117. **Ohta, K., Tatsuki, S., Uchiumi, K., Kurihara, M., and Fukami, J.,** Structures of sex pheromones of rice stem borer, *Agric. Biol. Chem.,* 40, 1897, 1976.

118. **McDonough, L. M. and Kamm, J. A.,** Sex pheromone of the cranberry girdler, *Chrysoteuchia topiaria* (Zeller) (Lepidoptera: Pyralidae), *J. Chem. Ecol.,* 5, 211, 1979.

119. **Kamm, J. A. and McDonough, L. M.,** Field tests with the sex pheromone of the cranberry girdler, *Environ. Entomol.,* 8, 773, 1979.

120. **Kamm, J. A. and McDonough, L. M.,** Synergism of the sex pheromone of the cranberry girdler, *Environ. Entomol.,* 9, 795, 1980.

121. **Kuwahara, Y., Hara, H., Ishii, S., and Fukami, H.,** The sex pheromone of the Mediterranean flour moth, *Agric. Biol. Chem.,* 35, 447, 1971.

122. **Levinson, H. Z. and Buchelos, C. Th.,** Surveillance of storage moth species (Pyralidae, Gelechiidae) in a flour mill by adhesive traps with notes on the pheromone-mediated behaviour of male moths, *Z. Angew. Entomol.,* 92, 233, 1981.

123. **Coffelt, J. A., Vick, K. W., Sonnet, P. E., and Doolittle, R. E.,** Isolation, identification, and synthesis of a female sex pheromone of the navel orange worm *Amylois transitella* (Lepidoptera: Pyralidae), *J. Chem. Ecol.,* 5, 955, 1979.

124. **Kuwahara, Y., Kitamura, C., Takahashi, S., Hara, H., Ishii, S., and Fukami, H.,** Sex pheromone of the almond moth and the Indian meal moth: *cis*-9,*trans*-12-tetradecadienyl acetate, *Science,* 171, 801, 1971.

125. **Brady, U. E., Tumlinson, J. H., Brownlee, R. G., and Silverstein, R. M.,** Sex stimulant and attractant in the Indian meal moth and in the almond moth, *Science,* 171, 802, 1971.

126. **Brady, U. E.,** Isolation, identification and stimulatory activity of a second component of the sex pheromone system (complex) of the female almond moth, *Cadra cautella* (Walker), *Life Sci.,* 13, 227, 1973.

127. **Hoppe, T. and Levinson, H. Z.,** Befallserkennung und Populationsüberwachung vorratsschädlicher Motten (Phycitinae) in einer Schokoladenfabrik mit Hilfe pheromonbeköderter Klebefallen, *Anz. Schaedlingskd. Pflanz. Umweltschutz,* 52, 177, 1979.

128. **Brady, U. E. and Daley, R. C.,** Identification of a sex pheromone from the female raisin moth, *Cadra figulilella, Ann. Entomol. Soc. Am.,* 65, 1356, 1972.

129. **Brady, U. E. and Nordlund, D. A.,** *cis*-9,*trans*-12-Tetradecadien-1-yl acetate in the female tobacco moth, *Ephestia elutella* (Hübner) and evidence for an additional component of the sex pheromone, *Life Sci.,* 10, 797, 1971.

130. **Sato, R., Yaginuma, K., and Kumakura, M.,** Sex attractant of the stone leekminer, *Acrolepia alliella* Semenv et Kuznesov: Z-11-hexadecenyl acetate, Z-11-hexadecenal and Z-11-hexadecenol, *Jpn. J. Appl. Entomol. Zool.,* 23, 115, 1979.

131. **Ando, T., Kuroko, H., Nakagaki, S., Saito, O., Oku, T., and Takahashi, N.,** Multi-component sex attractants in systematic field tests of male Lepidoptera, *Agric. Biol. Chem.,* 45, 487, 1981.

132. **Underhill, E. W., Steck, W., Chisholm, M. D., Worden, H. A., and Howe, J. A. G.,** A sex attractant for the cottonwood corn borer, *Aegeria tibialis* (Lepidoptera: Sesiidae), *Can. Entomol.,* 110, 495, 1978.

133. **Nielsen, D. G., Purrington, F. F., and Shambaugh, G. F.,** EAG and field responses of sesiid males to sex pheromones and related compounds, in Pheromones of the Sesiidae, Neal, J. W., Jr., Ed., USDA, SEA, Agric. Res. Results, ARR-NE-6, Washington, D.C., 1979, 11.

134. **Campbell, R. L.,** A sex pheromone for *Vespamima sequoisa, Bull. Oregon Entomol. Soc.,* 61, 499, 1976.

135. **Sharp, J. L., McLaughlin, J. R., James, J., Eichlin, T. D., and Tumlinson, J. H.,** Seasonal occurrence of male Sesiidae in North Central Florida determined with pheromone trapping methods, *Fla. Entomol.,* 61, 245, 1978.

136. **Nielsen, D. G., Purrington, F. F., Tumlinson, J. H., Doolittle, R. E., and Yonce, C. E.,** Response of male clearwing moths to caged virgin females, female extracts and synthetic sex attractants, *Environ. Entomol.,* 4, 451, 1975.

137. **Duckworth, W. D. and Eichlin, T. D.,** A new species of clearwing moth from south central Texas, *Pan-Pacific Entomol.,* 53, 175, 1977.

138. **Greenfield, M. D. and Karandinos, M. G.,** A new species of *Paranthrene* (Lepidoptera, Sesiidae), *Proc. Entomol. Soc. Washington,* 81, 499, 1979.

139. **Nielsen, D. G. and Purrington, F. F.,** Sex attractants: a new warning system to time clearwing borer control practices, *Ornamental Plants — 1978: A Summary of Research, Ohio Agric. Res. Dev. Center Res. Circ.,* 236, 7, 1978.

140. **Levinson, H. Z. and Levinson, A. R.,** Trapping of storage insects by sex and food attractants as a tool of integrated control, in *Chemical Ecology: Odour Communication in Animals,* Ritter, F. J., Ed., Elsevier, Amsterdam, 1979, 327.

141. **Underhill, E. W., Arthur, A. P., Chisholm, M. D., and Steck, W. F.,** Sex pheromone components of the sunflower moth, *Homoeosoma electellum:* Z-9,E-12-tetradecadienol and Z-9-tetradecenol, *Environ. Entomol.,* 8, 740, 1979.

142. **Coffelt, J. A., Sower, L. L., and Vick, K. W.,** Quantitative analysis of identified compounds in pheromone gland rinses of *Plodia interpunctella* and *Ephestia cautella* at different time of day, *Environ. Entomol.,* 7, 502, 1978.

143. **Vick, K. W., Coffelt, J. A., Mankin, R. W., and Soderstrom, E. L.,** Recent development in the use of pheromones to monitor *Plodia interpunctella* and *Ephestia cautella,* in *Management of Insect Pests with Semiochemicals, Concepts and Practice,* Mitchell, E. R., Ed., Plenum Press, New York, 1981, 119.

144. **Konno, Y., Arai, K., Sekiguchi, K., and Matsumoto, Y.,** (E)-10-hexadecenal, a sex pheromone component of the yellow peach moth, *Dichocrosis punctiferalis* Guenée (Lepidoptera: Pyralidae), *Appl. Entomol. Zool.,* 17, 207, 1982.

145. **Ando, T., Saito, O., Arai, K., and Takahashi, N.,** (Z)- and (E)-12-tetradecenyl acetates: sex pheromone components of oriental corn borer (Lepidoptera: Pyralidae), *Agric. Biol. Chem.,* 44, 2643, 1980.

146. **Klun, J. A., Bier-Leonhardt, B. A., Schwarz, M., Litsinger, J. A., Barrion, A. T., Chiang, H. C., and Jiang, Z.,** Sex pheromone of the Asian corn borer moth, *Life Sci.,* 27, 1603, 1980.

147. **Cheng, Z.-Q., Xiao, J.-C., Huang, X.-T., Chen, D.-L., Li, J.-Q., He, Y.-S., Huang, S.-R., Luo, Q.-C., Yang, C.-M., and Yang, T.-H.,** Sex pheromone components isolated from China corn borer, *Ostrinia furnacalis* Guenée (Lepidoptera: Pyralidae), (E)- and (Z)-12-tetradecenyl acetates, *J. Chem. Ecol.,* 7, 841, 1981.

148. **Klun, J. A. and Brindley, T. A.,** cis-11-Tetradecenyl acetate, a sex stimulant of the European corn borer, *J. Econ. Entomol.,* 63, 779, 1970.

149. **Kochansky, J., Cardé, R. T., Liebherr, J., and Roelofs, W. L.,** Sex pheromones of the European corn borer (*Ostrinia nubilalis*) in New York, *J. Chem. Ecol.,* 1, 225, 1975.

150. **Klun, J. A., Haynes, K. F., Bierl-Leonhardt, B. A., Birch, M. C., and Plimmer, J. R.,** Sex pheromone of the female artichoke plume moth, *Platyptilia carduidactyla, Environ. Entomol.,* 10, 763, 1981.

151. **Kochansky, J., Tette, J., Taschenberg, E. F., Cardé, R. T., Kaissling, K.-E., and Roelofs, W. L.,** Sex pheromone of the moth, *Antheraea polyphemus, J. Insect Physiol.,* 21, 1977, 1975.

152. **Henderson, H. E. and Warren, F. L.,** Sex-pheromones. cis-Dec-5-en-1-yl 3-methylbutanoate as the pheromone from the pine emperor moth (*Nudaurelia cytherea cytherea* Fabr.), *J. Chem. Soc. Chem. Commun.,* p. 686, 1972.

153. **Henderson, H. E., Warren, F. L., Aucustyn, O. P. H., Burger, B. V., Schneider, D. F., Boshoff, P. R., Spies, H. S. C., and Geertsema, H.,** Isolation and structure of the sex pheromone of the moth, *Nudaurelia cytherea cytherea, J. Insect Physiol.,* 19, 1257, 1973.

154. **Starratt, A. N., Dahm, K. H., Allen, N., Hildebrand, J. G., Payne, T. L., and Röller, H.,** Bombykal, a sex pheromone of the sphinx moth, *Manduca sexta, Z. Naturforsch.,* 34c, 9, 1979.

155. **Persoons, C. J., Ritter, F. J., Hainaut, D., and Demoult, J. P.,** Sex pheromone of the false codling moth *Cryptophlebia* (= *Argyroploce) leucotreta* (Lepidoptera: Tortricidae). trans-8-Dodecenyl acetate, a corrected structure, *Meded. Fac. Landbouwwet. Rijksuniv. Gent,* 41, 937, 1976.

156. **Persoons, C. J., Ritter, F. J., and Nooyen, W. J.,** Sex pheromone of the false codling moth *Cryptophlebia leucotreta* (Lepidoptera: Tortricidae). Evidence for a two-component system, *J. Chem. Ecol.,* 3, 717, 1977.

157. **Angelini, A., Descoins, C., Lhoste, J., Trijau, J.-P., and Zagatti, P.,** Essai de nouvelles formulations d'attractifs de synthèse pour le piégeage sexuel de *Cryptophlebia leucotreta* Meyr (Lepidoptera), *Cotton Fibres Trop.,* 36, 259, 1981.

158. **Roelofs, W. L., Comeau, A., and Selle, R.,** Sex pheromone of the oriental fruit moth, *Nature (London),* 224, 723, 1969.

159. **Cardé, A. M., Baker, T. C., and Cardé, R. T.,** Identification of a four-component sex pheromone of the female oriental fruit moth, *Grapholitha molesta* (Lepidoptera: Tortricidae), *J. Chem. Ecol.,* 5, 423, 1979.

160. **Frérot, B., Priesner, E., and Gallois, M.,** A sex attractant for the green budworm moth, *Hedya nubiferana, Z. Naturforsch.,* 34c, 1248, 1979.

161. **Roelofs, W., Comeau, A., Hill, A., and Millicevic, G.,** Sex attractant of the codling moth: characterization with electroantennogram technique, *Science,* 174, 297, 1971.

162. **Beroza, M., Bierl, B. A., and Moffitt, H. R.,** Sex pheromones: (E,E)-8,10-dodecadien-1-ol in the codling moth, *Science,* 183, 89, 1974.

163. **Roelofs, W. L., Kochansky, J., Cardé, R., Arn, H., and Rauscher, S.,** Sex attractant of the grape vine moth, *Lobesia botrana, Mitt. Schweiz. Entomol. Ges.,* 46, 71, 1973.

164. **Roelofs, W. L., Tette, J. P., Taschenberg, E. F., and Comeau, A.,** Sex pheromone of the grape berry moth: identification by classical and electroantennogram methods, and field tests, *J. Insect Physiol.,* 17, 2235, 1971.

165. **Smith, R. G., Daterman, C. D., Daves, G. D., Jr., McMurtrey, K. D., and Roelofs, W. L.,** Sex pheromone of the European pine shoot moth: chemical identification and field tests, *J. Insect Physiol.,* 20, 661, 1974.

166. **Hill, A. S., Berisford, C. W., Brady, U. E., and Roelofs, W. L.,** Nantucket pine tip moth, *Rhyacionia frustrana:* identification of two sex pheromone components, *J. Chem. Ecol.,* 7, 517, 1981.

167. **Hill, A. S., Berisford, C. W., Brady, U. E., and Roelofs, W. L.,** Sex pheromone of the pitch pine tip moth, *Rhyacionia rigidana, Environ. Entomol.,* 5, 959, 1976.

168. **Roelofs, W. L., Hill, A. S., Berisford, C. W., and Godbee, J. F.,** Sex pheromone of the subtropical pine tip moth, *Rhyacionia subtropica, Environ. Entomol.,* 8, 894, 1979.

169. **Arn, H., Schwarz, C., Limacher, H., and Mani, E.,** Sex attractant inhibitors of the codling moth *Laspeyresia pomonella* L., *Experientia,* 30, 1142, 1974.

170. **Arn, H., Städler, E., and Rauscher, S.,** The electroantennographic detector — a selective and sensitive tool in the gas chromatographic analysis of insect pheromones, *Z. Naturforsch.,* 30c, 722, 1975.

171. **Roelofs, W. L., Cardé, R., Benz, G., and Salis, G.,** Sex attractant of the larch bud moth found by electroantennogram method, *Experientia*, 27, 1438, 1971.

172. **Bjostad, L. B., Taschenberg, E. F., and Roelofs, W. L.,** Sex pheromone of the woodbine leafroller moth, *Sparganothis* sp., *J. Chem. Ecol.*, 6, 797, 1980.

173. **Bjostad, L. B., Taschenberg, E. F., and Roelofs, W. L.,** Sex pheromone of the chokecherry leafroller moth, *Sparganothis directana, J. Chem. Ecol.*, 6, 487, 1980.

174. **Tamaki, Y., Noguchi, H., Yushima, T., and Hirano, C.,** Two sex pheromones of the smaller tea tortrix: isolation, identification, and synthesis, *Appl. Entomol. Zool.*, 6, 139, 1971.

175. **Tamaki, Y., Noguchi, H., Sugie, H., Sato, R., and Kariya, A.,** Minor components of the female sex-attractant pheromone of the smaller tea tortrix moth (Lepidoptera: Tortricidae): isolation and identification, *Appl. Entomol. Zool.*, 14, 101, 1979.

176. **Tamaki, Y., Noguchi, H., Sugie, H., Kariya, A., Arai, S., Ohba, M., Terada, T., Suguro, T., and Mori, K.,** Four-component synthetic sex pheromone of the smaller tea tortrix moth: field evaluation of its potency as an attractant for male moth, *Jpn. J. Appl. Entomol. Zool.*, 24, 221, 1980.

177. **Ritter, F. J.,** Some recent developments in the field of insect pheromones, *Meded. Fac. Landbouwwet. Rijksuniv. Gent*, 36, 874, 1971.

178. **Meijer, G. M., Ritter, F. J., Persoons, C. J., Minks, A. K., and Voerman, S.,** Sex pheromones of summerfruit tortrix moth, *Adoxophyes orana:* two synergistic isomers, *Science*, 175, 1469, 1972.

179. **Guerin, P. M., Arn, H., Buser, H. R., and Charmillot, P. J.,** The sex pheromone of *Adoxophyes orana* — preliminary findings from a reinvestigation, November 16, 1981, Les Colloques d l'INRA, Paris, 1982, 267.

180. **Tamaki, Y., Noguchi, H., Yushima, T., Hirano, C., Honma, K., and Sugawara, H.,** Sex pheromone of the summerfruit tortrix: isolation and identification, *Kontyu*, 39, 338, 1971.

181. **McDonough, L. M., Hoffmann, M. P., Bierl-Leonhard, B. A., Smithhisler, C. L., Bailey, J. B., and Davis, H. G.,** Sex pheromone of the avocado pest, *Amorbia cuneana* (Walsingham) (Lepidoptera: Tortricidae): structure and synthesis, *J. Chem. Ecol.*, 8, 255, 1982.

182. **Sugie, II., Yaginuma, K., and Tamaki, Y.,** Sex pheromone of the Asiatic leafroller, *Archippus breviplicanus* Walsingham (Lepidoptera: Tortricidae): isolation and identification, *Appl. Entomol. Zool.*, 12, 69, 1977.

183. **Roelofs, W., Hill, A., Cardé, R., Tette, J., Madsen, H., and Vakenti, J.,** Sex pheromones of the fruittree leafroller moth, *Archips argyrospilus, Environ. Entomol.*, 3, 747, 1974.

184. **Cardé, R. T., Cardé, A. M., Hill, A. S., and Roelofs, W. L.,** Sex pheromone specificity as a reproductive isolating mechanism among the sibling species *Archips argyrospilus* and *Archips mortuanus* and other sympatric tortricine moths (Lepidoptera: Tortricinae), *J. Chem. Ecol.*, 3, 71, 1977.

185. **Roelofs, W. L., Tamhankar, A. J., Comeau, A., Hill, A. S., and Taschenberg, E. F.,** Moth activity periods and identification of the sex pheromone of the uglynest caterpillar, *Archips cerasivoranus, Ann. Entomol. Soc. Am.*, 73, 631, 1980.

186. **Persoons, C. J., Minks, A. K., Voerman, S., Roelofs, W. L., and Ritter, F. J.,** Sex pheromones of the moth, *Archips podana:* isolation, identification and field evaluation of the two synergistic geometrical isomers, *J. Insect Physiol.*, 20, 1184, 1974.

187. **Roelofs, W., Hill, A., Cardé, A., Cardé, R., Madsen, H., and Vakenti, J.,** Sex pheromone of the European leafroller *Archips rosanus, Environ. Entomol.*, 5, 362, 1976.

188. **Miller, J. R., Baker, T. C., Cardé, R. T., and Roelofs, W. L.,** Reinvestigation of oak leafroller sex pheromone components and the hypothesis that they vary with diet, *Science*, 192, 140, 1976.

189. **Hill, A. S., Cardé, R. T., Kido, H., and Roelofs, W. L.,** Sex pheromones of the orange tortrix moth *Argyrotaenia citrana, J. Chem. Ecol.*, 1, 215, 1975.

190. **Roelofs, W. L. and Arn, H.,** Sex attractant of the red-banded leafroller moth, *Nature (London)*, 219, 513, 1968.

191. **Roelofs, W., Hill, A., and Cardé, R.,** Sex pheromone components of the redbanded leafroller, *Argyrotaenia velutinana* (Lepidoptera: Tortricidae), *J. Chem. Ecol.*, 1, 83, 1975.

192. **Weatherston, J., Roelofs, W., Comeau, A., and Sanders, C. J.,** Studies of physiologically active arthropod secretions. X. Sex pheromone of the eastern spruce budworm, *Choristoneura fumiferana* (Lepidoptera: Tortricidae), *Can. Entomol.*, 103, 1741, 1971.

193. **Sanders, C. J. and Weatherston, J.,** Sex pheromone of the eastern spruce budworm (Lepidoptera: Tortricidae): optimum blend of *trans-* and *cis-*11-tetradecenal, *Can. Entomol.*, 108, 1285, 1976.

194. **Priesner, E., Bogenschütz, H., and Arn, H.,** A sex attractant for the European fir budworm moth, *Choristoneura murinana, Z. Naturforsch.*, 35c, 390, 1980.

195. **Burghardt, G., Knauf, W., Bestmann, H. J., Vostrowsky, O., and Koschantzky, K. H.,** Synthetische Sexual-Pheromone und deren Inhibitoren für *Choristoneura murinana* Hbn. (Tannentribwickler, Lepidoptera: Tortricidae), *Anz. Schaedlingskd. Pflanz. Umweltschutz*, 53, 49, 1980.

196. **Roelofs, W. L. and Tette, J. P.,** Sex pheromone of the oblique-banded leafroller, *Nature (London)*, 226, 1172, 1970.

197. **Hill, A. S. and Roelofs, W. L.,** Sex pheromone components of the obliquebanded leafroller moth, *Choristoneura rosaceana, J. Chem. Ecol.,* 5, 3, 1979.

198. **Minks, A. K., Roelofs, W. L., Ritter, F. J., and Persoons, C. J.,** Reproductive isolation of two tortricid moth species by different ratios of a two-component sex attractant, *Science,* 180, 1073, 1973.

199. **Minks, A. K., Roelofs, W. L., Schuurmans-van dijk, E., Persoons, C. J., and Ritter, F. J.,** Electroantennogram responses of two tortricid moths using two-component sex pheromones, *J. Insect Physiol.,* 20, 1659, 1974.

200. **Biwer, G., Descoins, C., Gallois, M., Priesner, E., Chambon, J.-P., Genestier, G., and Martinez, M.,** Etude de la sécrétion phéromonale de la femelle vierge de la Tordeuse des céréales: *Cnephasia pumicana* Zeller (Lépidoptère: Tortricidae), *Ann. Zool. Ecol. Anim.,* 10, 129, 1977.

201. **Lalanne-Cassou, B. and Frérot, B.,** Analyse de la sécrétion phéromonale de la tordeuse sud-africaine de l'Oeillet: (*Epichoristodes acerbella*) (Walker) (Lépidoptère, Tortricidae), *C. R. Acad. Sci. Paris, Ser. D,* 291, 845, 1980.

202. **Kochansky, J. P., Roelofs, W. L., and Sivapalan, P.,** Sex pheromone of the tea tortrix moth (*Homona coffearia* Neitner), *J. Chem. Ecol.,* 4, 623, 1978.

203. **Noguchi, H., Tamaki, Y., and Yushima, T.,** Sex pheromone of the tea tortrix moth: isolation and identification, *Appl. Entomol. Zool.,* 14, 225, 1979.

204. **Tonini, C., Cassani, G., Piccardi, P., Maini, S., Castellari, P. L., and Pasqualini, E.,** Identification of *Pandemis cerasana* sex pheromone components, November 16, 1981, Les Colloques de l'INRA, Paris, 1982, 395.

205. **Tonini, C., Cassani, G., Piccardi, P., Maini, S., Castellari, P. L., and Pasqualini, E.,** Sex pheromone components of the leafroller moth, *Pandemis cerasana, J. Insect Physiol.,* 28, 443, 1982.

206. **Frérot, B., Gallois, M., and Einhorn, J.,** La phéromone sexuelle produite par la female vierge de *Pandemis heparana* (Den. et Schiff.) (Lépidoptère: Tortricidae, Tortricinae), *C. R. Acad. Sci. Paris Ser. D,* 288, 1611, 1979.

207. **Frérot, B., Gallois, M., Lettere, M., Einhorn, J., Michelot, D., and Descoins, C.,** Sex pheromone of *Pandemis heparana* (Den. and Schiff.) (Lepidoptera: Tortricidae), *J. Chem. Ecol.,* 8, 663, 1982.

208. **Roelofs, W., Cardé, A., Hill, A., and Cardé, R.,** Sex pheromone of the threelined leafroller, *Pandemis limitata, Environ. Entomol.,* 5, 649, 1976.

209. **Roelofs, W. L., Lagier, R. F., and Hoyt, S. C.,** Sex pheromone of the moth, *Pandemis pyrusana, Environ. Entomol.,* 6, 353, 1977.

210. **Hill, A. S., Cardé, R. T., Bode, W. M., and Roelofs, W. L.,** Sex pheromone components of the variegated leafroller moth, *Platynota flavedana, J. Chem. Ecol.,* 3, 369, 1977.

211. **Hill, A., Cardé, R., Comeau, A., Bode, W., and Roelofs, W.,** Sex pheromones of the tufted apple bud moth (*Platynota idaeusalis*), *Environ. Entomol.,* 3, 249, 1974.

212. **Hill, A. S. and Roelofs, W. L.,** Sex pheromone components of the omnivorous leafroller moth, *Platynota stultana, J. Chem. Ecol.,* 1, 91, 1975.

213. **Benn, M. H., Galbreath, R. A., Holt, V. A., and Young, H.,** A sex attractant for *Sperchia intractana* Walker (Lepidoptera: Tortricidae) in New Zealand, *Z. Naturforsch.,* 36c, 178, 1981.

214. **Tamaki, Y., Kawasaki, K., Yamada, H., Koshihara, T., Osaki, N., Ando, T., Yoshida, S., and Kakinohana, H.,** (Z)-11-hexadecenal and (Z)-11-hexadecenyl acetate: sex pheromone components of the diamondback moth (Lepidoptera: Plutellidae), *Appl. Entomol. Zool.,* 12, 208, 1977.

215. **Koshihara, T., Yamada, H., Tamaki, Y., and Ando, T.,** Field attractiveness of the synthetic sex pheromone of the diamondback moth, *Plutella xylostella* (L.), *Appl. Entomol. Zool.,* 13, 138, 1978.

216. **Nesbitt, B. F., Beevor, P. S., Hall, D. R., Lester, R., Sternlicht, M., and Goldenberg, S.,** Identification and synthesis of the female sex pheromone of the citrus flower moth, *Prays citri, Insect Biochem.,* 7, 355, 1977.

217. **Campion, D. G., McVeigh, L. J., Polyrakis, J., Michaelaklis, S., Stravrakis, G. N., Beevor, P. S., Hall, D. R., and Nesbitt, B. F.,** Laboratory and field studies of the female sex pheromone of the olive moth, *Prays oleae, Experientia,* 35, 1146, 1979.

218. **Renou, M., Descoins, C., Priesner, E., Gallois, M., and Lettere, M.,** Le tétradécène-7 Z al-1, constituant principal de la sécrétion phéromonale de la teigne de l'olivier: *Prays oleae* Bern. (Lépidoptère, Yponomeutidae), *C. R. Acad. Sci. Paris Ser. D,* 288, 1559, 1979.

219. **Clearwater, J. R.,** Chemistry and function of a pheromone produced by the male of the southern armyworm *Pseudaletia separata, J. Insect Physiol.,* 18, 781, 1972.

220. **Hirai, K., Shorey, H. H., and Gaston, L. K.,** Competition among courting male moths: male-to-male inhibitory pheromone, *Science,* 202, 644, 1978.

221. **Hirai, K.,** Behavioral function of male scent in moths (Lepidoptera: Noctuidae), *Bull. Chugoku Natl. Agric. Exp. Stn.,* E16, 1, 1980.

222. **Jacobson, M., Adler, V. E., Kishaba, A. N., and Priesner, E.,** 2-Phenylethanol, a presumed sexual stimulant produced by the male cabbage looper moth, *Trichoplusia ni, Experientia,* 32, 964, 1976.

223. **Pliske, T. E. and Eisner, T.,** Sex pheromone of the queen butterfly: biology, *Science,* 164, 1170, 1969.
224. **Meinwald, J., Meinwald, Y. C., and Mazzocchi, P. H.,** Sex pheromone of the queen butterfly: chemistry, *Science,* 164, 1174, 1969.
225. **Schneider, D. and Seibt, U.,** Sex pheromone of the queen butterfly: electroantennogram responses, *Science,* 164, 1173, 1969.
226. **Grula, J. W., McChesney, J. D., and Taylor, O. R., Jr.,** Aphrodisiac pheromones of the sulfur butterflies *Colias eurytheme* and *Colias philodice* (Lepidoptera: Pieridae), *J. Chem. Ecol.,* 6, 241, 1980.
227. **Dahm, K. H., Meyer, D., Finn, W. E., Reinhold, V., and Röller, H.,** The olfactory and auditory mediated sex attraction in *Achrois grisella* (Fabr.), *Naturwissenschaften,* 58, 265, 1971.
228. **Kuwahara, Y.,** Isolation and identification of male-secreted possible sex pheromone from a pyralid moth, *Aphomia gularis* Zeller (Pyralidae: Lepidoptera), *Appl. Entomol. Zool.,* 15, 478, 1980.
229. **Kunesch, G., Zagatti, P., Lallemand, J. Y., Debal, A., and Vigneron, J. P.,** Structure and synthesis of the wing pheromone of the male African sugarcane borer: *Eldana saccharina* (Wlk.) (Lepidoptera, Pyralidae), *Tetrahedron Lett.,* 22, 5271, 1981.
230. **Zagatti, P.,** Comportment sexuel de la pyrale de la canne à sucre *Eldana saccharina* (Wlk.) lié à deux phéromones émises par le male, *Behaviour,* 78, 81, 1981.
231. **Kunesch, G. and Zagatti, P.,** Male sex pheromones of the African sugarcane borer: *Eldana saccharina* Wlk: identification and behaviour, Les Colloques de l'INRA, Paris, 1982, 281.
232. **Röller, H., Biemann, K., Bjerke, J. S., Norgard, D. W., and McShan, W. H.,** Sex pheromones of pyralid moths. I. Isolation and determination of the sex-attractant of *Galleria mellonella* L., *Acta Entomol. Bohemoslov.,* 65, 208, 1968.
233. **Leyrer, R. L. and Monroe, R. E.,** Isolation and identification of the scent of the moth, *Galleria mellonella,* and a re-evaluation of its sex pheromone, *J. Insect Physiol.,* 19, 2267, 1973.
234. **Baker, T. C., Nishida, R., and Roelofs, W. L.,** Close-range attraction of female oriental fruit moths to male hairpencils' herbal essence, *Science,* 214, 1359, 1981.
235. **Nishida, R., Baker, T. C., and Roelofs, W. L.,** Hairpencil pheromone components of male oriental fruit moths, *Grapholitha molesta, J. Chem. Ecol.,* 8, 947, 1982.
236. **Edgar, J. A., Cokrum, P. A., and Carrodus, B. B.,** Male scent organ chemicals of the vine moth, *Phalaenoides glycinae* Lew. (Agaristidae), *Experientia,* 35, 861, 1979.
237. **Aplin, R. T. and Birch, M. C.,** Identification of odorous compounds from male Lepidoptera, *Experientia,* 26, 1193, 1970.
238. **Bestmann, H. J., Vostrowsky, O., and Platz, H.,** Pheromone. XII. Männchenduftstoffe von Noctuiden (Lepidoptera), *Experientia,* 33, 874, 1977.
239. **Aplin, R. T. and Birch, M. C.,** Pheromones from the abdominal brushes of male noctuid Lepidoptera, *Nature (London),* 217, 1167, 1968.
240. **Clearwater, J. R.,** Pheromone metabolism in male *Pseudaletia separata* (Walk.) and *Mamestra configurata* (Walk.) (Lepidoptera: Noctuidae), *Comp. Biochem. Physiol.,* 50B, 77, 1975.
241. **Birch, M. C., Grant, G. G., and Brady, U. E.,** Male scent brush of *Peridroma saucia:* chemistry of secretion, *Ann. Entomol. Soc. Am.,* 69, 491, 1976.
242. **Hirai, K.,** Male scent emitted by armyworms, *Psudaletia unipuncta* and *P. separata* (Lepidoptera: Noctuidae), *Appl. Entomol. Zool.,* 15, 310, 1980.
243. **Grant, G. G., Brady, U. E., and Brand, J. M.,** Male armyworm scent brush secretion, identification and electroantennogram study of major components, *Ann. Entomol. Soc. Am.,* 65, 1224, 1972.
244. **Zagatti, P., Kunesch, G., and Morin, N.,** La vanilline, constituant majoritaire de la sécrétion aphrodisiaque émise par les androconies du mále de la pyrale de la canne à sucre: *Eldana saccharina* (Wlk.) (Lépidoptère, Pyralidae, Galleriinae), *C. R. Acad. Sci. Paris,* 292, 633, 1981.
245. **Meinwald, J., Boriack, C. J., Schneider, D., Boppré, M., Wood, W. F., and Eisner, T.,** Volatile ketones in the hairpencil secretion of danaid butterflies (*Amauris* and *Danaus*), *Experientia,* 30, 721, 1974.
246. **Petty, R. L., Boppre, M., Schneider, D., and Meinwald, J.,** Identification and localization of volatile hairpencil components in male *Amauris ochlea* butterflies (Danaidae), *Experientia,* 33, 1324, 1977.
247. **Edgar, J. A., Culvenor, C. C. J., and Smith, L. W.,** Dihydropyrrolizine derivatives in the "hairpencil" secretions of Danaid butterflies, *Experientia,* 27, 761, 1971.
248. **Meinwald, J., Thompson, W. R., and Eisner, T.,** Pheromones. VII. African monarch: major components of the hairpencil secretion, *Tetrahedron Lett.,* 38, 3485, 1971.
249. **Bellas, T. E., Brownless, R. G., and Silverstein, R. M.,** Isolation, tentative identification, and synthesis. Studies of the volatile components of the hairpencil secretion of the monarch butterfly, *Tetrahedron Lett.,* 30, 2267, 1974.
250. **Meinwald, J. and Meinwald, Y. C.,** Structure and synthesis of the major components in the hairpencil secretion of a male butterfly, *Lycorea ceres ceres* (Cramer), *J. Am. Chem. Soc.,* 88, 1305, 1966.
251. **Meinwald, J., Meinwald, Y. C., Wheeler, J. W., Eisner, T., and Brower, L. P.,** Major components in the exocrine secretion of a male butterfly (*Lycorea*), *Science,* 151, 583, 1966.

252. **Lundgren, L. and Bergström, G.,** Wing scents and scent-released phases in the courtship behavior of *Lycaeides argyrognomon* (Lepidoptera: Lycaenidae), *J. Chem. Ecol.,* 1, 399, 1975.

253. **Hayashi, N., Kuwahara, Y., and Komae, H.,** The scent scale substances of male *Pieris* butterflies (*Pieris melete* and *Pieris napi*), *Experientia,* 34, 684, 1978.

254. **Kuwahara, Y.,** Scent scale substances of male *Pieris melete* Ménétriés (Pieridae: Lepidoptera), *Appl. Entomol. Zool.,* 14, 350, 1979.

255. **Bergström, G. and Lundgren, L.,** Androconical secretion of three species of butterflies of the genus *Pieris* (Lep., Pieridae), *Zool. Suppl.,* 1, 67, 1973.

256. **Arn, H., Baltensweiler, W., Bues, R., Buser, H. R., Esbjerg, P., Guerin, P., Mani, E., Rauscher, S., Szöcs, G., and Toth, M.,** Refining lepidopteran sex attractants, Les Colloques de l'INRA, Paris, 1982, 261.

257. **Yamaoka, R., Ueno, T., and Fukami, H.,** Micro structure determination of Lepidopteran sex pheromones using GC-MS, Proc. 29th Ann. Conf. Mass Spectrometry and Applied Topics, Minneapolis, May 24, 1981, 7.

258. **Silk, P. J., Tan, S. H., Wiesner, C. J., Ross, R. J., and Lonergan, G. C.,** Sex pheromone chemistry of the eastern spruce budworm, *Choristoneura fumiferana, Environ. Entomol.,* 9, 640, 1980.

259. **McDonough, L. M., Kamm, J. A., and Biel-Leonhardt, B. A.,** Sex pheromone of the armyworm, *Pseudaletia unipuncta* (Haworth) (Lepidoptera: Noctuidae), *J. Chem. Ecol.,* 6, 565, 1980.

260. **Farine, J.-P., Frérot, B., and Isart, J.,** Facteurs d'isolement chimique dans la sécrétion phéromonale de deux Noctuelles Hadeninae: *Mamestra brassicae* (L.) et *Pseudaletia unipuncta* (Haw.), *C. R. Acad. Sci. Paris,* 292(III), 101, 1981.

261. **Kuwahara, Y. and Yen, L. T. M.,** Identification of bombykol in *Bombyx mandarina* females (Lepidoptera: Bombycidae), *Appl. Entomol. Zool.,* 14, 114, 1979.

262. **Steck, W., Underhill, E. W., Chisholm, M. D., and Bailey, B. K.,** (Z)-5-tetradecenyl acetate, a trapping synergist for the major sex pheromone of the spotted cutworm, *Amathes c-nigrum, Entomol. Exp. Appl.,* 31, 328, 1982.

263. **Doolittle, R. E., Tagestad, A., and McKnight, M. E.,** Trapping carpenterworms and aspen carpenterworms with sex attractants in North Dakota, *Environ. Entomol.,* 5, 267, 1976.

264. **Takahashi, S., Kawaradani, M., Sato, Y., and Sakai, M.,** Sex pheromone components of *Leucania separata* Walker and *Leucaniua loreyi* Duponchel, *Jpn. J. Appl. Entomol. Zool.,* 23, 78, 1979.

265. **Tumlinson, J. H., Mitchell, E. R., Browner, S. M., and Lindquist, D. A.,** A sex pheromone for the soybean looper, *Environ. Entomol.,* 1, 466, 1972.

266. **Berger, R. S.,** Isolation, identification, and synthesis of the sex attractant of the cabbage looper, *Trichoplusia ni, Ann. Entomol. Soc. Am.,* 59, 767, 1966.

267. **Bjostad, L. B., Gaston, L. K., and Shorey, H. H.,** Temporal pattern of sex pheromone release by female, *Trichoplusia ni, J. Insect Physiol.,* 26, 493, 1980.

268. **Yushima, T., Tamaki, Y., Kamano, S., and Oyama, M.,** Field evaluation of a synthetic sex pheromone, "litlure", as an attractant for males of *Spodoptera litura* (F.) (Lepidoptera: Nocotuidae), *Appl. Entomol. Zool.,* 9, 147, 1974.

269. **Sato, Y., Takahashi, S., Sakai, M., and Kodama, T.,** Attractiveness of the synthetic sex pheromone to the males of the armyworm, *Leucania separata* Walker and the lorey leafworm, *Leucania loryei* Duponchel (Lepidoptera: Noctuidae), *Appl. Entomol. Zool.,* 15, 334, 1980.

270. **Nielsen, D. G., Purrington, F. F., Campbell, R. L., Wilmot, T. R., Capizzi, J., and Tumlinson, J. H.,** Sex attractants for sequoia pitch moth and strawberry crown moth, *Environ. Entomol.,* 7, 544, 1978.

271. **Yaginuma, K., Kumakura, M., Tamaki, Y., Yushima, T., and Tumlinson, J. H.,** Sex attractant for the cherry tree borer, *Synanthedon hector* Butler (Lepidoptera: Sesiidae), *Appl. Entomol. Zool.,* 11, 266, 1976.

272. **Tamaki, Y., Yushima, T., Oda, M., Kida, K., Kitamura, K., Yabuki, S., and Tumlinson, J. H.,** Attractiveness of 3,13-octadecadienyl acetates for males of clearwing moths, *Jpn. J. Appl. Entomol. Zool.,* 21, 106, 1977.

273. **Duckworth, W. D. and Eichlin, T. D.,** Two new species of clearwing moths (Sesiidae) from eastern North America clarified by sex pheromones, *J. Lepid. Soc.,* 31, 191, 1977.

274. **Voerman, S., Minks, A. K., Vanwetswinkel, G., and Tumlinson, J. H.,** Attractivity of 3,13-octadecadien-1-ol acetates to the male clearwing moth *Synanthedon myopaeformis* (Borkhausen) (Lepidoptera, Sesiidae), *Entomol. Exp. Appl.,* 23, 301, 1978.

275. **Capek, M.,** Vplyv orientacie feromonovych lapakov na ich atractivitu pre mnisku vel'kohlavu (*Lymantiria dispar* L.), *Lesnictvi,* 25, 301, 1979.

276. **Daterman, G. E., Robbins, R. G., Eichlin, T. D., and Pierce, J.,** Forest Lepidoptera attracted by known sex attractants of western spruce budworms, *Choristoneura* spp. (Lepidoptera: Tortricidae), *Can. Entomol.,* 109, 875, 1977.

277. **Hrdy, I., Marek, J., and Krampl, F.,** Sexual pheromone activity of 8-dodecenyl and 11-tetradecenyl acetates for males of several Lepidopteran species in field trials, *Acta Entomol. Bohemoslov.,* 76, 65, 1979.

278. **Ando, T., Yoshida, S., Tatsuki, S., and Takahashi, N.,** Sex attractants for male Lepidoptera, *Agric. Biol. Chem.,* 41, 1485, 1977.

279. **Struble, D. L. and Lilly, C. E.,** An attractant for the beet webworm, *Loxosotege sticticalis* (Lepidoptera: Pyralidiae), *Can. Entomol.,* 109, 261, 1977.

280. **Kovaleva, A. S., Ivanov, L. L., Pyratnova, I. B., Kovalev, B. G., Minyailo, V. A., and Minylio, A. K.,** Results of testing synthetic compounds for their ability to attract insects in nature, *Khemoretseptsiya Nasekomykh,* 2, 159, 1975.

281. **Roelofs, W. L. and Comeau, A.,** Lepidopterous sex attractants discovered by field screening tests, *J. Econ. Entomol.,* 63, 969, 1970.

282. **Mitchell, E. R. and Tumlinson, J. H.,** Attractant for males of *Spodoptera dolichos* (Lepidoptera: Noctuidae), *Ann. Entomol. Soc. Am.,* 66, 917, 1973.

283. **Oetting, R. D.,** Immature stages and biology of *Chionodes psiloptera* a pest in bluegrass seed fields (Lepidoptera, Gelechiidae), *Pan-Pacific Entomol.,* 53, 258, 1977.

284. **Roelofs, W. L. and Comeau, A.,** Sex attractants in Lepidoptera, Proc. 2nd Entomol. Congr. Pest Chem., IUPAC, Tel Aviv, 1971.

285. **Rothschild, G. H. L.,** Attractants for monitoring *Pectinophora scutigera* and related species in Australia, *Environ. Entomol.,* 4, 983, 1975.

286. **Nagano, M., Sato, Y., Sakai, M., and Fujiwara, H.,** Cross-attractiveness of synthetic sex pheromones of lepidopterous species, *J. Takeda Res. Lab.,* 38, 209, 1979.

287. **Biwer, G., Lalanne-Cassou, B., Descoins, C., and Samain, D.,** Sur le piegeage sexuel de *Sterrha biselata* (Hufn.) (Lepidoptera, Geometridae, Sterrhinae) par l'acetoxy-1 dodecadiene-7*E*,9*Z* pheromone sexualle de *Lobesia botrana* (Schiff.) (Lepidoptera, Tortricidae, Olethreutinae), *C. R. Acad. Sci. Paris, Ser. D,* 280, 1469, 1975.

288. **Voerman, S. and Herrebout, W. M.,** A sex attractant for the leaf miner moth *Lithocolletis corylifoliella* and its influence on that of *L. blancardella* (Lep., Gracillariidae), *Entomol. Exp. Appl.,* 23, 96, 1978.

289. **Roelofs, W. L., Reissig, W. H., and Weires, R. W.,** Sex attractant for the spotted tentiform leaf miner moth, *Lithocolletis blancardella, Environ. Entomol.,* 6, 373, 1977.

290. **Daterman, G. E., Peterson, L. J., Robbins, R. G., Sower, L. L., Daves, G. D., Jr., and Smith, R. G.,** Laboratory and field bioassay of the douglas-fir tussock moth pheromone, (*Z*)-6-heneicosen-11-one, *Environ. Entomol.,* 5, 1187, 1976.

291. **Grant, G. G.,** Interspecific pheromone responses of tussock moths and some isolating mechanisms of eastern species, *Environ. Entomol.,* 6, 739, 1977.

292. **Beroza, M., Katagiri, K., Iwata, Z., Ishizuka, H., Suzuki, S., and Bierl, B. A.,** Disparlure and analogues as attractants for two Japanese Lymantriid moths, *Environ. Entomol.,* 2, 966, 1973.

293. **Beroza, M., Punjapi, A. A., and Bierl, B. A.,** Disparlure and analogues as attractants for *Lymantria obfuscata, J. Econ. Entomol.,* 66, 1215, 1973.

294. **Grant, G. G. and Frech, D.,** Disruption of pheromone communication of the rusty tussock moth, *Orgyia antiqua* (Lepidoptera, Lymantriidae), with (*Z*)-6-heneicosen-11-one, *Can. Entomol.,* 112, 221, 1980.

295. **Steck, W. F., Underhill, E. W., Bailey, B. K., and Chisholm, M. D.,** Trace co-attractants in synthetic sex lures for 22 noctuid moths, *Experientia,* 38, 94, 1982.

296. **Steck, W., Underhill, E. W., Chisholm, M. D., Bailey, B. K., Loeffler, J., and Devlin, C. G.,** Sex attractants for males of 12 moth species found in western Canada, *Can. Entomol.,* 109, 157, 1977.

297. **Steck, W. F., Chisholm, M. D., Bailey, B. K., and Underhill, E. W.,** Moth sex attractants found by systematic field testing of 3-component acetate-aldehyde candidate lures, *Can. Entomol.,* 111, 1263, 1979.

298. **Underhill, E. W., Chisholm, M. D., Steck, W.,** Olefinic aldehydes as constituents of sex attractants for noctuid moths, *Environ. Entomol.,* 6, 333, 1977.

299. **Steck, W., Underhill, E. W., and Chisholm, M. D.,** Structure-activity relationships in sex attractants for north American noctuid moths, *J. Chem. Ecol.,* 8, 731, 1982.

300. **Struble, D. L. and Swailes, G. E.,** A sex attractant for adult males of the pale western cutworm, *Agrotis orthogonia* (Lepidoptera: Noctuidae), *Can. Entomol.,* 110, 769, 1978.

301. **Chisholm, M. D., Underhill, E. W., and Steck, W. F.,** Field trapping of the diamondback moth *Plutella xylostella* using synthetic sex attractants, *Environ. Entomol.,* 8, 516, 1979.

302. **Ando, T., Yoshida, S., Tatsuki, S., and Takahashi, N.,** Lepidopterous sex attractants, *Agric. Biol. Chem.,* 39, 1163, 1975.

303. **Steck, W. F., Underhill, E. W., Chisholm, M. D., and Gerber, H. S.,** Sex attractant for male alfalfa looper moths, *Autographa californica* (Speyer), *Environ. Entomol.,* 98, 373, 1979.

304. **Butler, L. I., Halfhill, J. E., McDonough, L. M., and Butt, B. A.,** Sex attractant of the alfalfa looper *Autographa californica* and the celery looper *Anagrapha falcifera* (Lepoidoptera: Noctuidae), *J. Chem. Ecol.,* 3, 65, 1977.

305. **McLaughlin, J. R., Mitchell, E. R., Beroza, M., and Bierl, B. A.,** Effect of *E*-*Z* concentration of 7-dodecenyl acetate on captures of four noctuid species in pheromone traps, *J. Georgia Entomol. Soc.,* 10, 338, 1975.

306. **Steck, W. F., Underhill, E. W., Chisholm, M. D., and Gerber, H. S.,** Sex attractant for male alfalfa looper moths, *Autographa californica, Environ. Entomol.,* 8, 373, 1979.

307. **Hall, D. R. and Read, J. S.,** Sex attractants for two zygaenid moths, *J. Entomol. Soc. South Afr.,* 42, 115, 1979.

308. **Ghizdavu, I., Hodosan, F., Popovici, N., Botar, A., and Barbas, A.,** A sex attractant for *Autographa gamma* L. Moth., *Rev. Roum. Biol. Ser. Biol. Anim.,* 24, 87, 1979.

309. **Steck, W., Underhill, E. W., Bailey, B. K., and Chisholm, M. D.,** A sex attractant for male moths of the glassy cutworm *Crymodeos devastator* (Brace): a mixture of Z-11-hexadecen-1-yl acetate, Z-11-hexadecenal and Z-7-dodecen-1-yl acetate, *Environ. Entomol.,* 6, 270, 1977.

310. **Kondratyev, Y. A., Lebedeva, E. V., and Pyatnova, Y. B.,** Progress in insect pheromone investigation, *Proc. VIII Int. Congr. Plant Protection Moscow,* 3, 44, 1975.

311. **Chisholm, M. D., Steck, W., and Underhill, E. W.,** Effects of additional double bonds on some olefinic moth sex attractants, *J. Chem. Ecol.,* 6, 203, 1980.

312. **Steck, W. F., Bailey, B. K., Underhill, E. W., and Chisholm, M. D.,** A sex attractant for the great dart, *Eurois occulta:* a mixture of (Z)-9-tetradecen-1-ol acetate and (Z)-11-hexadecen-1-ol acetate, *Environ. Entomol.,* 5, 523, 1976.

313. **Steck, W. F., Underhill, E. W., Chisholm, M. D., and Gerber, H. S.,** Sex attractant for male alfalfa looper moths, *Autographa californica* (Speyer), *Environ. Entomol.,* 98, 373, 1979.

314. **Struble, D. L. and Swailes, G. E.,** A sex attractant for the adult males of the cutworm *Euxoa auxiliaris:* a mixture of Z-5-tetradecenyl acetate and E-7-tetradecenyl acetate, *Environ. Entomol.,* 6, 719, 1977.

315. **Struble, D. L.,** Modification of the attractant blend for adult males of the army cutworm, *Euxoa auxiliaris* (Grote), and the development of an alternate 3-component attractant blend for this species, *Environ. Entomol.,* 10, 167, 1981.

316. **Underhill, E. W., Steck, W. F., Byers, J. R., and Chisholm, M. D.,** Z-5-decenyl acetate, a sex attractant for three closely related species, *Euxoa declarata,* Euxoa campestris and *Euxoa rockburnei, Can. Entomol.,* 113, 245, 1981.

317. **Byers, J. R., Underhill, E. W., Chisholm, M. D., and Teal, P. E.,** Biosystematics of the genus *Euxoa* Lepidoptera Noctuidae 15. Sex pheromone cross attractancy among the 3 closely related species of the *Euxoa declarata* group, *Can. Entomol.,* 113, 235, 1981.

318. **Struble, D. L. and Swailes, G. E.,** A sex attractant for adult males of the pale western cutworm, *Agrotis orthogonia* (Lepidoptera: Noctuidae), *Can. Entomol.,* 110, 769, 1978.

319. **Struble, D. L., Swailes, G. E., and Ayre, G. L.,** A sex attractant for males of the dark-sided cutworm, *Euxoa messoria* (Lepidoptera: Noctuidae), *Can. Entomol.,* 109, 975, 1977.

320. **Struble, D. L., Swailes, G. E., Steck, W. F., Underhill, E. W., and Chisholm, M. D.,** A sex attractant for *Leucania commoides* Gn.: a mixture of Z-9-tetradecen-1-yl acetate, Z-11-hexadecen-1-yl acetate and Z-11-hexadecen-1-ol, *Can. Entomol.,* 109, 1393, 1977.

321. **Steck, W., Bailey, B. K., Chisholm, M. D., and Underhill, E. W.,** A sex attractant for males of the cutworm *Euxoa pleuritica* (Lepidoptera: Noctuidae), *Can. Entomol.,* 110, 775, 1978.

332. **Struble, D. L., Steck, W. F., Swailes, G. E., Chisholm, M. D., Underhill, E. W., and Lilly, C. E.,** Two tow-component sex attractant blends for adult males of the striped cutworm, *Euxoa tessellata* (Lepidoptera: Noctuidae), *Can. Entomol.,* 114, 359, 1982.

323. **Steck, W. F., Struble, D. L., Lilly, C. E., Chisholm, M. D., Underhill, E. W., and Swailes, G. E.,** A sex attractant for males of the early cutworm, *Euxoa tristicula* (Lepidoptera: Noctuidae), *Can. Entomol.,* 111, 337, 1979.

324. **Voerman, S.,** Synthesis, purification, and field screening of potential insect sex pheromones, in *Chemical Ecology: Odour Communication in Animals,* Ritter, F. J., Ed., Elsevier/North-Holland, Amsterdam, 1979, 353.

325. **Roelofs, W., Cardé, A., Hill, A., and Cardé, R.,** Sex pheromone of the threelined leafroller, *Pandemis limitata, Environ. Entomol.,* 5, 649, 1976.

326. **Hill, A. S. and Roelofs, W. L.,** Two sex attractants for male speckled green fruitworm moths, *Orthosia hibisci* Guenée (Lepidoptera: Noctuidae), *J. N.Y. Entomol. Soc.,* LXXXVI, 296, 1978.

327. **Priesner, E., Bogenschütz, H., Altenkirch, W., and Arn, H.,** A sex attractant for pine beauty moth, *Panolis flammea, Z. Naturforsch.,* 33c, 1000, 1978.

328. **Struble, D. L., Swailes, G. E., Steck, W. F., Underhill, E. W., and Chisholm, M. D.,** A sex attractant for the adult males of variegated cutworm, *Peridroma saucia, Environ. Entomol.,* 5, 988, 1976.

329. **Steck, W., Underhill, E. W., and Chisholm, M. D.,** Trace components in lepidopterous sex attractants, *Environ. Entomol.,* 9, 583, 1980.

330. **Priesner, E.,** Sex attractant system in *Polia pisi* L. (Lepidoptera: Noctuidae), *Z. Naturforsch.,* 35c, 990, 1980.

331. **Kamm, J. A., McDonough, L. M., and Smithhisler, C. L.,** Sex attractant for *Protagrotis obscura* a pest of grass grown for seed, *Environ. Entomol.,* 11, 118, 1982.

332. **Weatherston, J., Davidson, L. M., and Simonini, D.,** Attractants for several male forest Lepidoptera, *Can. Entomol.,* 106, 781, 1974.
333. **Arsura, E., Capizzi, A., Piccardi, P., and Spinelli, P.,** (Z)-9-tetradecen-1-ol and (Z)-9-tetradecenyl acetate: a potent attractant system for male *Sesamia cretica* Led. (Lep., Noctuidae), *Experientia,* 33, 1423, 1977.
334. **Campion, D. G.,** Some observations on the use of pheromone traps as a survey tool for *Spodoptera littoralis, Centre Overseas Pest Res. Misc. Rep.,* 4, 1972.
335. **Blair, B. W. and Tannock, J.,** A possible pheromone for *Spodoptera triturata* (Walker) (Lepidoptera: Noctuidae), *Rhod. J. Agric. Res.,* 15, 225, 1977.
336. **Blair, B. W. and Tannock, J.,** A further note on the possible pheromone for *Spodoptera triturata* (Walker)(Lepidoptera, Noctuidae), *Rhod. J. Agric. Res.,* 16, 221, 1978.
337. **Campion, D. G.,** Sex pheromones and their uses for control of insects of the genus *Spodoptera, Meded. Fac. Landbouwwet. Rijksuniv. Gent,* 40, 283, 1975.
338. **Miyailo, V. A., Kovalev, B. G., Kirov, E. I., and Minyailo, A. K.,** On the attractiveness of disparlure, the sex pheromone of the gypsy moth *Porthetria dispar* (Lepidoptera: Orgyidae), for males of *Zanclognatha lunaris* (Lepidoptera, Noctuidae), *Zool. Zh.,* 56, 309, 1977.
339. **Adler, V. E., Jacobson, M., Edmiston, J. F., and Fleming, M. H.,** (Z,E)-9,12-tetradecadien-1-ol acetate: a potent attractant for male *Ephestiodes infimella, J. Econ. Entomol.,* 69, 706, 1976.
340. **Klun, J. A. and Robinson, J. F.,** Olfactory discrimination in the European corn borer and several pheromonally analogous moths, *Ann. Entomol. Soc. Am.,* 65, 1337, 1972.
341. **Stockel, J. and Anglade, P.,** Influence de la concentration en isomeres *cis* et *trans* dans la pheromone sexuelle d'*Ostrinia nubilalis* Hb. sur le comportement d'orientation de cette espece et sur celui de *Pyrausta aurata* Sc., *C. R. Acad. Sci. Paris, Ser. D,* 285, 61, 1977.
342. **Anglade, P.,** Emploi de pheromones sexuelles synthetiques pour l'attraction des males de la pyrale du mais (*O. nubilalis* Hbn.), *Rev. Zool. Agric. Pathol. Veg.,* 73, 37, 1974.
343. **Brennan, B. M.,** Attraction of *Ereunetis simulans* (Lepidoptera: Tineidae) to the sex attractant of peach tree borer, *Sanninoidea exitiosa* (Lepidoptera: Sessidae), *Proc. Hawaii Entomol. Soc.,* 22, 457, 1977.
344. **Ando, T., Kuroko, H., Nakagaki, S., Saito, O., Oku, T., and Takahashi, N.,** Two-component sex attractants for male moths of the subfamily Tortricinae (Lepidoptera), *Agric. Biol. Chem.,* 42, 1081, 1978.
345. **Frérot, B., Descoins, C., Lalanne-Cassou, B., Saglio, P., and Beauvais, F.,** Essais de piégeage sexuel des Lépidoptères Tortricidae des vergers de pommiers par des attractifs de synthése, *Ann. Zool. Ecol. Anim.,* 11, 617, 1979.
346. **Tagesdta, A. D.,** Spruce budworm detection in North Dakota shelterbelts and nurseries with a synthetic attractant, *North Dakota Farm Res.,* 33, 17, 1975.
347. **Roelofs, W. L. and Cardé, R. T.,** Sex pheromones in the reproductive isolation of lepidopterous species, in *Pheromones,* Birch, M. C., Ed., American Elsevier, New York, 1974, 96.
348. **Sziraki, G.,** Examination on Tortricid moths trapped by synthetic attractants (Lepidoptera), *Folia Entomol. Hung.,* 31, 259, 1978.
349. **Weatherston, J., Hedin, A. F., Ruth, D. S., MacDonald, L. M., Leznoff, C. C., and Fyles, T. M.,** Chemical and field studies on the sex pheromones of the corn and seed moths *Barbara colfaxiana* and *Laspeyresia youngana, Experientia,* 33, 723, 1977.
350. **Milaire, H. G.,** Role et portee du piegeage avec des substances attractives sexuelles dans les cultures fruitleres, *Pomol. Fruit Grow. Soc. Annu. Rep.,* 17, 13, 1975.
351. **Sanders, C. J., Daterman, G. E., Sheppard, R. F., and Cerezke, H.,** Sex attractants for two species of western spruce budworm, *Choristoneura biennis* and *C. viridis* (Lepidoptera, Tortricidae), *Can. Entomol.,* 106, 157, 1973.
352. **Weatherston, J., Percy, J. E., and MacDonald, L. M.,** Field testing of *cis*-11-tetradecenal as attractant or synergist in Tortricinae, *Experientia,* 32, 178, 1976.
353. **Burghardt, G., Knauf, W., Bestmann, H. J., Vostrowsky, O., and Koschatzky, K. H.,** Synthetische sexual-pheromone und deren inhibitoren für *Choristoneura muriana* Hbn. (Tannentriewickler, Lepidoptera: Tortricidae), *Anz. Schaedlingskd. Pflanz. Umweltschutz,* 53, 49, 1980.
354. **Jacobson, M. and Jennings, D. T.,** Attraction of *Rhyacionia neomexicana* (Dyar) to synthetic pheromones, *Environ. Sci. Health,* 13, 429, 1978.
355. **Frilli, F.,** L'impiego di attrattivi sessuali di sintesi per il controllo di *Cydia molesta, Entomoligica,* 10, 31, 1974.
356. **Beauvais, F., Biwer, G., and Charmillot, P. J.,** Essais de piegeage sexuel de *Grapholitha janthinana* Dup. par un melange d'acetoxy-1 dodecene 8Z et 8E (8Z DDA et 8E DDA), *Ann. Zool. Ecol. Anim.,* 9, 457, 1977.
357. **Alford, D. V.,** Observations on the specificity of pheromone-baited traps for *Cydia funebrana* (Treitschke) (Lepidoptera: Tortricidae), *Bull. Entomol. Res.,* 68, 97, 1978.

358. **Weatherston, J., Grant, G. G., MacDonald, L. M., Frech, D., Werner, R. A., Leznoff, C. C., and Fyles, T. M.,** Attraction of various tortricine moths to blends containing *cis*-11-tetradecenal, *J. Chem. Ecol.,* 4, 543, 1978.

359. **Greenway, A. R. and Wall, C.,** Attractant lures for males of the pea moth, *Cydia nigricana* (F.) containing (*E*)-10-dodecen-1-yl acetate and (*E,E*)-8,10-dodecadien-1-yl acetate, *J. Chem. Ecol.,* 7, 563, 1981.

360. **Minks, A. K., Voerman, S., and Vrie, van de,** A sex attractant for the cherry-bark tortrix moth, *Enarmonia formosana, Entomol. Exp. Appl.,* 19, 301, 1976.

361. **Chambon, J. P. and d'Aguilar, J.,** Remarques sur la specificite de quelques pheromones sexuelles de synthese: analyse faunistque des Tortricides en vergers, *Ann. Zool. Ecol. Anim.,* 6, 423, 1974.

362. **Bartell, R. J. and Lawrence, L. A.,** Reduction in responsiveness of male light-brown apple moths, *Epiphyas postivittana* to sex pheromone following pulsed pre-exposure to pheromone components, *Physiol. Entomol.,* 2, 89, 1977.

363. **Sower, L. L., Daterman, G. E., Sartwell, C., and Cory, H. T.,** Attractants for the western pine shoot borer, *Eucosma sonomana* and *Rhyacionia zozana* determined by field screening, *Environ. Entomol.,* 8, 265, 1979.

364. **Gentry, C. R., Beroza, M., and Blythe, J. L.,** Pecan bud moth: captures in Georgia in traps baited with the pheromone of the oriental fruit moth, *Environ. Entomol.,* 4, 227, 1975.

365. **Davis, H. G. and McDonough, L. M.,** Sex attractant for the filbertworm, *Melissopus latiferranus* (Walsingham), *Environ. Entomol.,* 10, 390, 1981.

366. **Guerin, P. M., Arn, H., Blaser, C., and Lettéré, M.,** Z,E-8,10-dodecadien-1-ol, attractant for male *Pammene rhediella, Ent. Exp. Appl.,* in press.

367. **Dix, M. E., Tagestad, A., and Jacobson, M.,** Synthetic attractants for male Lepidoptera, *Proc. North Central Branch Entomol. Soc. Am.,* 31, 29, 1977.

368. **Dix, M. E., Tagestad, A. D., and Jacobson, M.,** Trapping male *Petrova metallica* (Busck) with synthetic attractants, *Proc. North Central Branch Entomol. Soc. Am.,* 32, 63, 1977.

369. **Kinjyo, M., Sugie, H., and Tamaki, Y.,** Sex pheromone of the sugarcane borer, *Tetramoera schistaceana* Snellen. I. Isolation and identification, *Proc. 23rd Ann. Meet. Jpn. Soc. Appl. Entomol. Zool. Fukuoka,* Jpn. Soc. Appl. Entomol. Zool., 1979, 80.

370. **Knauf, W., Bestmann, H. J., Koschatzky, K. H., Suss, J., and Vostrowsky, O.,** Untersuchungen über die Lockwirkung synthetischer sex-pheromone bei *Tortrix viridana* (Eichenwickler) und *Panolis flammea* (Kieferneule), *Z. Angew. Entomol.,* 88, 307, 1979.

371. **Greenway, A. R. and Wall, C.,** Sex attractant lures for pea moth, *Rep. Rothamsted Exp. Stn. 1979,* 1, 117, 1980.

372. **Benz, G. and Salis, G.,** Use of synthetic sex attractant of larch bud moth *Zeiraphera diniana* (Gn.) in monitoring traps under different conditions, and antagonistic action of *cis*-isomer, *Experientia,* 29, 729, 1973.

373. **Baltensweiler, W., Priesner, E., Arn, H., and Delucchi, V.,** Unterschiedliche Sexuallockstoffe bei Larchen- und Arvenform des Grauen Larchenwicklers (*Zeiraphera diniana* Gn., Lep. Tortricidae), *Mitt. Schweiz. Entomol. Ges.,* 51, 133, 1979.

374. **Vrkoc, J., Skuhravy, V., and Baltensweiler, W.,** Freilanduntersuchungen zur Sexullockstoff-Reaction der Fichtenform des grauen Lärchenwicklers, *Zeiraphera diniana* Gn. (Lep., Tortricidae), *Anz. Schaedlingskd. Pflanz. Umweltschutz,* 52, 129, 1979.

375. **Ritter, F. J. and Persoons, C. J.,** Recent development in insect pheromone research, in particular in the Netherlands, *Netherlands J. Zool.,* 25, 261, 1975.

376. **Herrebout, W. N.,** Sex attractants for *Yponomeuta* species, *Netherlands J. Zool.,* 28, 276, 1978.

377. **McDonough, L. M., Kamm, J. A., George, D. A., Smithhisler, C. L., and Voerman, S.,** Sex attractant for the western lawn moth, *Tehama bonifatella* Eulst, *Environ. Entomol.,* 11, 711, 1982.

378. **Löfstedt, C., Van der Pers, J. N. C., Lofqvist, J., Lanne, B. S., Appelgren, M., Bergström, G., and Thelin, B.,** Sex pheromone components of the turnip moth, *Agrotis segetum:* chemical identification, electrophysiological evaluation and behavioral activity, *J. Chem. Ecol.,* 8, 1305, 1982.

379. **Honda, K. and Kawatoko, M.,** Exocrine substances of the white cabbage butterfly, *Pieris rapae crucivora* (Lepidoptera: Pieridae), *Appl. Entomol. Zool.,* 17, 325, 1982.

380. **Honda, K.,** Odor of a Papilionid butterfly: odoriferous substances emitted by *Atrophaneura alcinous alcinous* (Lepidoptera: Papilionidae), *J. Chem. Ecol.,* 6, 867, 1980.

381. **Bartell, R. J. and Bellas, T. E.,** Evidence for naturally occurring, secondary compounds of the codling moth female sex pheromone, *J. Aust. Entomol. Soc.,* 20, 197, 1981.

382. **Cory, H. T., Daterman, G. E., Daves, G. D., Jr., Sower, L. L., Shepherd, R. F., and Sanders, C. J.,** Chemistry and field evaluation of the sex pheromone of western spruce budworm, *Choristoneura occidentalis* Freeman, *J. Chem. Ecol.,* 8, 339, 1982.

383. **Novák, L., Toth, M., Balla, J., and Szántay, Cs.,** Sex pheromone of the cabbage armyworm, *Mamestra brassicae:* isolation, identification and stereocontrolled synthesis, *Acta Chim. Acad. Sci. Hung.,* 102, 135, 1979.

384. **Arai, K., Ando, T., Sakurai, A., Yamada, H., Koshihara, T., and Takahashi, N.**, Identification of the female sex pheromone of the cabbage webworm, *Agric. Biol. Chem.*, 46, 2395, 1982.

385. **Roelofs, W. L., Hill, A. S., Linn, C. E., Meinwald, J., Jain, S. C., Herbert, H. J., and Smith, R. F.**, Sex pheromone of the winter moth, a geometrid with unusually low temperature precopulatory responses, *Science*, 217, 657, 1982.

386. **Hart, E. R., Klun, J. A., and Weatherby, J. C.**, Field tests of the sex pheromone for the yellow-headed fireworm, *Acleris minuta* (Robinson), *Abstr. Natl. Entomol. Soc. Am. Meet. San Diego*, 1981, 77.

387. **Arn, H., Priesner, E., Bogenschutz, H., Buser, H. R., Struble, D. L., Rauscher, S., Voerman, S.**, Sex pheromone of *Tortrix viridana*: (Z)-11-tetradecenyl acetate as the main component, *Z. Naturforsch.*, 34c, 1281, 1979.

388. **Wall, C., Greenway, A. R., and Burt, P. E.**, Electroantennographic and field responses of the pea moth, *Cydia nigricana*, to synthetic sex attractants and related compounds, *Physiol. Entomol.*, 1, 151, 1976.

389. **Chisholm, M. D., Steck, W. F., Underhill, E. W., and Palaniswamy, P.**, Field trapping of diamondback moth *Plutella xylostella* using an improved four-component sex attractant blend, *J. Chem. Ecol.*, 9, 113, 1983.

390. **Roelofs, W. L. and Comeau, A.**, Sex attractants in Lepidoptera, in *Chemical Releasers in Insects*, Tahori, A. S., Ed., 1971, 91.

391. **Granges, J. and Baggiolini, M.**, Une pheromone sexuelle synthetique attractive pour le carpocapse des prunes (*Grapholitha funebrana* Tr. (Lep., Tortricidae)), *Rev. Suisse Vitic. Arboric.*, 3, 93, 1971.

392. **Mitchell, E. R., Sugie, H., and Tumlinson, J. H.**, *Spodoptera exigua*: capture of feral males in traps baited with blends of pheromone components, *J. Chem. Ecol.*, 9, 95, 1983.

393. **Arn, H., Delley, B., Baggiolini, M., and Charmillot, P. J.**, Communication disruption with sex attractant for control of the plum fruit moth, *Grapholitha funebrana*: a two-year field study, *Entomol. Exp. Appl.*, 19, 139, 1976.

394. **Schneider, D., Boppré, M., Zweig, J., Borsley, S. B., Bell, T. W., Meinwald, J., Hansen, K., and Diehl, E. W.**, Scent organ development in *Creatonotos* moths: regulation by pyrrolizidine alkaloids, *Science*, 215, 1264, 1982.

395. **Priesner, E., Bogenschtz, H., and Arn, H.**, A sex attractant for the European fir budworm moth, *Choristoneura murinana*, *Z. Naturforsch.*, 35c, 390, 1980.

396. **Struble, D. L. and Swailes, G. E.**, A sex attractant for the clover cutworm, *Scotogramma trifolii* (Rottenberg), a mixture of Z-11-hexadecen-1-ol acetate and Z-11-hexadecen-1-ol, *Environ. Entomol.*, 4, 632, 1975.

397. **Tamaki, Y., Ohsawa, T., Yushima, T., and Noguchi, H.**, Sex pheromone and related compounds secreted by the virgin females of *Spodoptera litura* (F.), *Jpn. J. Appl. Entomol. Zool.*, 20, 81, 1976.

398. **Kwangtung Institute of Analysis**, Isolation, identification, and synthesis of the sex pheromone of *Argyloploce (Eucosma) schistaceana* Snellen (Lepidoptera: Tortricidae), *Kexue Tongbao (K'o Hsueh T'ung Pao)*, 23, 438, 1978.

399. **Ando, T., Koshihara, T., Yamada, H., Vu, M. H., Takahashi, N., and Tamaki, Y.**, Electroantennogram activities of sex pheromone analogues and their synergistic effect on field attraction in the diamondback moth, *Appl. Entomol. Zool.*, 14, 362, 1979.

400. **Stevens, R. E., Sartwell, C., Koerber, T. W., Daterman, G. E., Sower, L. L., and Powell, J. A.**, Western *Rhyacionia* (Lepidoptera: Tortricidae, Olethreutinae) pine tip moths trapped using synthetic sex attractants, *Can. Entomol.*, 112, 591, 1980.

401. **Wu, D.-M. and Li, G.-L.**, EAG analysis and field attractant effect of sex pheromone of the striped sugarcane borer, *Acta Entomol. Sinica*, 25, 358, 1982.

402. **Chisholm, M. D., Underhill, E. W., and Steck, W. F.**, Field trapping of the diamondback moth *Plutella xylostella* using synthetic sex attractants, *Environ. Entomol.*, 8, 516, 1979.

403. **Sarmiento, R., Beroza, M., and Tardif, J. G. R.**, Activity of compounds related to disparlure, the sex attractant of the gypsymoth, *J. Econ. Entomol.*, 65, 665, 1972.

404. **Plimmer, J. R., Schwalbe, C. P., Paszek, E. C., Bierl, B. A., Webb, R. E., Marumo, S., and Iwaki, S.**, Contrasting effectiveness of (+) and (−) enantiomers of disparlure for trapping native populations of gypsy moth in Massachusetts, *Environ. Entomol.*, 6, 518, 1977.

405. **Miller, J. R., Mori, K., and Roelofs, W. L.**, Gypsy moth field trapping and electroantennogram studies with pheromone entiomers, *J. Insect Physiol.*, 23, 1447, 1977.

406. **Priesner, E.**, A new approach to insect pheromone specificity, *Olfaction Taste*, 3, 235, 1969.

407. **Gaston, L. K., Payne, T. L., Takahashi, S., and Shorey, H. H.**, Correlation of chemical structure and sex pheromone activity in *Trichoplusia ni* (Noctuidae), *Olfaction Taste*, 4, 167, 1971.

408. **Voerman, S., Minks, A. K., and Goewie, E. A.**, Specificity of the pheromone system of *Adoxophyes orana* and *Clepsis spectrana*, *J. Chem. Ecol.*, 1, 423, 1975.

409. **Cardé, R. T. and Roelofs, W. L.**, Attraction of redbanded leafroller moths, *Argyrotaenia velutinana*, to blends of (Z)-11- and (E)-11-tridecenyl acetate, *J. Chem. Ecol.*, 3, 143, 1977.

410. **Negishi, T., Ishiwatari, T., Uchida, M., and Asano, S.,** A new synergist, 11-methyl-*cis*-9,12-trideca-dienyl acetate for the smaller tea tortrix sex attractant, *Appl. Entomol. Zool.,* 14, 478, 1979.

411. **Birch, M. C.,** Aphrodisiac pheromones in insects, in *Pheromones,* Birch, M. C., Ed., American Elsevier, New York, 1974, 115.

412. **Hagan, D. V. and Brady, U. E.,** Absence of detectable 2-phenylethanol in *Trichoplusia ni,* a reported pheromone of males, *J. Georgia Entomol. Soc.,* 16, 192, 1981.

413. **Schneider, D., Boppré, M., Schneider, H., Thompson, W. R., Boriack, C. J., Petty, R. L., and Meinwald, J.,** A pheromone precursor and its uptake in male *Danaus* butterflies, *J. Comp. Physiol.,* 97, 245, 1975.

414. **Jones, I. F. and Berger, R. S.,** Incorporation of (1-^{14}C)acetate into *cis*-7-dodecen-1-ol acetate, a sex pheromone in the cabbage looper (*Trichoplusia ni*), *Environ. Entomol.,* 7, 666, 1978.

415. **Bjostad, L. B. and Roelofs, W. L.,** Sex pheromone biosynthesis from radiolabelled fatty acids in the redbanded leafroller moth, *J. Biol. Chem.,* 256, 7936, 1981.

416. **Schmidt, S. P. and Monroe, R. E.,** Biosynthesis of the wax moth sex attractants, *Insect Biochem.,* 6, 377, 1976.

417. **Inoue, S. and Hamamura, Y.,** The biosynthesis of "bombykol", sex pheromone of *Bombyx mori, Proc. Jpn. Acad.,* 48, 323, 1972.

418. **Yamaoka, R. and Hayashiya, K.,** Daily changes in the characteristic fatty acid, (Z)-11-hexadecenoic acid, of the pheromone gland of the silkworm pupa and moth, *Bombyx mori* L. (Lepidoptera: Bombycidae), *Jpn. J. Appl. Entomol. Zool.,* 26, 125, 1982.

419. **Bjostad, L. B., Wolf, W. A., and Roelofs, W. L.,** Total lipid analysis of the sex pheromone gland of the redbanded leafroller moth, *Argyrotaenia velutinana,* with reference to pheromone biosynthesis, *Insect Biochem.,* 11, 73, 1981.

420. **Tumlinson, J. H., Teal, R. E. A., and Heath, R. R.,** Chemical ethology: a holistic approach to the study of insect pheromones, *Proc. 1st Japan/USA Symp. on IPM, Tukuba (Japan),* Japan Plant Protection Assoc., Tokyo, 1981, 19.

421. **Tamaki, Y.,** Sex pheromones, in *Comprehensive Insect Physiology, Biochemistry and Pharmacology,* Vol. 9, Kerkut, G. A. and Gilbert, L. I., Eds., Pergamon Press, New York, 1985, 145.

422. **Cardé, R. T. and Webster, R. P.,** Endogenous and exogenous factors controlling insect sex pheromone production and responsiveness, particularly among the Lepidoptera, in *Regulation of Insect Development and Behaviour,* Kloza, M., Ed., Tech. Univ. Wroclaw, Karpacz, 1981, 977.

423. **Sower, L. L., Shorey, H. H., and Gaston, L. K.,** Sex pheromones of noctuid moths. XXI. Light: dark cycle regulation and light inhibition of sex pheromone release by females of *Trichoplusia ni, Ann. Entomol. Soc. Am.,* 63, 1090, 1970.

424. **Sower, L. L., Shorey, H. H., and Gaston, L. K.,** Sex pheromones of noctuid moths. XXV. Effects of temperature and photoperiod on circadian rhythms of sex pheromone release of females of *Trichoplusia ni, Ann. Entomol. Soc. Am.,* 64, 488, 1971.

425. **Bjostad, L. B., Gaston, L. K., and Shorey, H. H.,** Temporal pattern of sex pheromone release by female *Trichoplusia ni, J. Insect Physiol.,* 26, 493, 1980.

426. **Sugie, H., Yamazaki, S., and Tamaki, Y.,** On the origin of the sex pheromone components of the smaller tea tortric moth, *Adoxophyes fasciata* Walsingham (Lepidoptera: Tortricidae), *Appl. Entomol. Zool.,* 11, 371, 1976.

427. **Tamaki, Y.,** Rearing of *Adoxophyes orana* and *Homona magnanima* on simplified artificial food, *Jpn. J. Appl. Entomol. Zool.,* 10, 46, 1966.

428. **Arn, H., Esbjerg, P., Bues, R., Toth, M., Szocs, G., Guerin, P., and Rauscher, S.,** Field attraction of *Agrotis segetum* males in four European countries to mixtures containing three homologous acetates, *J. Chem. Ecol.,* 9, 267, 1983.

429. **Buser, H.-R., Rauscher, S., and Arn, H.,** Sex pheromone of *Lobesia botrana:* (*E,Z*)-7,9-dodecadienyl acetate in the female grape vine moth, *Z. Naturforsch.,* 29c, 781, 1974.

430. **Biwer, G. and Descoins, C.,** Approche d'un mecanisme d'isolement sexuel entre quatre especes de Tortricicae du genre *Grapholitha, C. R. Acad. Sci. Paris,* 286, 875, 1978.

431. **Hirai, K.,** Directional flow of male scent released by *Pseudaletia separata* Walker (Lepidoptera: Noctuidae) and its repellent effect on adults and larvae of four noctuid and one phycitine moth, *J. Chem. Ecol.,* 8, 1263, 1982.

432. **Hirai, K.,** Observation on the function of male scent brushes and mating behavior in *Leucania separata* W. and *Mamestra brassicae* L. (Lepidoptera, Noctuidae), *Appl. Entomol. Zool.,* 12, 347, 1977.

433. **Clearwater, J. R.,** personal communication.

434. **Hall, D. R., Beevor, P. S., Cork, A., Nesbitt, B. F., and La Croix, E. A. S.,** Laboratory and field studies of the female sex pheromones of *Cryptophlebia batrachopa* and *Cryptophlebia leucotreta,* pests of macadamia in Malawi, *Entomol. Exp. Appl.,* in press.

435. **Steck, W. F., Underhill, E. W., Bailey, B. K., and Chisholm, M. D.,** (*Z*)-7-tetradecenal, a seasonally dependent sex pheromone of the w-marked cutworm, *Spaelotis clandestina* (Harris) (Lepidoptera: Noctuidae), *Environ. Entomol.,* 11, 1119, 1982.

436. **Rothschild, G. H. L., Nesbitt, B. F., Beevor, P. S., Cork, A., Hall, D. R., and Vickers, R. A.,** Studies of the female sex pheromone of the native budworm, *Heliothis punctiger, Entomol. Exp. Appl.,* 31, 395, 1982.

437. **Wong, J. W., Palaniswamy, P., Underhill, E. W., Steck, W. F., and Chisholm, M. D.,** Novel sex pheromone components from the fall cankerworm moth, *Alsophila pometaria, J. Chem. Ecol.,* 10, 463, 1984.

438. **Underhill, E. W., Palaniswamy, P., Abrams, S. R., Bailey, B. K., Steck, W. F., and Chisholm, M. D.,** Triunsaturated hydrocarbones, sex pheromone components of *Caenurgina erechtea, J. Chem. Ecol.,* in press.

439. **Kassang, G., Schneider, D., and Schäfer, W.,** The silkworm moth *Bombyx mori:* presence of the (*E,E*)-stereoisomer of bombykol in the female pheromone gland, *Naturwissenschaften,* 65, 337, 1978.

440. **Capizzi, A., Tonini, C., Arsura, E., Guglielmetti, G., Massardo, P., and Piccardi, P.,** Sex pheromone components of the European goat moth, *Cossus cossus, J. Chem. Ecol.,* 9, 191, 1983.

441. **Zagatti, P., Bosson, G. A., Etienne, J., Brenière, J., Descoins, C., and Gallois, M.,** Phéromone sexuelle de *Chilo zacconius* Blesz., foreur du Riz en Afrique (Lépidoptère, Pyralidae), *C. R. Acad. Sci. Paris Ser. III,* 296, 85, 1983.

442. **Priesner, E. and Bogenschütz, H.,** Weitere Feldversuche zur Spezifität synthetischer Sexuallockstoffe bei *Choristoneura murinana, Z. Angew. Entomol.,* 94, 331, 1982.

443. **Benn, M. H., Galbreath, R. A., Holt, V. A., Young, H., Down, G., and Priesner, E.,** The sex pheromone of the silver Y moth *Chrysodeixis eriosoma* (Doubleday) in New Zealand, *Z. Naturforsch.,* 37c, 1130, 1982.

444. **Priesner, E., Altenkirch, W., Baltensweiler, W., and Bogenschütz, H.,** Evaluation of (*Z*)-5-decen-1-ol as an attractant for male larch casebearer moths, *Coleophora laricella, Z. Naturforsch.,* 37c, 953, 1982.

445. **Schwarz, M., Klun, J. A., Leonhardt, B. A., and Johnson, D. T.,** (*E,Z*)-2,13-octadecadien-1-ol acetate. A new pheromone structure for sesiid moths, *Tetrahedron Lett.,* 24, 1007, 1983.

446. **Struble, D. L.,** Sex pheromone components of *Euxoa drewseni:* chemical identification, electrophysiological evaluation, and field attractancy tests, *J. Chem. Ecol.,* 9, 335, 1983.

447. **Chisholm, M. D., Steck, W. F., Underhill, E. W., and Palaniswamy, P.,** Field trapping of diamondback moth *Plutella xylostella* using an improved four-component sex attractant blend, *J. Chem. Ecol.,* 9, 113, 1983.

448. **Steck, W. F., Underhill, E. W., Bailey, B. K., and Chisholm, M. D.,** A 4-component sex attractant for male moths of the armyworm, *Pseudaletia unipuncta, Entomol. Exp. Appl.,* 32, 302, 1982.

449. **Teal, P. E. A., McLaughlin, J. R., Tumlinson, J. H., and Rush, R. R.,** A sex attractant for *Platysenta videns* (Gn.) similar to the sex pheromone of *Heliothis virescens* (F.), *Fla. Entomol.,* 65, 583, 1982.

450. **Mitchell, E. R., Sugie, H., and Tumlinson, J. H.,** Rubber septa as dispensers for the fall armyworm sex attractant, *J. Environ. Sci. Health,* A18, 463, 1983.

451. **Toth, M.,** Male produced pheromone of *Mamestra brassicae* (L.) (Lepidoptera: Noctuidae): its possible role in courtship behaviour and study of other effects, *Acta Phytopathol. Acad. Sci. Hung.,* 17, 123, 1982.

452. **Klun, J. A., Leonhardt, B. A., Lopez, J. D., Jr., and Lachance, L. E.,** Female *Heliothis subflexa* (Lepidoptera: Noctuidae) sex pheromone: chemistry and congeneric comparisons, *Environ. Entomol.,* 11, 1084, 1982.

453. **Meyer, W. L., Debarr, G. L., Berisford, C. W., Barber, L. R., and Roelofs, W. L.,** Identification of the sex pheromone of the webbing coneworm moth, *Dioryctria disclusa* (Lepidoptera: Pyralidae), *Environ. Entomol.,* 11, 986, 1982.

454. **Kamm, J. A., McDonough, L. M., and Gustin, R. D.,** Armyworm (Lepidoptera: Noctuidae) sex pheromone: field tests, *Environ. Entomol.,* 11, 917, 1982.

455. **Neal, J. W., Jr., Klun, J. A., Bier-Leonhardt, B. A., and Schwarz, M.,** Female sex pheromone of *Choristoneura parallela* (Lepidoptera: Tortricidae), *Environ. Entomol.,* 11, 893, 1982.

456. **Underhill, E. W., Rogers, C. E., Chisholm, M. D., and Steck, W. F.,** Monitoring field populations of the sunflower moth, *Homoeosoma electellum* (Lepidoptera: Pyralidae), with its sex pheromone, *Environ. Entomol.,* 11, 681, 1982.

457. **Meinwald, J., Chalmer, A. M., Pliske, T. E., and Eisner, T.,** Pheromones. III. Identification of *trans,trans*-10-hydroxy-3,7-dimethyl-2,6-decadienoic acid as a major component in "hairpencil" secretion of the male monarch butterfly, *Tetrahedron Lett.,* 47, 4895, 1968.

458. **Meinwald, J., Chalmers, A. M., Pliske, T. E., and Eisner, T.,** Identification and synthesis of *trans,trans*-3,7-dimethyl-2,6-decadien-1,10-dioic acid, a component of the pheromonal secretion of the male monarch butterfly, *Chem. Commun.,* p. 86, 1969.

459. **Edgar, J. A., Culvenor, C. C. J., and Pliske, T. E.,** Isolation of a lactone, structurally related to the esterifying acids of pyrrolizidine alkaloidis, from the costal fringes of male Ithomiinae, *J. Chem. Ecol.,* 2, 263, 1976.

460. **Edgar, J. A., Culvenor, C. C. J., and Robinson, G. S.,** Hairpencil dihydropyrrolizines of Danainae from the New Herbrides, *J. Aust. Entomol. Soc.,* 12, 144, 1973.

461. **Edgar, J. A. and Culvenor, C. C. J.,** Pyrrolizidine ester alkaloid in danaid butterflies, *Nature (London),* 248, 614, 1974.

462. **Grant, G. G. and Slessor, K. N.,** Sex pheromone of the maple leaf roller *Cenopis acerivorana. (Lepidoptera: Tortricidae), Can. Entomol.,* 115, 189, 1983.

463. **Grant, G. G., Frech, D., MacDonald, L., and Doyle, B.,** Effect of additional components on a sex attractant for the oak leaf shredder, *Croesia semipurpurana* (Lepidoptera: Tortricidae), *Can. Entomol.,* 113, 449, 1981.

464. **Rothschild, G. H. L.,** Monitoring of pinkspotted bollworm, *Pectinophora scutigera* (Holdaway) (Lepidoptera: Gelechiidae) with sex attractant traps, *J. Aust. Entomol. Soc.,* 22, 161, 1983.

465. **Jain, S. C., Dussourd, D. E., Conner, W. E., Eisner, T. E., Guerrero, A., and Meinwald, J.,** Polyene pheromone components from an arctiid moth (*Utetheisa ornatrix*): characterization and synthesis, *J. Org. Chem.,* 48, 2266, 1983.

466. **Arn, H., Roehrich, R., Descoins, C., and Rauscher, S.,** Performance of five sex attractant formulations for the grape moth, *Eupoecilia ambiguella* Hb., in European vineyards, *Mitt. Schweiz. Entomol. Ges.,* 52, 45, 1979.

467. **Saglio, P., Descosins, C., Gallois, M., Lettere, M., Jaouen, D., and Mercier, J.,** Etude de la phéromone sexuelle de la Cochylis de la vigne *Eupoecilia (Clysia) ambiguella* Hb., Lépidoptère Tortricoidea, Cochylidae, *Ann. Zool. Ecol. Anim.,* 9, 533, 1977.

468. **Bestman, H. J., Brosche, T., Koschantzky, K. H., Michaelis, K., Platz, H., Roth, K., Süβ, J., Vostrowsky, O., and Knauf, W.,** Pheromone — XLII, 1,3,6,9-Nonadecatetraen, das Sexualpheromon des Frostspanners, *Operophtera brumata* (Geometridae), *Tetrahedron Lett.,* 23, 4007, 1982.

469. **Einhorn, J., Lallemand, J.-Y., Zagatti, P., Gallois, M., Virelizier, H., Riom, J., and Menassieu, P.,** Isolement et identification de la phéromone sexuelle attractive de *Hyphantria cunea* (Drury) (Lépidoptère, Arctiidae), *C. R. Acad. Sci. Paris Ser. II,* 294, 41, 1982.

470. **Bosson, G. A. and Gallois, M.,** Analyse de la sécrétion phéromonale émise par les femelles vierges de la mineuse des pousses de l'acajou: *Hypsipyla robusta* (Moore) (Lépidoptère, Pyralidae, Phycitinae), *C. R. Acad. Sci. Paris Ser. III,* 294, 819, 1982.

471. **Zagatti, P., Lalanne-Cassou, B., Descoins, C., and Gallois, M.,** Données nouvelles sur la sécrétion phéromonale de *Cryptophlebia leucotreta* (Meyr.) (Lepidoptera, Tortricidae), *Agronomie,* 3, 75, 1983.

472. **Frerot, B., Renou, M., Gallois, M., and Descoins, C.,** Un attractif sexuel pour la tordeuse des bourgeons: *Archips xylosteana* L. (Lepid., Tortricidae, Tortricinae), *Agronomie,* 3, 173, 1983.

473. **Chambon, J.-P. and Biwer, G.,** A Propos de la capture d'un Tortricidae nouveau pour la France: *Laspeyresia exquisitana* Rebel, *Bull. Soc. Entomol. France,* 83, 211, 1978.

474. **Decamps, C., Merle, P., Gourio, C., and Luquet, G.,** Attraction d'espèces du genre *Zygaena* F. par des substances phéromonales de Tordeuses (Lepidoptera, Zygaenidae et Tortricidae), *Ann. Soc. Entomol. France,* 17, 441, 1981.

475. **Renou, M. and Decamps, C.,** Les phéromones sexuelles des Zygènes (Lepidoptera, Zygaenidae): données d'électroantennographie, *C. R. Acad. Sci. Paris Ser. III,* 295, 623, 1982.

476. **Birch, M. C.,** Male abdominal brush-organs in British noctuid moths and their values as a taxonomic character, (Part I), *Entomologist,* 105, 185; (Part II), *Entomologist,* 105, 233, 1972.

477. **Culvenor, C. C. J. and Edgar, J. A.,** Dihydropyrrolidine secretions associated with coremata of *Utetheisa* moth (family Arctiidae), *Experientia,* 28, 627, 1972.

478. **Struble, D. L. and Richards, K. W.,** Identification of sex pheromone components of the female driedfruit moth, *Vitula edmandsae serratilineella,* and a blend for attraction of female moths, *J. Chem. Ecol.,* 9, 785, 1983.

479. **Bellas, T. E., Bartell, R. J., and Hill, A.,** Identification of two components of the sex pheromone of the moth, *Epiphyas postvittana* (Lepidoptera, Tortricidae), *J. Chem. Ecol.,* 9, 503, 1983.

480. **Honma, K., Kawasaki, K., and Tamaki, Y.,** Sex-pheromone activities of synthetic 7-alken-11-ones for male peach fruit moths, *Carposina niponensis* Walsingham (Lepidoptera: Carposinidae), *Jpn. J. Appl. Entomol. Zool.,* 22, 87, 1978.

481. **Shirasaki, S., Yamada, M., Sato, R., Yaginuma, K., Kumakura, M., and Tamaki, Y.,** Field tests on attractiveness of the synthetic sex pheromone of the peach fruit moth, *Carposina niponensis* Walsingham (Lepidoptera: Carposinidae), *Jpn. J. Appl. Entomol. Zool.,* 23, 240, 1979.

482. **Wakamura, S.,** Sex attractant pheromone of the turnip moth, *Agrotis segetum* Denis et Schiffermüller (Lepidoptera: Noctuidae): effects of the geometrical isomers on the attractant activity of the synthetic sex pheromone, *Appl. Entomol. Zool.,* 16, 496, 1981.

483. **Tatsuki, S., Kurihara, M., Usui, K., Ohguchi, Y., Uchiumi, K., Fukami, J., Arai, K., Yabuki, S., and Tanaka, F.,** Sex pheromone of the rice stem borer, *Chilo suppressalis* (Walker) (Lepidoptera: Pyralidae): the third component, Z-9-hexadecenal, *Appl. Entomol. Zool.,* 18, 443, 1983.

484. **Noguchi, H., Tamaki, Y., Arai, S., Shimoda, M., and Ishikawa, I.,** Field evaluation of synthetic sex pheromone of the oriental tea tortrix moth, *Homona magnanima* Diakonoff (Lepidoptera: Tortricidae), *Jpn. J. Appl. Entomol. Zool.,* 25, 170, 1981.

485. **Paker, S. P., Ed.,** *Synopsis and Classification of Living Organisms,* Vol. 2, McGraw-Hill, New York, 1982, 1232.

486. **Roelofs, W. L.,** Developing the potential of lepidopterous pheromones in insect control, in *Environmental Protection and Biological Forms of Control of Pest Organisms,* Lundholm, B. and Stackerud, M., Eds., *Ecol. Bull. (Stockholm),* 31, 25, 1980.

487. **Roelofs, W. L.,** Attractive and aggregating pheromones, in *Semiochemicals: Their Role in Pest Control,* Nordlund, D. A., Jones, R. L., and Lewis, W. J., Eds., John Wiley & Sons, New York, 1981, 215.

488. **Silverstein, R. M.,** Pheromones: background and potential for use in insect pest control, *Science,* 213, 1326, 1981.

489. **Mitchell, E. R., Ed.,** *Management of Insect Pests with Semiochemicals: Concepts and Practice,* Plenum Press, New York, 1981, 507.

490. **Campion, D. G. and Nesbitt, B. F.,** Lepidopteran sex pheromones and pest management in developing countries, *Trop. Pest Manage.,* 27, 53, 1981.

491. **Bartell, R. J.,** Mechanisms of communication disruption by pheromone in the control of Lepidoptera: a review, *Physiol. Entomol.,* 7, 353, 1982.

492. **Tamaki, Y.,** Insect sex pheromones and integrated pest management: problems and perspectives, in *IUPAC Pesticide Chemistry: Human Welfare and the Environment,* Miyamoto, J., Ed., IUPAC, Oxford, 1983, 37.

493. **Nakamura, K. and Tamaki, Y.,** *Sex Pheromone and Insect Pest Management: Experiments and Feasibility,* Kokin-Shoin, Tokyo, 1983, 202.

494. **Minks, A. K. and DeJong, D. J.,** Determination of spraying date for *Adoxophyes orana* by sex pheromone traps and temperature recordings, *J. Econ. Entomol.,* 68, 729, 1975.

495. **Riedl, H., Croft, B. A., and Howitt, A. J.,** Forecasting codling moth phenology based on pheromone trap catches and physiological-time models, *Can. Entomol.,* 108, 449, 1976.

496. **Tanaka, F. and Yabuki, S.,** Forecasting the oriental fruit moth, *Grapholitha molesta* Busk. emergence time on the pheromone trap method by the estimate of temperature, *Jpn. J. Appl. Entomol. Zool.,* 22, 162, 1978.

497. **Madsen, H. F. and Davis, W. W.,** A progress report on the use of female-baited traps as indicators of codling moth populations, *J. Entomol. Soc. Br. Columb.,* 68, 11, 1971.

498. **Madsen, H. F., Myburgh, A. C., Rust, D. J., and Bosman, I. P.,** Codling moth (Lepidoptera: Olethreutidae): correlation of male sex attractant trap captures and injured fruit in South African apple and pea orchards, *Phytophylactica,* 6, 185, 1974.

499. **Riedl, H. and Croft, B. A.,** A study of pheromone trap catches in relation to codling moth (Lepidoptera: Olethreutidae) damage, *Can. Entomol.,* 106, 525, 1974.

500. **Nakasuji, F. and Kiritani, K.,** Estimating the control threshold density of the tobacco cutworm *Spodoptera litura* (Lepidoptera: Noctuidae) on a corm crop, taro, by means of pheromone traps, *Protect. Ecol.,* 1, 23, 1978.

501. **Knipling, J. S. and McGuire, J. U., Jr.,** Population models to test theoretical effects of sex attractants used for insect control, *USDA Inf. Bull.,* 308, 2, 1966.

502. **Roelofs, W. L., Glass, E. H., Tette, J., and Comeau, A.,** Sex pheromone trapping for red-banded leafroller control: theoretical and actual, *J. Econ. Entomol.,* 63, 1162, 1970.

503. **Nakamura, K. and Kawasaki, K.,** The active space of the *Spodoptera litura* (F.) sex pheromone and the pheromone component determining this space, *Appl. Entomol. Zool.,* 12, 162, 1977.

504. **Teich, I., Waters, R. M., Jacobson, M., and Neumark, S.,** Reducing the number of insecticidal treatments by mass trapping of *Spodoptera littoralis* in safety belts around host crops in Israel, *J. Environ. Sci. Health,* A12, 291, 1977.

505. **Teich, I., Neumark, S., Jacobson, M., Klug, J., Shani, A., and Waters, R. M.,** Mass trapping of Egyptian cotton leafworm (*Spodoptera littoralis*) and large-scale synthesis of prodlure, in *Chemical Ecology: Odour Communication in Animals,* Ritter, F. J., Ed., Elsevier, Amsterdam, 1979, 343.

506. **Kovalev, B. G., Nedopekina, S. F., Cheknov, M. I., Osetsimsky, B. I., Lebedeva, K. V., and Bocharova, N. I.,** Issledovanie polovogo feromona, *Tortrix viridana, Khim. Prir. Soedin.,* 645, 1981.

507. **Yanagisawa, K.,** Use of sex pheromone for the control of *Spodoptera litura, Shokubutsu-Boeki (Plant Protection),* 33, 406, 1979.

508. **Shimada, K.,** Sex pheromone mass trapping for control of the smaller tea tortrix, *Appl. Entomol. Zool.,* 24, 81, 1980.

509. **Negishi, T., Ishiwatari, T., Asano, S., and Fujikawa, H.,** Trapping for the smaller tea tortrix control, *Appl. Entomol. Zool.,* 15, 113, 1980.

510. **Beroza, M.,** Insect attractants are taking hold, *Agric. Chem.,* 15, 37, 1960.
511. **Shorey, H. H., Gaston, L. K., and Saario, C. A.,** Sex pheromones of noctuid moths. XIV. Feasibility of behavioural control by disrupting pheromone communication in cabbage loopers, *J. Econ. Entomol.,* 60, 1541, 1967.
512. **Rothschild, G. H. L.,** Mating disruption of lepidopterous pests: current status and future prospects, in *Management of Insect Pests with Semiochemicals: Concepts and Practice,* Mitchell, E. R., Ed., Plenum Press, New York, 1981, 207.
513. **Brooks, T. W., Doane, C. C., and Staten, R. T.,** Experience with the first commercial pheromone communication disruptive for suppression of an agricultural insect pest, in *Chemical Ecology: Odour Communication in Animals,* Ritter, F. J., Ed., Elsevier, Amsterdam, 1979, 375.
514. **Doane, C. C. and Brooks, T. W.,** Research and development of pheromones for insect control with emphasis on the pink bollworm, in *Management of Pest Insects with Semiochemicals: Concepts and Practice,* Mitchell, E. R., Ed., Plenum Press, New York, 1981, 285.
515. **Ohtaishi, M. and Horikawa, T.,** Sex pheromones and the control of the smaller tea tortrix moth and the tea tortrix moth (Lepidoptera: Tortricidae), *Proc. 1st Japan/USA Symp. on IPM, Tsukuba (Japan),* 1981, 41.
516. **Tamaki, Y., Noguchi, H., Sugie, H., Ohtaishi, M., Horikawa, T., and Ohba, M.,** Simultaneous disruption of pheromonal communication of the two tortricid moths in the tea plantation, in *IUPAC Pesticide Chemistry: Human Welfare and the Environment,* Miyamoto, J., Ed., Pergamon Press, Oxford, 1983, 103.
517. **Tamaki, Y., Noguchi, H., and Sugie, H.,** Selection of disruptants of pheromonal communication in both the smaller tea tortrix moth and the tea tortrix moth (Lepidoptera: Tortricidae), *Jpn. J. Appl. Entomol. Zool.,* 27, 124, 1983.
518. **Vrkoc, J., Konyukhov, V. P., and Kovalev, B. G.,** Isolation and identification of main components of the pheromone complex of the moth *Agrotis exclamationis* (Lepidoptera, Noctuidae), *Acta Entomol. Bohemoslov.,* 80, 184, 1983.
519. **Kovalev, B. G., Nedopekina, S. F., and Konstantin, E. N.,** Sex attractant of females of gamma moth, *Insect Chemoreception,* Vol. 5, Mokslas, Vilnius, U.S.S.R., 1980, 43.
520. **Kovalev, B. G., Bolgar, T. S., and Konjukhov, V. P.,** Identification of *Amathes c-nigrum* L. sex pheromone component (Lepidoptera, Noctuidae), *Insect Chemoreception,* Vol. 5, Mokslas Vilnius, U.S.S.R., 1980, 48.
521. **Kovalev, B. G., Nedopekina, S. F., and Kost, A. N.,** Identifikatsiya polovogo feromona samok ogorodnoi sovki, *Mamestra oleracia* L. (Lepidoptera: Noctuidae), *Dokl. Akad. Nauk USSR,* 249, 370, 1979.
522. **Kovalev, B. G., Nedopekina, S. F., Lebedeva, K. V., and Kost, A. N.,** Vydclenie i identifikatsiya polovogo feromona samok kapustnoi sovki, *Bioorgan. Khim.,* 5, 912, 1979.

PHEROMONES OF THE COLEOPTERA

Hans Jürgen Bestmann and Otto Vostrowsky

INTRODUCTION

With the isolation and identification of (10*E*,12*Z*)-10,12-hexadecadienol, bombykol, the sex pheromone of the silk worm moth, *Bombyx mori* (Lepidoptera), by Butenandt and co-workers,[1] an exciting and promising method for a specific control of insect pests arose. This discovery established a new chemical basis for the understanding of insect behavior and an integral part of regulatory biology and thus a technology for manipulating insects. However, many investigations on the use of pheromones in pest management have done much to dispel this early optimism, almost certainly based on wrong or incomplete information or on misconceptions. In spite of this sometimes justified skepticism on the potential of pheromones for control, a better understanding of the subject is coming about by interdisciplinary collaboration between biologists and chemists.

Before starting to describe the chemical structures and the biological activity of pheromones, a number of terms have to be defined, designating the various kinds of chemical communication between insect individuals.[2] Chemicals which are delivering a message are called messenger substances or *semiochemicals*.[3] Among the semiochemicals which affect individuals other than the producing organism are those which have an effect on other species (interspecific effect), called *allelochemicals*,[4] and are divided into those which give an adaptive advantage to the receiver (*kairomones*[5]) or to the emitter (*allomones*[5]). Kairomones e.g., are substances that enable predators to locate their prey, host scents, and phago-stimulants whereas allomones are again divided into those with antagonistic effect, such as the defensive secretions of many arthropods, repellents, warning substances, feeding deterrents, antibiotics and poisons, and those with mutualistic effects, such as flower scents, attractants between symbionts, etc. Intraspecific semiochemicals, i.e., those which are sent and received by individuals of the same species, are called pheromones.[6,7] (For a critical comment on the definitions of signal substances and pheromones see Reference 8.)

Pheromones are classified, according to the response they elicit, into first, primer pheromones,[9,10] chemicals inducing delayed, lasting responses, possibly changing the endocrine state of the receiver and therewith its receptiveness for later signals. The second class is the releaser pheromones,[9,10] chemical stimuli that trigger an immediate and reversible change in the behavior of the recipient. The latter group, which is discussed below, includes for insects the female and male sex attractants, aggregation or recruitment pheromones, trail marking substances, and alarm pheromones. In a chemically relevant sense we arrange in this group also substances which influence this type of chemical communication, e.g., pheromone inhibitors, repellents, synergists, and mating disruptants, even when they are totally synthetic.

Sex pheromones are secreted by one sex to attract the other as an initial part of the mating process or to elicit responses of precopulatory and mating behavior. Sex attractants or *parapheromones,* chemicals often found through a screening program, are attractive to one sex (producing increased trap captures over captures in unbaited traps) and may also elicit additional responses similar to those evoked by the natural pheromonal emission. An attractant also produces a significant increase of trap captures, but is sex-unspecific and does not produce any other behavioral response. In some species, particularly among beetles (Coleoptera), the pheromones may be secreted by one or by both sexes and may also attract both sexes and therefore serve more than one function. These compounds are called *aggregation* or *recruitment pheromones*. *Trail pheromones* are chemicals used by foraging insects to lay a trail between a rich food source and the nest. This chemical cue guides other insects to the food location or to a new nesting site. *Alarm pheromones* are used by many social insects to alert other members of their species, causing aggressive behavior and alarm reactions, but may have other functions also in different concentrations.

A pheromone inhibitor reduces trap catches when exposed in a trap baited with an attractant, a pheromonal extract, or live insects; a mating disruptant is a chemical which, when disseminated throughout a test area, prevents intersexual chemical information and thus interferes with the ability of insects to locate the mates.

The pheromonal behavior response may be due to an individual substance, or, as is often the case, to a mixture of chemicals. In the latter, the total mixture is the pheromone or the pheromone complex, and its individual components are termed pheromone components. In some pheromone complexes all the components are necessary for eliciting a maximum response and quite often a distinct ratio of substances is required. If the total effect is greater than the sum of effects of the individual components, the pheromone is termed synergistic. Thus, a *synergist* is a compound which may not be attractive at all, but causes a significant increase in trap captures when exposed with an attractant.

Nonpheromonal chemicals, also releasing behavioral reactions and therefore potential agrochemicals for insect pest management, are the *kairomones,* produced by insects, which attract or stimulate a parasite or predator of that insect. Other substances of that kind are the larval attractants, oviposition stimulants, feeding stimulants, food baits, and repellents.

The distinction between the categories mentioned above is sometimes quite arbitrary. Thus, a compound designated as a synergist only, may ultimately prove to be an obligatory ingredient of a multicomponent pheromone complex. Many of these substances may have different affects on insects, sometimes depending on their concentration. Even a species-characteristic sex attractant can act as an inhibitor in high doses; the pheromones of some insects also serve as kairomones in that they attract predators.

The literature in the field of insect chemistry is increasing almost exponentially. At the time of this writing it certainly will have reached some 15,000 scientific papers published. A number of excellent reviews have been published in the form of books, periodicals, and articles in journals, dealing with insect pheromones.[11-32]

THE CHEMICAL STRUCTURE OF COLEOPTERAN PHEROMONES

Isolation

The structural elucidation of a pheromone, similar for all types of semiochemicals, can be divided into isolation and purification steps, the actual fractionation operations, different techniques of identification, the structural proof by synthesis, and the final test of the synthesis product for its biological activity.

For the isolation of a sex pheromone the storage and release rate of the pheromone by the insect should be considered. If it is occurring in relatively large quantities ready for release, it can be isolated by common extraction methods, by rinsing the pheromone gland segments, or by a distillation of suspensions or homogenates of insect parts. If released in small concentrations only or slowly over an extended period, it can be collected from the insect's surrounding air. In the first case the whole insects or the corresponding pheromone gland segments can be used for extraction and the cell tissue is often destroyed by grinding, mercerizing, or homogenizing by ultrasonic vibration. The solvents used can be decisive for the extract quality. Sometimes it is recommended to choose a rather inefficient solvent which extracts just enough of the pheromone desired, but reduces the accumulation of other impurities, e.g., lipids, fatty acids etc.

Rinsing the male dried bean beetle, *Acanthoscelides obtectus,* Horler isolated the sex pheromone methyl (2*E*)-2,4,5-tetradecatrienoate[33] and Henzell and Lowe found phenol in ether washings of female *Costelytra zealandica* abdomens.[34]

The odorous substances of Dufour glands of insects were isolated by vacuum distillation in a short-path distillation apparatus[35] or by condensing the volatile components in cooled capillaries.[36,37] By "sweep-codistillation" with nitrogen from the injection port of a gas chromatograph Buser and Arn isolated two lepidopteran sex pheromones.[38]

An isolation of volatiles from the air surrounding the insects is sometimes preferred to extraction, often saving further purification steps. Moreover, the pheromone emission can be taken from the insect containers by passing air over the insects. The pheromone-loaded airstream is usually passed into a conventional cold or cryogenic (liquid nitrogen) trap[39] or through an absorbent such as Porapak Q[40] or Tenax[41] (c.f., isolation of bark beetle pheromones[42-44]). Alternatively, the filter paper linings of the cages containing the insects were extracted for pheromone isolation, as well as parts of the insects kept in contact with GC-support material or adsorber material. (Methods for pheromone collection and isolation have been reviewed in References 45 to 50.)

Fractionation and Purification

Numerous methods of fractionation and purification have been used in the structure elucidation of pheromones. Due to the small amount of material available chromatographic techniques are the methods of choice. Among these, gel permeation chromatography (GPC, also known as steric exclusion chromatography, gel filtration, or gel chromatography) is often used in the first stage of purification. Column chromatography as a form of liquid chromatography using adsorbents like silica gel, aluminum oxide, and Florisil (silica gel and magnesium oxide) in ratios of 100:1 relative to the mixture to be separated, is often applied in pheromone isolation. Using eluents of increasing polarity, crude mixtures can be fractionated according to the polarities of each of the components. Since many pheromones possess an olefinic structure, silver nitrate impregnated adsorbents (usually 10 to 25% $AgNO_3$, Adsorbil CABN, Hi-Florisil-Ag, etc.) have proved useful in the separation of geometrical isomers.

The conventional technique of column chromatography has in recent years been more and more displaced by high performance liquid chromatography (HPLC). This method, approaching the efficiency of gas chromatographic separation and using automatic instrumentation[51] is finding increased use in insect chemistry. Thus, the four components of the pheromone complex of the boll weevil *Anthonomus grandis* were detected in the frass (feces) of feeding males by a rapid HPLC method with dual UV detection.[52] Using reversed-phase HPLC and a polar mobile phase containing $AgNO_3$, rapid separations of geometrical isomers of synthetic pheromones were achieved.[53]

Thin-layer chromatography (TLC) because of its small material demand, can also be a versatile tool for separations of relatively complex mixtures as well as for separation of (*Z*)- and (*E*)-olefins, using silver-loaded TLC plates. The method of high performance thin layer chromatography (HPTLC),[54] developed in the past few years, because of new techniques of charging the mobile phase, improvement in sample application, its high precision of separation, fast analysis time, and a high standard of reproducibility also promises to become a common method of separation in pheromone identification.

Because most semiochemicals are volatiles, GC is almost universally used in pheromone studies for the fractionation of mixtures and collection of the various components. For the condensation of the separated volatiles, cooled glass capillaries, capillary tubes with a temperature gradient, stationary gas chromatographic phases, glass wool, or powdered graphite are used or the eluted substances are introduced into cooled solvents. The quantitative gas chromatographic separation of enantiomers of the beetle pheromone chalcograne succeeded by means of complexation with a Ni(II)-salt.[55] (Fractionation and purification have been reviewed in detail.)[45,46,48,56-59]

Identification

The common techniques of spectroscopic and chromatographic identification, determination of physicochemical data, and the comparison with synthetic compounds can be used for elucidation of the chemical structure of pheromones. Although these substances have rather simple structures, the identification is no easy matter, since the amount produced by an individual is frequently at the nanogram level or even less. Not long ago, Tumlinson et al.[60] required 4.5 million boll weevils (*Anthonomus grandis*) to get enough material for the identification. With the improvement of instrumentation and the elaboration of ingenious microtechniques in the field of natural product chemistry, rapid progress in pheromone identification has been made in the past few years. Since analytical procedures for pheromone identification have been reviewed (c.f. References 45, 46, 49, 59, 61, and 62) only a few examples of these different techniques will be listed here for illustration.

Spectroscopic Methods

IR (infrared) spectra are useful in the determination of certain functional groups of unknown compounds. With micro IR cells, micro KBr disks, beam condensers, Fourier-transform IR spectrometers, GC-IR combinations, and computer accumulation, spectra can be obtained in the microgram scale[63] or even at the nanogram level.[64] Thus, Riley et al.[65] used IR spectroscopy for the identification of a termite trail pheromone, taking spectra of 25-μg samples, Persoons used 7- to 10-μg samples for the identification of the cockroach pheromone, periplanone B,[66] and Tumlinson et al.[67] similar quantities for those of two moth species. Williams et al.[68] took IR spectra from $CDCl_3$ solutions of the sex pheromone components of the male lesser grain borer, *Rhyzopertha dominica*, using 10-$\mu\ell$ cavity cells.

UV (ultraviolet) spectroscopy has found application for the identification of insect pheromones only in those cases when a UV absorbing chromophore was present in the molecular structure investigated. The more highly sensitive UV detectors are used in liquid chromatographic separations for the recognition of UV absorbing molecules during the fractionation of pheromones.

NMR (nuclear magnetic resonance) spectroscopy in the form of ^{1}H-proton and ^{13}C-carbon resonance spectroscopy certainly is the most widely used analytical technique in organic chemistry. Until recently the smallest sample size with which reliable proton NMR spectra could be obtained was the 50- to 100-μg scale (micro NMR sample tubes and computer accumulation of spectra). Tumlinson and Heath[61] have recently reported good spectra from only 2 μg of a pheromone component of the boll weevil. Byrne et al.[69] determined two structural elements in the beetle pheromone sulcatol. Chiral shift reagents have been used for the determination of the enantiomeric composition of several chiral pheromone bicyclic ketals,[70] esters,[68] and alcohols.[71]

α-Methoxy-α-trifluoromethylphenylacetyl derivatives (MTPA) of these alcohols were also used for the analysis of the enantiomeric composition.[71-73]

The required large sample size of more than 1 mg limits the applicability of ^{13}C-NMR spectroscopy in pheromone identification. However, it has proved useful in the determination of optical purity of synthetic α-multistriatin[74] and of the stereochemistry of bicyclic ketal pheromones.[75] Nishida et al.[76] determined the structure of a cockroach pheromone (38-mg samples) by means of ^{13}C-spectroscopy.

On some chiral pheromones optical rotatory dispersion (ORD) measurements[74] have been carried out as well as measurements of classical optical rotation. It should be pointed out here that the sign of rotation of a chiral substance can depend upon the solvent used: thus, (1S,4S,5S)-*cis*-verbenol is levorotatory in chloroform[77] and dextrorotatory in methanol and acetone.[73,78]

The determination of optical activity of chiral beetle pheromones is carried out indirectly in most cases. Usually the pure enantiomers are stereospecifically synthesized and the biological activity subsequently determined by bioassay. In some cases the optically active compounds are reacted with an enantiomeric pure reagent giving rise to diastereomeric products with different physical and chemical properties. Thus, by treating NMR samples with chiral shift reagents (see above), a different chemical shift of the ^{1}H- as well as the ^{13}C-resonance bands can be obtained in NMR spectroscopy. Similar results can be expected in taking spectra of diastereomeric reaction products (e.g., MTPA derivatives of alcohols). A difference in physical properties can also mean different retention behavior of diastereomeric compounds in chromatography. Consequently, these substances can be separated chromatographically. Another technique is the separation of enantiomers using chiral mobile (LC) or stationary (LC, HPLC, TLC, and GLC) phases.

The most frequently used analytical technique in pheromone chemistry and analysis of natural products is mass spectroscopy (MS). Modern mass spectrometers, magnetic sector instruments, as well as quadrupole or multipole instruments, already give spectra from ~10 ng or less of sample. In pheromone studies dealing with volatile compounds, the spectrometer is usually interfaced with a gas chromatograph (GC-MS combination) and connected with a data system. This device allows continuous recording, storage and delayed output of spectra, control of the mass spectrometer and regulation of scanning parameters, automatic registration of retention time, calculation of the signal areas, background subtraction, and recalling of spectra at a later time. In the case of electron impact ionization (EI spectra), the molecular weight of the compound analyzed can often be obtained; the fragmentation pattern from which the chemical structure, the class of compound, and the presence of functional groups can also be deduced. Using an ionized reagent gas for ionizating the sample (chemical ionization, CI spectra) results in a stable molecular ion (giving the molecular weight) and a spectrum with a small portion of decay products only. Field ionization (FI) or field desorption (FD) mass spectrometry is especially useful for evidence on unstable or nonvolatile compounds.

Apart from taking the mass spectra of actual pheromones sometimes the spectra of reaction products have been obtained for identification. In these cases the hydrogenolysis of sites of unsaturation, ozone cleavage of double bonds, and degradative hydrogenation for the determination of the carbon skeleton of the compounds investigated were the methods mostly used. Insect pheromones have also been identified with special mass spectrometry techniques such as mass chromatography or mass fragmentography (single or multiple ion detection). Using the first method, Beroza et al.[79] proved the presence of (8*E*,10*E*)-8,10-dodecadienol (codlemone) in a poorly purified extract of female codling moths *Laspeyresia pomonella*, and Bestmann et al.[80,81] proved the presence of (Z)-5-decenyl acetate in the female turnip moth *Agrotis segetum*. With the latter technique Buser and Arn[38] demonstrated the presence of (7*E*,9*Z*)-7,9-dodecadienyl acetate and codlemone in the sex pheromone extracts of the European grape berry moth *Lobesia botrana* and the codling moth *L. pomonella*, respectively, and Hendry et al.[82] proved the presence of the pheromone components of the oak leaf roller *Archips semiferanus*.

Chromatographic Methods

Knowledge of exact retention data also helps in the structure elucidation of insect pheromones. Thus, the retention time of a compound in GC i.e., the time of elution from the separating column, which is determined on polar and nonpolar GC phases, gives information with respect to polarity, molecular size, and the presence of functional groups in the compound analyzed. The use of the Kovats system of GC retention indices (RI),[83,84] which represent net retention time relative to the net retention time of an *n*-alkane expressed by the number of carbon atoms times 100 (e.g., hexane RI = 600, dodecane RI = 1200), considerably improves this analytical technique. Since lepidopterous sex pheromones are mostly long chain unsaturated compounds with one single functional group, this method has been especially helpful with the IR analysis.[85]

A solid injection technique for studying bark beetle pheromones was developed by Bridges and Guinn.[86] With this technique a whole beetle is placed directly into the injection port of a gas chromatograph, and the volatiles are vaporized for 1 min. The amount of volatile substances, thus obtained, was not different from those obtained by extraction with solvent.[86]

A single-insect aeration apparatus was designed for the determination of the pheromone release of the furniture carpet beetle *Anthrenus flavipes*. Tenax was selected as the adsorption agent. The pheromone (Z)-3-decenoic acid was quantitatively analyzed as its pentafluoro-benzyl derivative by electron-capture GC detection.[41]

Comparison of retention behavior of known and unknown compounds by TLC and HPLC can be used to establish identity. Argentation chromatography (using silver salt impregnated thin-layer plates) has often been applied to distinguish between geometrical isomers and thus to determine double bond configuration.

Reaction GC[87-91] gives the retention time of unknown substances as well as information about their chemical structures. By performing simple chemical reactions either before or after the separation of the analysis mixture, a signal of a certain compound can be eliminated from the chromatogram and a new signal arises. One common technique is to use hydrogen as a carrier gas and an appropriate catalyst in a precolumn. On injection of an unsaturated compound the substance may undergo hydrogenation, thus giving signals for the corresponding saturated compound, which quite often can be analyzed more easily. This method had been used in the structure elucidation of the pheromone of the curcullionid *Anthonomus grandis*.[92]

Carbon skeleton chromatography[93,94] utilizes hydrogenolytic conditions at elevated temperature and strips off all functional groups of the compounds investigated, thus giving the completely reduced parent hydrocarbons having the same carbon skeletons. With this analytic technique the structure of the bark beetle pheromone brevicomin could be determined from the final hydrogenation product *n*-nonane and additional spectral data[95] as well as the structure of the pheromone of the smaller European elm bark beetle *Scolytus multistriatus*.[96] Reaction or subtraction loops, connected to the GC column may "subtract" (retain) compounds possessing certain functional groups.[97,98] These loops can contain various chemicals for specific reactions thus aiding functional group analysis even for minute samples. The structure of the gypsy moth (*Lymantria dispar*) sex pheromone was provided in part when it was subtracted by phosphoric acid known to remove epoxides.[99]

Microchemical Methods

Chemical reactions performed on a micro scale are often used for the structure elucidation of pheromones, especially when spectral and chromatographic data are not sufficient for the analysis of the unknown compound.

Hydrogenation by GC (see above) or in solution is often used for the determination of the number of olefinic bonds. Double bond positions are usually determined by micro-ozonolysis and examination of the ozone cleavage products,[62,66,100,101] as well as by epoxidation,[102-104] glycol formation and derivatization,[105,106] and by methoxymercuration[107-109] and subsequent mass spectrometry.

The nature of functional groups in a pheromone can be analyzed by a number of functional group tests and determination of the loss (and possibly the subsequent recovery) of biological activity can be achieved in a bioassay.[45,48,62,110]

Table 1 shows a series of simple test reactions with which the presence or absence of a certain class of compounds can be proved in pheromone extracts. These group tests can be carried out in the form of a spot reaction with a corresponding TLC fraction as well as in reaction loops during gas chromatographic separations.

Electroantennogram, Olfactometer, and Field Screening Methods

The most commonly used technique for insect pheromone activity determination is the

Table 1
FUNCTIONAL GROUP TESTS AND BIOLOGICAL ACTIVITY DETERMINATION OF REACTION PRODUCTS BY BIOASSAY

Functional group of pheromone	Chemical reaction	Activity
Double bond	Hydrogenolysis,[a] ozonolysis, KMnO$_4$, perbenzoic acid, bromination	−
	Reaction of bromination products with Zn	+
Ester	Hydrolysis, LiAlH$_4$ reduction	−
Acetate	Acetylation of hydrolysis products in the neutral fraction	+
Methyl ester	Esterification of the acidic hydrolysis products with diazomethane	+
Carboxylic acid	Extraction with alkali, esterification with diazomethane	−
	Acidification of alkaline extracts, hydrolysis of methylated products	+
	LiAlH$_4$ or NaBH$_4$ reduction, ZnO reaction[a]	−
	Oxidation of reduction products with CrO$_3$/H$_2$SO$_4$	+
Alcohol	Acetylation and subsequent hydrolysis	+ −
	Boric acid reaction loop[a]	−
	Phthalic anhydride and subsequent hydrolysis	+ −
	Phenyl or naphthyl isocyanate	−
Aldehyde	LiAlH$_4$, NaBH$_4$, semicarbazide, 2,4-dinitrophenylhydrazone, hydroxylamine	−
	Benzidine subtraction loop[a], o-dianisidine subtraction loop[a], Girard-T-Girard-P-reagent	−
	Subsequent acidification with H$_2$SO$_4$	+
Ketone	LiAlH$_4$ reduction	−
	Subsequent oxidation with CrO$_3$ derivatives	+
Acetal	Hydrolysis with mineral acid	−
Epoxide	Treatment with dilute acid, periodic acid, phosphoric acid subtraction loop[a]	−
Amine	Extraction with acids	−
	neutralizing the acidic layer	+

[a] Used in the form of GC-loops in reaction gas chromatography.

electroantennogram (EAG).[111,112] This technique, the results of which are largely independent of environmental factors, is widely used for the determination of chain length, functional groups, and position and configuration of double bonds of the pheromone. With it the slow cell potentials of the antennal sensilla are measured electrophysiologically during stimulation with odorous compounds and the form and size of the response amplitude evaluated. (Roelofs et al.[113] obtained significant response to the (*E*)-7- and (*Z*)-9-isomers of dodecenyl acetates in the analysis of a moth pheromone, and consequently correctly predicted (7*E*,9*Z*)-7,9-dodecadienyl acetate to be the sex attractant.) Systematic EAG investigations may result in certain structure activity relationships, as was shown in Lepidoptera;[114] from these investigations pheromone structures have quite often been deduced.

Traditional bioassay techniques are the observation of stereotyped behavioral responses of the insects to specific test substances, observation of wing fluttering, trail following, anemotactic orientation etc., and an olfactometer test. An olfactometer[115] is a sort of a scent analyzer, consisting of chambers or tubes, differently loaded with stimulus chemicals, thus forcing the test insects to give behavioral answers like excitement, attraction, escaping reaction etc. A more modern type of olfactometer is a wind tunnel, developed for flying insects only.

With the increasing number of pheromones identified we gather knowledge about variations of chemical structure of insect pheromones with respect to order and family of insects. With this knowledge we may apply traps, baited with synthetic chemicals, to be tested in the field or in certain infestation areas for attractivity, pheromone inhibition, or repellency. But it should be noted that attraction to a particular synthetic compound does not constitute rigorous proof for a pheromone or pheromone component of that species.

SYNTHESIS

Before starting a detailed description of the syntheses of beetle pheromones we would like to focus the reader's attention to the chemical structure of these compounds and to their occurrence in the different insect species. In Table 2 the chemical names, referring to the nomenclature of organic chemistry, trivial names, synonyms, structural formulas as well as the occurrence in the insects are presented. (The identification and laboratory and field bioassays of beetle pheromones, scheduled according to the different beetle families and species, and the corresponding literature references are shown in detail in Table 5.) In the central column of Table 2 the chemical syntheses of the target molecules are found. The various syntheses are arranged according to the preparation of (1) racemic mixtures and (2) certain stereoisomers in the case of optically active signal molecules; a key reaction step or the starting material is given together with the corresponding literature reference.

Since in Table 2 the various syntheses of coleopteran pheromones are reviewed extensively, we intend to present here a few synthesis examples only, which are outstanding either through using special preparation techniques or because of the chosen target molecule. And since many beetle pheromones are chiral compounds we will concentrate on stereoselective syntheses in this section.

(*R*)(−)-14-methyl-(*Z*)-8-hexadecenal **XI** is the genuine sex pheromone of the dermestid species *Trogoderma inclusum* and *T. variabile*, isolated and identified by Cross et al.[276] without assigning the absolute configuration. Mori et al.[226] synthesized the (*R*)-enantiomer as well as the (*S*)-configured isomer, starting from (*R*)(+)-citronellol **I** (Scheme 1). The key intermediate was the optically active olefin (*R*)-**VI** which was converted into the decyne **VIII,** coupled with **IX,** hydrolyzed, (*Z*)-selectively partially hydrogenated, and oxidized to the (*R*)-aldehyde pheromone, (*R*)(−)-trogodermal, **XI**[226] (Scheme 1).

The synthesized (*R*)-isomer showed about 250-fold greater biological activity compared with its antipode when tested on *Trogoderma inclusum*.[226] Rossi and Carpita[228] prepared the (*S*)-**XI**-isomer by oxidizing the corresponding alcohol (*S*)-**X**, as well as the *trans*-compound (*S*)(+)-14-methyl-(*E*)-8-hexadecenal; both enantiomeric forms of the (*E*)-aldehyde were synthesized extending the synthesis given in Scheme 1.[232]

The allenic ester methyl (*R,E*)(−)-2,4,5-tetradecatrienoate **XVIII** was found in male dried bean beetles *Acanthoscelides obtectus*[33] and for the first time synthesized by Pirkle and Boeder[219] according to Scheme 2. The authors described synthesis of **XVIII** starting from 3-cyanopropanal **XII.** Three steps led to a diastereomeric mixture of the carbamates **XV** which could be separated by liquid chromatography. The coupling reaction of the (*S,R*)-isomer **XV** and lithium di-*n*-octylcuprate **XVI** gave methyl (*R*)-4,5-tetradecadienoate **XVII** which was converted into the (*R,E*)-2,4,5-tetradecatrienoate **XVIII**[219] (Scheme 2).

Table 2
CHEMICAL NAMES, SYNONYMS, STRUCTURAL FORMULAS, SYNTHESES, AND OCCURRENCE OF BEETLE PHEROMONES

Chemical name, common name or synonym (numbers refer to Tables 3 and 4, column 1)	Synthesis, key reaction, starting compound, and literature reference	Identified in, and/or biological activity proven for species
	Bicyclic Ketals	
(1*R*,5*S*,7*R*) (+)-exo-7-ethyl-5-methyl-6,8-dioxabicyclo[3.2.1]octane exo-Brevicomin (1)	Racemate syntheses: stereoselective and non-stereoselective;[95,116] thermal rearrangement of an epoxyketone;[117] Diels-Alder reaction;[118] mixed Kolbe electrolysis;[119] Eschenmoser fragmentation of epoxy ketone;[120] Diels-Alder and photochemical reaction;[121] fragmentation reaction of 3-chloro-2-cyclohexen-1-one;[122] cyclization of ketoepoxides;[123] as an exo-, endo-mixture via dimerization of methylvinylketone;[124] starting from pent-4-yn-1-ol[126] (1*R*,5*S*,7*R*) (+): from (2*S*,3*S*)-D-(−)-tartaric acid;[125] in a six-step sequence starting from diethyl (*S*,*S*)-tartrate;[127] starting from a nona-3.8-dienulose derivative[128] (1*S*,5*R*,7*S*)(−): from the enantiomeric tartaric acid[125]	*Dendroctonus adjunctus, D. brevicomis, D. jeffreyi, D. ponderosae, Leperisinus varius*
endo-7-Ethyl-5-methyl-6,8-dioxabicyclo[3.2.1]octane endo-Brevicomin (2)	Racemate syntheses: byproduct of nonstereoselective syntheses of exo-brevicomin;[116,118] stereoselective synthesis[95,129] thermal rearrangement of a *trans*-epoxy ketone;[117] epoxidation of a pure (*E*)-olefinic ketone; generated by Birch reduction;[120] starting from an (*E*)-olefinic alcohol;[130] dimerization of butadiene;[131] dimerization of methylvinylketone;[124] starting from pent-4-yn-1-ol[126]	*Dendroctonus brevicomis, D. frontalis, D. ponderosae*
(1*S*,5*R*)(−)-1,5-dimethyl-6,8-dioxabicyclo[3.2.1]octane Frontalin (3)	Racemic form:[132] Diels-Alder reactions;[118,133] thermolysis of epoxy ketones;[134] photochemical synthesis;[135] thermal decomposition of bicyclo[3.2.1]endoperoxide;[136] starting from pent-4-yn-1-ol[126] (1*S*,5*R*)(−): by separation of diastereomeric intermediates;[137] starting from D-glucose;[138] from methyl-α-D-glucopyranoside[139] (1*R*,5*S*)(+): by diastereomer separation;[137] starting from methyl-α-D-glucopyranoside;[139] using organosilicon reagent, starting from (*R*)(−)-linalool[140]	*Dendroctonus adjunctus, D. brevicomis, D. frontalis, D. ponderosae, D. pseudotsugae, Ips grandicollis*
2-Ethyl-1,6-dioxaspiro[4.4]nonane Chalcograne (4)	Racemate synthesis: stereoisomeric mixture;[141] aldol condensation and cyclizing hydrogenation;[142] Grignard reaction of acetylyde with lactone[143] (2*R*,5*RS*): starting from (*R*)(+)-caprolactone;[144] from D-glucose;[145] from α-amino-*n*-butyric acid;[146] building-block approach;[147] (2*S*,5*RS*): starting from α-amino-*n*-butyric acid;[146] from D-glucose;[145] separation of enantiomers by complexation chromatography[55]	*Pityogenes chalcographus*

Table 2 (continued)
CHEMICAL NAMES, SYNONYMS, STRUCTURAL FORMULAS, SYNTHESES, AND OCCURRENCE OF BEETLE PHEROMONES

Chemical name, common name or synonym (numbers refer to Tables 3 and 4, column 1)	Synthesis, key reaction, starting compound, and literature reference	Identified in, and/or biological activity proven for species
(1S,2R,4S,5R)(−)-2,4-dimethyl-5-ethyl-6,8-dioxabicyclo[3.2.1]octane α-Multistriatin (5)	Racemic form: tosylate coupling;[96] Diels-Alder approach;[148] stereoselective *trans*-opening of an epoxide;[149] methyl cyanocuprate and enol phosphate;[150] detailed synthesis description;[151] stereocontrolled synthesis[152] Isomeric mixture, starting from (S)(+)-2-methyl-3-butenoic acid, separated by GLC;[74] from (R)(+)-citronellol;[153] All stereoisomers with 1S-configuration: from (R)(+)-glyceraldehyde acetonide[154] (1S,2R,4S,5R)(−): starting material D-glucose[155] (−)-α- and (−)-γ-Multistriatin by resolving diastereomers[156]	*Scolytus multistriatus, S. scolytus*
(−)-δ-Multistriatin (1S,2S,4S,5R)(−)-δ-Multistriatin (6)	Racemate synthesis: Diels-Alder reaction[148] (1R,2R,4R,5S)(+): as a mixture with α-isomer, GLC separation;[74] from citronellol;[153] stereoselective ring opening of an epoxy ester[157] (1S,2S,4S,5R)(−): GLC separation from isomeric mixture;[154] ring opening of an epoxy ester[157]	

Pheromones with Terpenoid Structure

(S)(−)-2-methyl-6-methylene-7-octen-4-ol (S)(−)-ipsenol (7)	Racemic form: dithiane alkylation reaction;[78,158] building block type approach;[159] Grignard type reaction;[160] thermolysis of a cyclobutene derivative;[161] sigmatropic rearrangement of an allenyl vinyl ether;[162] Grignard reagent from chloroprene;[163] via retro-Diels-Alder reaction;[164] orthoester Claisen rearrangement;[165] coupling reaction of a terminal allene;[166] application of organosilane chemistry[167] (S)(−)-ipsenol: chiral epoxides and chiral α-methylene-γ-lactones as intermediates;[168,169] chiral epoxides and chloroprene Grignard reagent[170] (R)(+)-ipsenol: see synthesis[168,170]	*Ips acuminatus, I. amitinus, I. calligraphus, I. cembrae, I. confusus, I. cribricollis, I. grandicollis, I. latidens, I. sexdentatus, I. typographus, Pityokteines curvidens*
(S)(+)-2-methyl-6-methylene-2,7-octadien-4-ol (S)(+)-ipsdienol (8)	Racemate synthesis: alkylation of dithiane;[78] Grignard type reaction;[162,165,171] starting from myrcene;[172] coupling reaction and pyrolysis;[173] coupling of terminal allene;[166] isoprenylation with organosilane reagent;[167] Grignard type reaction[174] (S)(+)-ipsdienol: starting from (R)(+)-malic acid, chiral epoxide as key intermediate;[170] from optically active verbenone[175] (R)(−)-ipsdienol: starting from optically active verbenone;[175] via chiral epoxide, starting from (R)(+)-glyceraldehyde acetonide[170,176]	*Dendroctonus valens, Ips acuminatus, I. amitinus, I. avulsus, I. bonanseai, I. calligraphus, I. cembrae, I. confusus, I. cribricollis, I. duplicatus, I. integer, I. interstitialis, I. knausi, I. pini, I. sexdentatus, I. typographus*

Table 2 (continued)
CHEMICAL NAMES, SYNONYMS, STRUCTURAL FORMULAS, SYNTHESES, AND OCCURRENCE OF BEETLE PHEROMONES

Chemical name, common name or synonym (numbers refer to Tables 3 and 4, column 1)	Synthesis, key reaction, starting compound, and literature reference	Identified in, and/or biological activity proven for species
(1R,2S)(+)-1-methyl-1-(2-hydroxy)-ethyl-2-isopropenylcyclobutane (1R,2S)(+)-grandisol (9)	Racemate: photochemical cycloaddition;[60,92] photocycloaddition of ethylene and cyclohexenone;[177] cycloaddition of ethylene and 4-hydroxybutenolactone;[178,179] photocycloaddition of ethylene and methylcyclopentenone and ozonide fragmentation;[180] isoprene dimerization;[181] intramolecular photocyclization;[182] α-oxycyclopropylcarbinyl rearrangement;[183,184] cycloaddition of epoxynitrile;[185] cyclization of δ-chloroester;[186] annelation of a lithium cyclopropyl phenyl sulfide;[187] organocuprate reaction[188] (1R,2S)(+): starting from (1S,5S)(−)-β-pinene;[189,190] resolution of a bicyclic keto acid[191] (1S,2R)(−): see Ref. 191; oxidation of racemic grandisol, resolution and subsequent reduction[192]	*Anthonomus grandis, Rhabdoscelus obscurus*
(1R,4S,5R,7R)(+)-3,3,7-trimethyl-2,9-dioxatricyclo[3.3.1.0⁴'⁷]nonane Lineatin (10)	Racemate syntheses: isomeric mixture via 2 + 2 photocycloaddition;[193] photoaddition and chromatographical separation;[194] photocycloaddition with 2,4,4-trimethyl-2-cyclopenten-1-one;[195] cycloaddition of dichloroketene to isoprene;[196] in low yield via three synthesis pathways;[197] carbene addition to an unsaturated cyclic acetal;[198] thermal intramolecular ene-allene cyclization[199] (1R,4S,5R,7R)(+): resolution of intermediates of synthesis[195,196,200] (1S,4R,5S,7S)(−): Refs. 195,196,200	*Trypodendron domesticum, T. lineatum, Xyloterus signatus*
(Z)-3,3-dimethyl-Δ¹·ᵝ-cyclohexaneethanol (11)	Starting from 3,3-dimethylcyclohexanone;[60,92,201-204] cylization of methyl γ-geranate[205]	*Anthonomus grandis, Curculio caryae, Rhabdoscelus obscurus*
(Z)-3,3-dimethyl-Δ¹·α-cyclohexaneacetaldehyde (12)	From dimethylcyclohexanone;[60,92,201-204] from methyl γ-geranate;[205] Vilsmeier reaction of isophorone;[206,207] [3,3]sigmatropic rearrangement of allylic thiocarbamate;[208] as an isomer mixture via substitution of cyclohexanone derivative[209]	*Anthonomus grandis, Rhabdoscelus obscurus*
(E)-3,3-dimethyl-Δ1,α-cyclohexaneacetaldehyde (13)	Starting material dimethylcyclohexanone;[60,92,201-204] from γ-geranate;[205] from isophorone;[206,207] substitution with (Z)-2-ethoxyvinyllithium[209]	*Anthonomus grandis, Rhabdoscelus obscurus*

Table 2 (continued)
CHEMICAL NAMES, SYNONYMS, STRUCTURAL FORMULAS, SYNTHESES, AND OCCURRENCE OF BEETLE PHEROMONES

Chemical name, common name or synonym (numbers refer to Tables 3 and 4, column 1)	Synthesis, key reaction, starting compound, and literature reference	Identified in, and/or biological activity proven for species
(1R,4S,5R)(+)-pinen-4-ol (+)-*trans*-verbenol (14)	(1R,4S,5R)(+): from (1R,5R)(+)-α-pinene, resolution of the 3β-acetoxyetienate;[72] from (+)-*trans*-verbenyl acetate;[210] (1S,4R,5S)(−): see Ref. 72	*Dendroctonus brevicomis, D. frontalis, D. jeffreyi, D. ponderosae, D. valens, Ips amitinus, I. avulsus, I. bonanseai, I. calligraphus, I. confusus, I. grandicollis, I. knausi, I. pini*
(1S,4S,5S)-2-pinen-4-ol (S)-*cis*-verbenol (15)	Stereospecific reductions of verbenone[78,158] (1R,4R,5R): conversion of *trans*-verbenol[73]	*Dendroctonus brevicomis, D. frontalis, D. jeffreyi, D. ponderosae, D. valens, Ips amitinus, I. avulsus, I. bonanseai, I. calligraphus, I. confusus, I. grandicollis, I. integer, I. knausi, I. pini, I. typographus*
(−)-α-Cubebene (16)	Component of essential oils, e.g., oil of cubeb (*Piper cubeba*) As a mixture together with β-cubebene from (−)-*trans*-caran-2-one[211] β-Cubebene: from (−)-*trans*-caran-2-one[211]	*Scolytus multistriatus, S. scolytus*
Verbenone (17)	Component of essential oils From α-pinene with lead tetraacetate[212]	*Hylotrupes bajulus, Dendroctonus brevicomis, D. frontalis, D. jeffreyi, D. ponderosae, Ips confusus, Orthotomicus erosus*

Olefinic Compounds

Methyl (R,E)(−)-2,4,5-tetradecatrienoate (18)	Racemic form: reductive elimination reaction of α-hydroxyacetylenes;[213] substitution of a 3-acetoxyalkyne with organocopper reagent;[214] coupling of acetylenic compounds and elimination;[215] coupling of lithium diallenylcuprate;[216] orthoester Claisen rearrangement;[217] preparation of the corresponding tetradecatrien-1-ol[218] (R,E)(−): separation of diastereomeric carbamates as key intermediates[219] (S,E)(+): see synthesis;[219] orthoester Claisen rearrangement[220] *cis*-Isomer: via vinylstannanes[221]	*Acanthoscelides obtectus*
(R,Z)(−)-14-methyl-8-hexadecen-1-ol (19)	Racemate syntheses: Wittig carbonyl olefination;[222,223] Cope rearrangement[224] (R,Z)(−): acetylenic route, starting from citronellol;[225,226] coelectrolysis of (R)(−)-4-methylhexanoic acid with dodecynoic acid derivative[227]	*Trogoderma inclusum, T. variabile*

Table 2 (continued)
CHEMICAL NAMES, SYNONYMS, STRUCTURAL FORMULAS, SYNTHESES, AND OCCURRENCE OF BEETLE PHEROMONES

Chemical name, common name or synonym (numbers refer to Tables 3 and 4, column 1)	Synthesis, key reaction, starting compound, and literature reference	Identified in, and/or biological activity proven for species
Methyl (*R,Z*)(−)-14-methyl-8-hexadecenoate (20)	(*S,Z*)(+): starting from (*S*)-2-methylbutanol;[228-230] from citronellol[226] Racemate: Wittig olefination[222,223] (*S,Z*)(+): starting from (*S*)-2-methylbutanol[229,230]	*Trogoderma granarium, T. inclusum, T. variabile*
(*R,Z*)-14-methyl-8-hexadecenal (21)	Racemate synthesis: oxidation of the corresponding alcohol (*R,Z*)(−): starting from citronellol;[225,226] ring opening reaction of β-methyl-β-propiolactone[231] (*S,Z*)(+): starting from (*S*)-2-methylbutanol;[228] from citronellol[226]	*Trogoderma granarium, T. inclusum, T. variabile*
(*E*)-14-methyl-8-hexadecen-1-ol (22)	(*R,E*)(−): starting from (*R*)(+)-citronellol;[225,232] (*S,E*)(+): from (*S*)-2-methylbutanol, acetylenic route;[228] from citronellol[232]	*Trogoderma glabrum*
(*R,E*)(−)-14-methyl-8-hexadecenal (23)	(*S,E*)(+): starting from (*S*)-2-methylbutanol;[228] from citronellol[232] (*R,E*)(−): from citronellol[225,232]	*Trogoderma glabrum, T. granarium*
(*Z*)-3-decenoic acid (24)	Partial hydrogenation of acetylene derivative[233,234]	*Anthrenus flavipes*
(3*E*,5*Z*)-3,5-tetradecadienoic acid Megatomic acid (25)	Grignard reaction, alkyne hydrogenation with Lindlar catalyst;[235,236] as a mixture with (*E,E*)-isomer via condensation reactions with malonic acid[237]	*Attagenus megatoma*
(3*Z*,5*Z*)-3,5-tetradecadienoic acid (26)	Acetylene synthesis with final GLC fractionation from the isomers[236]	*Attagenus elongatus*
(*S*)(+)-1-methylbutyl (*E*)-2-methyl-2-pentenoate Dominicalure 1 (27)	Racemate synthesis: hydrolysis of a cyanohydrin and esterification with 2-pentanol[68] (*S*)(+)-enantiomer: esterification with (*S*)(+)-2-pentanol[68] (*R*)(−)-enantiomer: same reaction with (*R*)(−)-2-pentanol[68]	*Rhyzopertha dominica*
(*S*)(+)-1-methylbutyl (E)-2,4-dimethyl-2-pentenoate Dominicalure 2 (28)	Racemate synthesis: hydrolysis of cyanohydrin of methylpentanone and subsequent esterification[68] (*S*)(+)-enantiomer: esterification with optically active pentanol[68] (*R*)(−)-enantiomer: sequence like (*S*)-form[68]	*Rhyzopertha dominica*

Table 2 (continued)
CHEMICAL NAMES, SYNONYMS, STRUCTURAL FORMULAS, SYNTHESES, AND OCCURRENCE OF BEETLE PHEROMONES

Chemical name, common name or synonym (numbers refer to Tables 3 and 4, column 1)	Synthesis, key reaction, starting compound, and literature reference	Identified in, and/or biological activity proven for species
	Other Structures	

2,3-Dihydro-2,3,5-trimethyl-6-(1-methyl-2-oxobutyl)-4*H*-pyran-4-one (29)

Mixture of two diastereomeric racemates: biomimetic synthesis via condensation reactions[238,239]
(2*S*)-isomer: asymmetric reduction with baker's yeast[240]

Stegium paniceum

Stegobinone
3-Methyl-2-cyclohexen-1-ol

(2*R*,3*S*,7*RS*) and (2*S*,3*R*,7*RS*):[241]
(R)(+): starting from 1,3-cyclohexanedione, resolution of the 3-iodo-2-cyclohexen-1-ol intermediate[242]
(*S*)(–): see synthesis of (*R*)-isomer[242]

Dendroctonus pseudotsugae

Seudenol (30)

(3*S*,4*S*)(–)-threo-4-methylheptan-3-ol (31)

Racemate synthesis: reduction of the corresponding ketone;[96] Grignard synthesis[243]
Racemic threo- and erythro-isomers from nerol and geraniol, respectively[244]
(3*R*,4*R*)(+): starting from methyl (R)(+)-citronellate;[244] from optically pure butenolide;[245] enzymatic resolution of intermediates[246]
(3*S*,4*S*)(–): from optically pure butenolide;[245] enzymatic resolution of intermediates[246]
(*S*)-4-diastereomeric mixture: reduction of (*S*)-4-methylheptan-3-one[247]

Scolytus multistriatus, S. scolytus

4-Hydroxydodecanoic acid lactone
Dodecan-4-olide (32)

(*S*)(–)-isomer: enantioselective reduction of alkynyl ketones[248]
(*R*)(+)-isomer: from optically active propargylic carbinols[249]

Bledius mandibularius, B. spectabilis

(*R*,*Z*)(–)-4-hydroxy-5-tetradecenoic acid lactone (tetradecen-4-olide)
(*R*,*Z*)(–)-5-(1-decenyl)dihydro-2(3*H*)-furanone (33)

Racemate: see synthesis of (*R*,*Z*)[250]
(*R*,*Z*)(–): starting from (*R*)(–)-glutamic acid, (*Z*)-stereoselective Wittig reaction;[251] resolution of intermediates, acetylenic approach;[252] by asymmetric reduction;[254] enantioselective reduction of alkynyl ketones;[248] in essentially 100% optical purity from 4-oxo-5-tetradecynoate;[255] synthesis of racemic and (*R*,*Z*)-form;[250] asymmetric reduction and simple resolution[256]
(*S*,*Z*)(+): see Refs. 250 to 252

Popillia japonica

(*R*)(+)-γ-caprolactone (34)

Racemate commercially available[253]
(*R*)(+): starting from (*S*)(+)-glutamic acid;[257,258] from optically active propargylic carbinol;[249] broad spectrum synthesis of enantiomeric lactones;[259] asymmetric reducing reagent;[254]
(*S*)(–): from (*R*)(–)-glutamic acid;[257,258] see Ref. 259

Trogoderma glabrum, T. granarium, T. inclusum, T. variabile

Table 2 (continued)
CHEMICAL NAMES, SYNONYMS, STRUCTURAL FORMULAS, SYNTHESES, AND OCCURRENCE OF BEETLE PHEROMONES

Chemical name, common name or synonym (numbers refer to Tables 3 and 4, column 1)	Synthesis, key reaction, starting compound, and literature reference	Identified in, and/or biological activity proven for species
(S)(−)-2,5-dimethyl-2-isopropyl-2,3-dihydro-furan (35)	Racemate synthesis: 2,5-hexanedione and isopropyl magnesium bromide[260,261] (S)(−)-isomer: from methyl-2-C-acetyl-4,6-O-benzylidene-α-D-ribohexopyranose[260]	*Hylecoetus dermestoides*
(4S,6S,7S)-4,6-dimethyl-7-hydroxynonan-3-one (36) Serricornin	Racemate: Michael addition via ketonitrile[262] Diastereomeric mixture[263,264] (4RS,6R,7R)-form: starting material methylmalic acid, resolution of intermediates[265] (4RS,6S,7S)-form: yeast reduction of a β-keto ester[266] (4RS,6R,7S)-form from β-methylmalate, see Ref. 265 (4S,6R,7R) isomer: multistep synthesis, starting from D-glucose[267] (4R,6R,7S)-isomer:[268] (4S,6R,7S)-isomer:[268] (4S,6S,7S)-isomer:[269]	*Lasioderma serricorne*
2,6-Diethyl-3,5-dimethyl-3,4-dihydro-2H-pyran Anhydroserricornin (37)	Racemate: Diels-Alder reaction, hydrogenation and elimination;[262] Michael addition and cyclization of δ-hydroxyketones[260,262] (6S,7S)-isomer: optical purity ≥85%[266]	*Lasioderma serricorne*
(S)(+)-6-methyl-5-hepten-2-ol (S)(+)-sulcatol (38)	(S)(+)-isomer: asymmetric reduction, employing baker's yeast[270]	*Platypus flavicornis, Gnathotrichus retusus, G. sulcatus*
8-Methyl-2-decanol propanoate (65)	Racemate synthesis[271] Synthesis of all four stereoisomers[272]	*Diabrotica virgifera virgifera*
10-Methyl-2-tridecanone (66)	Racemate synthesis[273] Preparation of the two enantiomers[274]	*Diabrotica undecimpunctata*
4,8-Dimethyldecanal (43)	Facile synthesis, also of analogs[275]	*Tribolium castaneum, T. confusum*

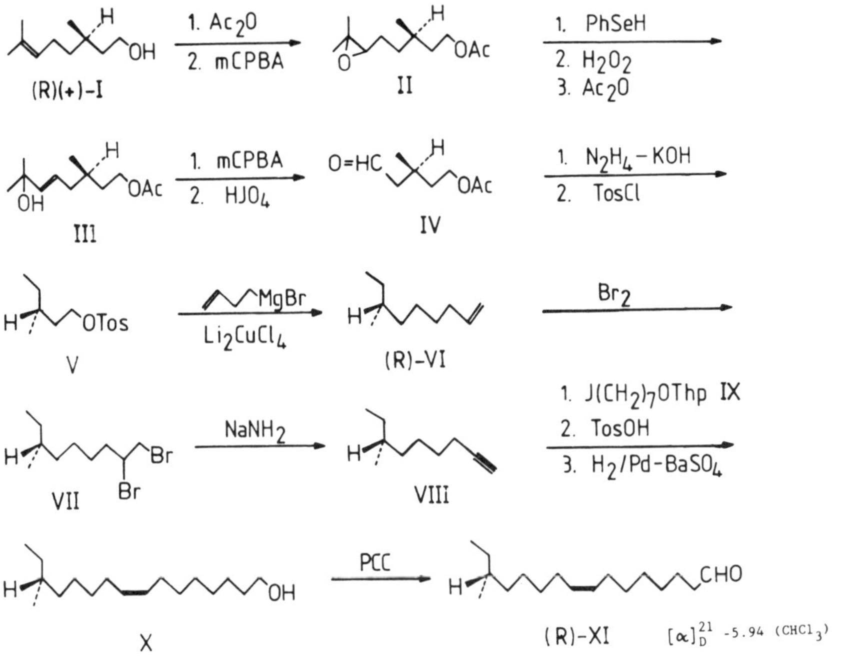

PCC pyridinium chlorochromate
m-CPBA m-chloroperbenzoic acid
Thp tetrahydropyranyl

Scheme 1.

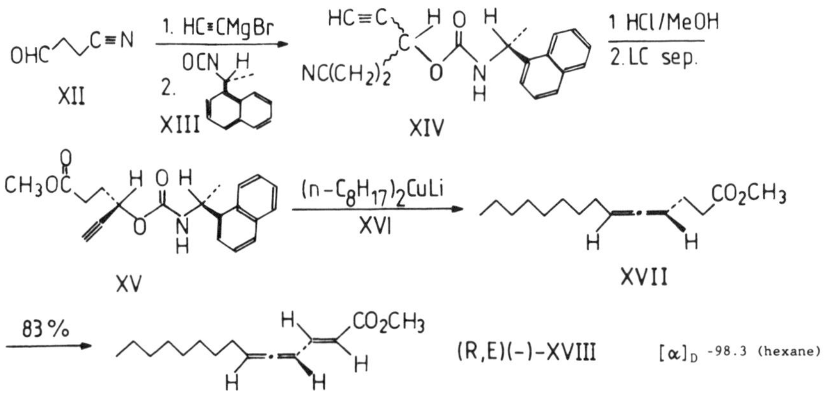

Scheme 2.

Using an orthoester Claisen rearrangement Mori et al.[220] recently synthesized (*S,E*) (+)-**XVIII.** This antipode of the natural pheromone had 125% of the (opposite) rotatory power of the pheromone, thus implying that the isolated natural substance was either chemically or enantiomerically impure.

Chalcograne, 2-ethyl-1,6-dioxaspiro[4.4]nonane **XXV,** is the aggregating pheromone of the bark beetle *Pityogenes chalcographus*;[141] the exact isomeric composition of the natural substance is still unknown. Whereas Smith et al.[144] obtained (2*R*,5*R*)- and (2*R*,5*S*)-chalcograne from optically active γ-caprolactone and separated the stereoisomers by preparative GLC, Mori prepared the four isomers (2*R*,5*R*)-**XXV,** (2*R*,5*S*)-**XXV,** (2*S*,5*R*)-**XXV,** and (2*S*,5*S*)-**XXV**[146] using the reaction of the dianion of **XXIII** with the chiral epoxides **XXIIa**

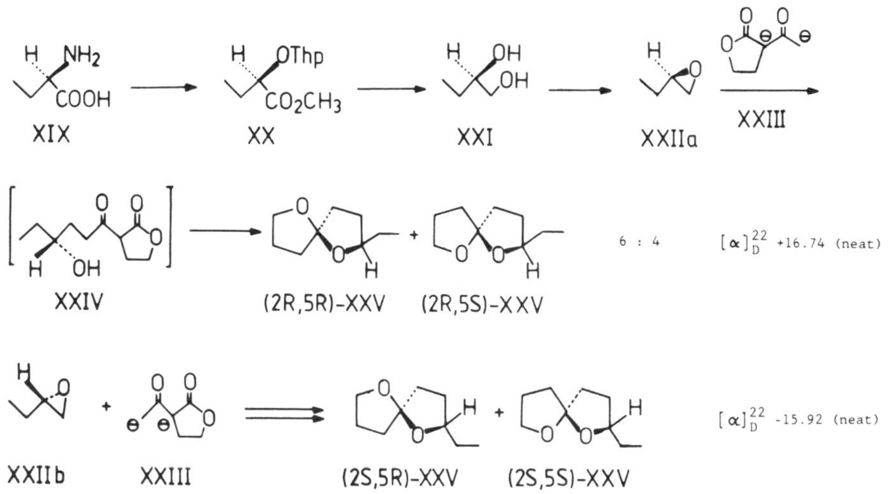

Scheme 3.

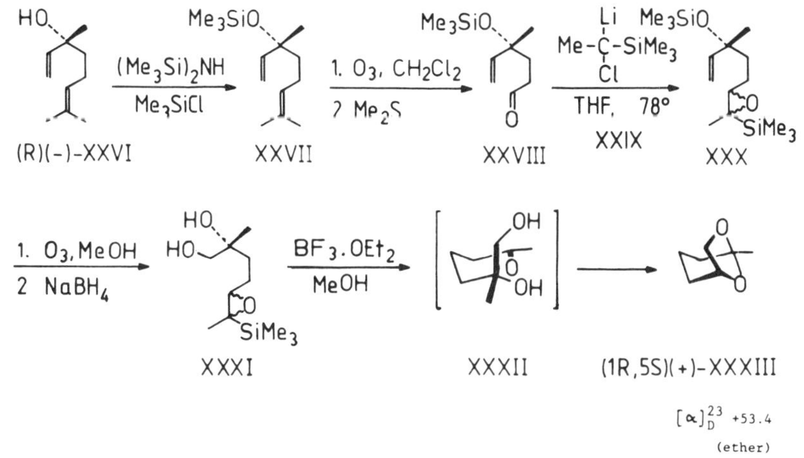

Scheme 4.

and **b**. This condensation yielded a 60:40 mixture of (2R,5R)- and (2R,5S)-**XXV** in the case of epoxide **XXIIa** and a 59:41 mixture of (2S,5R)- and (2S,5S)-**XXV**, respectively, with epoxide **XXIIb** (Scheme 3). The chiral epoxides **XXII** could be derived in seven steps from the enantiomers of optically active α-amino-*n*-butyric acid **XIX**. No separation of the mixture of the final products **XXV** was attempted since the naturally occurring pheromone itself was a mixture.

Another type of beetle pheromone with an intramolecular ketal linkage is frontalin, 1,5-dimethyl-6,8-dioxabicyclo[3.2.1]octane, which was extracted and identified from about 6500 hindguts of the male western pine beetle *Dendroctonus brevicomis*.[133] After a series of syntheses of the racemic form of the pheromone (c.f. Table 2), Mori reported a synthesis of the enantiomers (1S,5R)(−)-**XXXIII** and (1R,5S)(+)-**XXXIII** starting from an optically active lactonic acid, and reported biological activity for the (1S,5R)-isomer **XXXIII** only.[137] In 1976 two preparations of **XXXIII** were published[138,139] starting with D-glucose derivatives; whereas 2 years later Magnus and Roy described the preparation of the (1R,5S)-(+)-form of the pheromone **XXXIII** starting from (R)(−)-linalool **XXVI**, using a new organosilicon reagent **XXIX**[140] (Scheme 4).

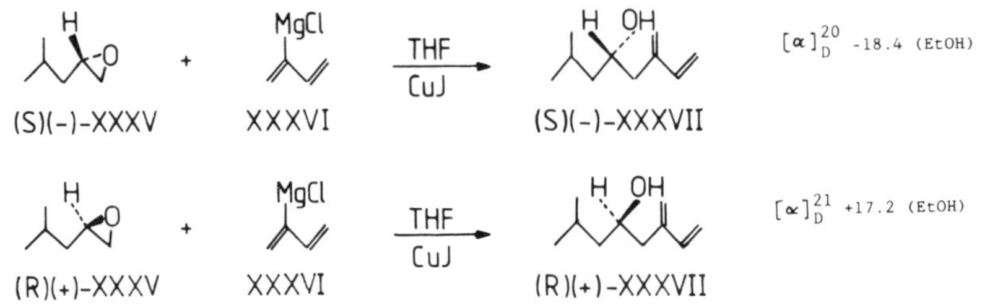

Scheme 5.

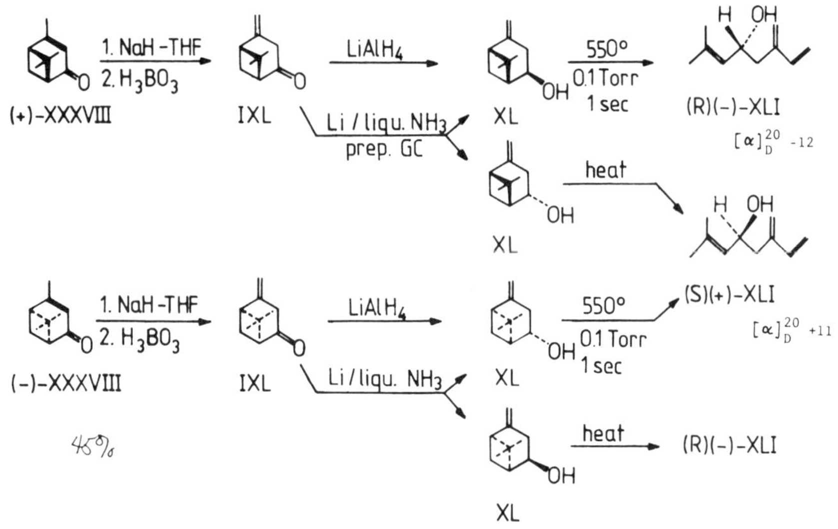

Scheme 6.

Ipsenol **XXXVII** and ipsdienol **XLI** were isolated by Silverstein et al.[158,277] from the frass of *Ips paraconfusus*. The natural ipsenol was levorotatory, $[\alpha]_D^{25} - 17.5° \pm 0.7$ (EtOH), whereas ipsdienol was dextrorotatory, $[\alpha]_D^{20} + 10° \pm 0.9$ (MeOH).[158,277] Mori's second synthesis of ipsenol **XXXVII** started with the chiral epoxides **XXXV,** which were derived from leucine **XXXIV**.[170] A simple coupling of **XXXV** with the Grignard reagent **XXXVI** prepared from chloroprene gave $(S)(-)$-ipsenol **XXXVII** and $(R)(+)$-**XXXVII,** respectively[170] (Scheme 5).

Optically active ipsdienol **XLI** was prepared in an elegant way by Ohloff and Giersch[175] starting with verbenone **XXXVIII**. Isomerization and reduction gave the four isomeric 2(10)-pinen-4-ols **XL** which were converted by flash pyrolysis into $(R)(-)$-ipsdienol **XLI** and $(S)(+)$-**XLI,** respectively. Since the two bridgehead chiral centers were destroyed during pyrolysis, the configuration of the hydroxy-substituted carbon was solely responsible for the geometry of the final products **XLI**[175] (Scheme 6).

(R,Z)-4-hydroxy-5-tetradecenoic acid lactone **XLVII** is the pheromone isolated from the female Japanese beetle *Popillia japonica*[251] and the synthesis of both enantiomers was originally described in the same paper.[251] The authors employed the Wittig reaction for the conversion of lactone aldehyde **XLV** with phosphorus ylide **XLVI** and obtained the final product **XLVII** in a highly optically pure state. The 10 to 15% contamination of **XLVII** with its corresponding (E)-isomer had to be removed by preparative HPLC and GLC. The enantiomers of the chiral lactones **XLIII** could be synthesized from $(R)(-)$-glutamic acid **XLII** and $(S)(+)$-glutamic acid, respectively[251] (Scheme 7).

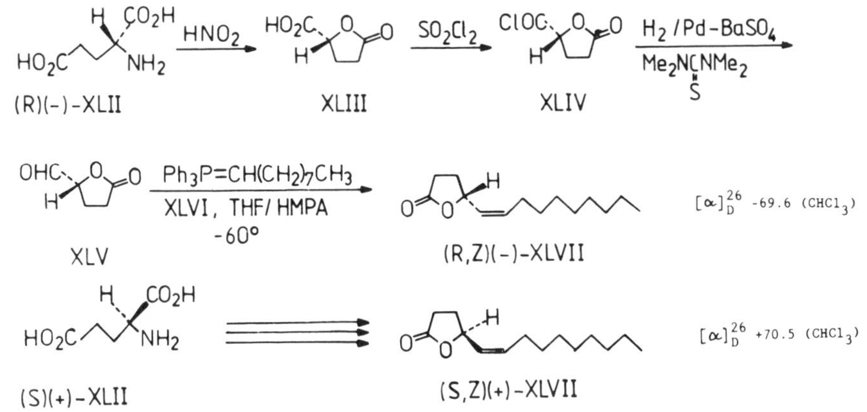

Scheme 7.

With these few examples of beetle pheromone syntheses we certainly do not claim to have presented completeness on this subject. However, these examples stand as representative for numerous other syntheses, showing the difficulties especially in the preparation of optically active substances and the ingenuity of chemists in sometimes forcing nature to react in one direction only.

Further details can be found in numerous papers and books reviewing the chemistry and synthesis of insect pheromones.[11,14-16,25,29,48,210,278-286]

PHYSICAL PROPERTIES AND ANALYTICAL AND SPECTRAL DATA OF COLEOPTERAN PHEROMONES

This section, presented in Tables 3 and 4, is related to the synthesis as well as to the identification of pheromones. It gives information on the physical properties such as melting and boiling points of the compounds, the elementary composition, the molecular weight, the refractive index (RI), and literature references (Table 3). In the case of chiral substances the sign and angle of rotation is given (Table 3). Table 4 gives spectral data including NMR spectra, IR and mass spectra, and literature references concerning analytical properties of the semiochemicals.

Chiral pheromones are often represented by one enantiomer only (Tables 3 and 4) which does not conclusively mean that this one isomer is the only one responsible for biological activity. Melting and boiling points are given in °C and the distillation pressure is given in mmHg. The superscripts of the RI n_D as well as of the angle of rotation $[\alpha]_D$ refer to the temperature. In most cases the sign and size of angle of rotation (for chiral compounds) are given for both enantiomers or for a series of stereoisomers in the case of more centers of chirality; the solvent is given in parentheses (Table 3). GC and other chromatographic retention data are indicated by a reference superscript, as well as elementary analyses, UV spectra, or ORD measurements (Table 3). In Table 4 NMR spectra, ^{1}H-proton resonance spectra, and ^{13}C-carbon spectra are found, giving the chemical shift of the resonance bands in δ-values. Furthermore, the solvent is given, and each signal is described according to its shape (e.g., singlet, doublet, triplet, multiplet, etc.) and the number of nuclei assigned is shown in parentheses. The IR data are wave numbers (cm^{-1}) and the amplitude of the absorption bands designated as strong (s), medium (m), weak (w), or shoulder (sh) in the usual way (the solvent is indicated in parentheses). The mass spectra are 70-eV electron impact spectra, unless stated otherwise. The last column of Tables 3 and 4 gives references to papers dealing with the spectral data of the compounds.

Table 3
PHYSICAL PROPERTIES AND CHROMATOGRAPHIC DATA ON COLEOPTERAN PHEROMONES

No.	Chemical name, common name, and synonym	Mol formula, mol wt	bp (°C/torr)	Refractive index (n_D) °C	$[\alpha]_D$ °C (solvent)	Additional analytical data[a]	Ref.
1	(1R,5S,7R)(+)-exo-7-Ethyl-5-methyl-6,8-dioxabicyclo[3.2.1]octane exo-Brevicomin	$C_9H_{16}O_2$ 158.24	95—100/110 70/20 (bath temp.) 86—88/50	1.4370^{27}	(1R,5S,7R): $+84.1^{26}$ (ether) (1S,5R,7S): -80.0^{24} (ether)	GC data[120,125,126]	116,120,125,126,287
2	endo-7-Ethyl-5-methyl-6,8-dioxabicyclo[3.2.1]octane endo-Brevicomin	$C_9H_{16}O_2$ 158.24	60/20 (bath temp.) 82—84/50			GC data[120,126]	116,120,126,131,288
3	(1S,5R)(−)-1,5-dimethyl-6,8-dioxabicyclo[3.2.1]octane Frontalin	$C_8H_{14}O_2$ 142.20	91/100 76—78/50	1.4386^{20} 1.4345^{27}	(1S,5R): -52.0^{27} (ether) (1R,5S): $+53.4^{23}$ (ether)	GC data[126,137] ORD data[137]	118,126,132,133,137
4	2-Ethyl-1,6-dioxaspiro[4.4]nonane Chalcograne	$C_9H_{16}O_2$ 156.23	62/15 (bath temp.)		(2R,5RS): $+13.5^{23}$ (CDCl₃) $+18.4^{20}$ (pentane) (2R,5R): -76^{25} (C₆D₆) (2R,5S): $+62^{25}$ (C₆D₆)	GC data[144,146,147]	141,142,144—147,290,291
5	(1S,2R,4S,5R)(−)-2,4-dimethyl-5-ethyl-6,8-dioxabicyclo[3.2.1]octane α-Multistriatin	$C_{10}H_{18}O_2$ 170.25	85—100/31 (bath temp.)	1.4488^{23}	-17.0^{23} (ether) -26.0 (hexane) -18.7 (hexane) -64.0^{24} (hexane)	GC data[154]	74,148,153—155,210
6	(1S,2S,4S,5R)(−)-δ-multistriatin	$C_{10}H_{18}O_2$ 170.25	85—100/31 (bath. temp.) 63—64/15	1.4488^{23}	-17.0^{23} (ether)		154
7	(S)(−)-2-methyl-6-methylene-7-octen-4-ol (S)(−)-Ipsenol	$C_{10}H_{18}O$ 154.24	63—64/15	1.4633^{24}	(S)(−): -18.4^{20} (EtOH) (R)(+): $+17.2^{21}$ (EtOH)	GC data[78,277] UV spectrum[78]	78,168,170,277
8	(S)(+)-2-methyl-6-methylene-2,7-octadien-4-ol	$C_{10}H_{16}O$ 152.23	55/0.15		(S)(+): $+11.9^{21}$ (MeOH)	GC data[277] UV spectrum[277]	170,175,176,277

	Name	Formula / M.W.	bp/mp / n	$[\alpha]$	Other data	Ref.
	(S)(−)-ipsdienol			$(R)(-)$: -12.0^{20} (MeOH)		
9	(1R,2S)(+)-1-methyl-1-(2-hydroxy)-ethyl-2-isopropenylcyclobutane (1R,2S)(+)-grandisol	$C_{10}H_{18}O$ 154.24		(1R,2S): $+19.6^{20}$ (hexane) (1S,2R): -20.0^{20} (hexane)	GC data[92] Analysis[92]	61,92,191
10	(1R,4S,5R,7R)(+)-3,3,7-trimethyl-2,9-dioxatricyclo[3.3.1.0⁴,⁷]nonane Lineatin	$C_{10}H_{16}O_2$ 168.24	110/53 (1R,4S,5R,7R): 1.4586[22] (1S,4R,5S,7S): 1.4588[22]	(1R,4S,5R,7R): $+84.8^{21.5}$ (CHCl₃) (1R,4R,5S,7S): -87.7^{22} (CHCl₃)	X-ray[196,200]	195,196,200
11	(Z)3,3-dimethyl-Δ¹·ᵝ-cyclohexaneethanol	$C_{10}H_{18}O$ 154.25			GC data[92] Analysis[92]	92
12	(Z)-3,3-dimethyl-Δ¹·ᵅ-cyclohexane-acetaldehyde	$C_{10}H_{16}O$ 152.24			GC data[92]	92
13	(E)-3,3-dimethyl-Δ¹·ᵅ-cyclohexane-acetaldehyde	$C_{10}H_{16}O$ 152.24			GC data[92] Analysis[92]	92
14	(1R,4S,5R)(+)-pinen-4-ol (+)-*trans*-verbenol	$C_{10}H_{16}O$ 152.24	88/9	1.4480[24]	(1R,4S,5R): $+141^{24}$ (CHCl₃) $+154$ (MeOH) (1S,4R,5S): -135^{24} (CHCl₃) -146 (MeOH)	72,210,292
15	(1S,4S,5S)-2-pinen-4-ol (S)-*cis*-Verbenol	$C_{10}H_{16}O$ 152.24	92/10 mp 64	1.4912[25]	(1S,4S,5S): -9.8^{22}(CHCl₃) $+11.4$ (MeOH) $+6.2$ (acetone) (1R,4R,5R): $+9.3^{22}$ (CHCl₃) -12.8 (MeOH) -6.8 (acetone) -20.0^{30} (CHCl₃)	GC data[277] UV spectrum[277] Pleasant menthol odor	73,277
16	(−)α-Cubebene	$C_{15}H_{24}$ 204.36		$+6.6$	UV spectrum[293] Reactions[293]	292,293
17	Verbenone	$C_{10}H_{14}O$ 150.22	103—104/16	1.4995		292
18	Methyl (R,E)(−)-2,4,5-tetra-decatrienoate	$C_{15}H_{24}O_2$ 236.36		$(S)(+)$: $+160^{20}$ (hexane)	GC data[33] R_F data[33] UV spectrum[33,214]	33,214,220,221

Table 3 (continued)
PHYSICAL PROPERTIES AND CHROMATOGRAPHIC DATA ON COLEOPTERAN PHEROMONES

No.	Chemical name, common name, and synonym	Mol formula, mol wt	bp (°C/torr)	Refractive index (n_D) °C	$[\alpha]_D$°C (solvent)	Additional analytical data[a]	Ref.
19	(R,Z)(−)-14-methyl-8-hexadecen-1-ol	$C_{17}H_{34}O$ 254.46	110/0.01 (bath temp.) (125—126/0.1) 132—133/0.2	(S)(+): 1.4568^{25} 1.4580^{25}	(R)(−): -5.43^{20} (CHCl₃) -5.27^{25} (CHCl₃) (S)(+): $+5.31^{25}$(CHCl₃)	GC data[228,229]	225—229
20	Methyl (R,Z)(−)-14-methyl-8-hexadecenoate	$C_{18}H_{34}O_2$ 282.47	(S)(+): 125—127/0.1	(S)(+): 1.4483^{25}	(S)(+): $+3.75$ (CHCl₃)	GC data[229]	229
21	(R,Z)-14-methyl-8-hexadecenal	$C_{17}H_{32}O$ 252.45	(R)(−): 116—117/ 0.15 (S)(+): 138—139/1	(R)(−): 1.4541 (S)(+): 1.4580^{25}	(R)(−): -5.90^{25} (ether) -5.94^{21} (CHCl₃) (S)(+): $+6.05^{25}$ (ether) $+6.02^{21}$ (CHCl₃)	Elementary analysis[228]	225,226,228,231
22	(E)-14-methyl-8-hexadecen-1-ol	$C_{17}H_{34}O$ 254.46	(R)(−): 118—119/0.07 125/0.1 (S)(+): 108—109/0.03	(R)(−): 1.5484^{25} (S)(+): 1.4584^{25}	(R)(−): -5.45^{25} (CHCl₃) -5.12 (CHCl₃) (S)(+): $+5.11^{25}$ (CHCl₃)	GC data[228]	225,227,228
23	(R,E)-14-methyl-8-hexadecenal	$C_{17}H_{32}O$ 252.45	(R)(−): 127/0.45 (S)(+): 127—129/0.5	(R)(−): 1.4532^{25} (S)(+): 1.4534^{25}	(R)(−): -5.04^{25} (ether) -5.99^{21} (S)(+): $+15.25^{25}$ (ether) $+6.18^{22}$	Elementary analysis[228]	225,228,231,232
24	(Z)-3-decenoic acid	$C_{10}H_{18}O_2$ 170.25	118—120/1.5 mp 1.4—4			GC data[233]	233,234,294
25	(3E,5Z)-3,5-tetradecadienoic acid Megatomic acid	$C_{14}H_{34}O_2$ 224.35				UV data[236] GC data[236] Elementary analysis[236]	236

No.	Name	Formula / MW	bp/mp	n_D	$[\alpha]$	Characterization	Ref.
26	(3Z,5Z)-3,5-tetradecadienoic acid	$C_{14}H_{24}O_2$ 224.35				GC data[236] UV spectrum[236]	236
27	(S)(+)-1-methylbutyl (E)-2-methyl-2-pentenoate Dominicalure 1	$C_{11}H_{22}O_2$ 186.29	92/10		(S)(+): +13.4[25] (ether) (R)(−): −29.7[25] (ether)	GC data[68]	68
28	(S)(+)-1-methylbutyl (E)-2,4-dimethyl-2-pentenoate Dominicalure 2	$C_{12}H_{22}O_2$ 198.31	94/10		(S)(+): +31.8[25] (ether) (R)(−): −30.1[25] (ether)	GC data[68]	68
29	2,3-Dihydro-2,3,5-trimethyl-6-(1-methyl-2-oxobutyl)-4H-pyran-4-one Stegobinone	$C_{13}H_{20}O_3$ 224.30	mp 52.5—53.5		(2S,3R,7RS): −129[23] (CHCl₃) (2R,3S,7RS): +121[22] (CHCl₃)	CD spectrum[241] GC data[241]	241
30	3-Methyl-2-cyclohexen-1-ol Seudenol	$C_7H_{12}O$ 112.17	(R)(+): 83—84/20.5 (S)(−): 82.5—83.5/21	(R)(±): 1.4818[20] (S)(−): 1.4817[20]		Elementary analysis[242]	242
31	(3S,4S)-threo-(−)-4-methylheptan-3-ol	$C_8H_{18}O$ 130.23	Racemate: 110/123 (3S,4S): 105—108/102 (3R,4R): 99—101/98	Racemate: 1.4264[23] (3S,4S): 1.4244[23] (3R,4R): 1.4244[23]	(3R,4R): 22.7[23.5] (hexane) (3S,4R): +13[23] (hexane) (3S,4S): −21.7[22] (hexane)	GC data[244,246] Analysis[244]	244,246,292
32	4-Hydroxydodecanoic acid lactone Dodecan-4-olide	$C_{14}H_{22}O_2$ 198.31	130/0.5 170—171/11 148/0.5	1.4522[20]	(S)(−): −28.7[22] (MeOH)	GC data[295]	248,249,294,295
33	(R,Z)(−)-4-hydroxy-5-tetradecenoic acid lactone (Z)-5-tetradecen-4-olide (R,Z)(−)-5-(1-decenyl)dihydro-2(3H)-furanone	$C_{12}H_{24}O_2$ 224.35	100—104/0.01 110/0.02 (bath temp.)		(R)(−): −69.6 (CHCl₃) (S)(+): +70.5 (CHCl₃)	GC data[256]	251,252,254,256
34	(R)(+)-γ-caprolactone	$C_6H_{10}O_2$ 114.15	107—109/17 60/0.05 (bath temp.)	1.4393[20]	(R)(+): +53.3 (MeOH) (S)(−): −53.2 (MeOH)		249,257,258,292

Table 3 (continued)

PHYSICAL PROPERTIES AND CHROMATOGRAPHIC DATA ON COLEOPTERAN PHEROMONES

No.	Chemical name, common name, and synonym	Mol formula, mol wt	bp (°C/torr)	Refractive index (n_D) °C	$[\alpha]_D$°C (solvent)	Additional analytical data[a]	Ref.
35	(S)(−)-2,5-dimethyl-2-isopropyl-2,3-dihydrofuran	$C_9H_{16}O$ 140.23			(S)(−): −1.1 (pentane)		260,261
36	(4S,6S,7S)-4,6-dimethyl-7-hydroxynonan-3-one Serricornin	$C_{11}H_{22}O_2$ 186.30			(4S,6R,7R)-acetate: +36.75[23] (hexane) (4RS,6R,7R)-acetate: +16.6[23] (MeOH)	UV spectrum[265] GC data[265]	261,262,264—267,269
37	2,6-Diethyl-3,5-dimethyl-3,4-dihydro-2H-pyran Anhydroserricornin	$C_{11}H_{20}O$ 168.28	58—64/15	1.4517[23]	(6S,7S): −61.5 (CHCl$_3$)	Elementary analysis[266] 1H—400 MHz[266]	261,262,266
38	(S)(+)-6-methyl-5-hepten-2-ol Sulcatol	$C_8H_{16}O$ 128.22	73/10 88/25	Racemate: 1.4481[20] 1.4515[11.5]	(S)(+): +14.4[23] (EtOH) (R)(−): −14.5[23] (EtOH)	GC data[69]	69,294
39	3-Hydroxy-3-methylbutan-2-one	$C_5H_{10}O_2$ 102.14	65/50 139/750	1.4158[20]		GC data[296] LC retention[296]	296
40	2,3-Dihydroxy-2-methylbutane	$C_5H_{12}O_2$ 104.15	65/5 93—95/24	1.4380[25]			296
41	Phenol	C_6H_6O 94.11	183/760 mp 43	1.5445		GC data[34]	34,294
42	Hexan-2-ol	$C_6H_{14}O$ 102.18	54.5/17 139.9/760	1.4156[20] 1.4141[25]			292,296

No.	Compound	Formula M.W.	b.p. (°C/mm)	n_D	Notes	Ref.
43	4,8-Dimethyldecanal	$C_{12}H_{24}O$ 184.32	92—93/6	1.4388^{20} 1.4367^{25} 1.4361^{21}	GC data[275,297]	275,297
44	1-Pentadecene $CH_2=CH(CH_2)_{12}CH_3$	$C_{15}H_{30}$ 210.41	102/2 133.5/10	1.4345^{20}		292
45	Hexadecane $CH_3(CH_2)_{14}CH_3$	$C_{16}H_{34}$ 226.45	117/2 155—158/15	1.4436^{20}		292,298
46	1-Heptadecene $CH_2=CH(CH_2)_{14}CH_3$	$C_{17}H_{34}$ 238.46	103/0.5 159/10	1.4099^{15}		292
47	Pentanoic acid $CH_3(CH_2)_3CO_2H$	$C_5H_{10}O_2$ 102.14	186/760	1.4188^{15}		292,298
48	Hexanoic acid $CH_3(CH_2)_4CO_2H$	$C_6H_{12}O_2$ 116.16	102/15 208/760	1.4149^{25}		292,298
49	Isopropyl dodecanoate Isopropyl laurate $CH_3(CH_2)_{10}CO_2CH(CH_3)_2$	$C_{15}H_{30}O_2$ 242.41			Kovats indices[299]	299
50	Isopropyl (Z)-5-dodecenoate $CH_3(CH_2)_5-\overset{H}{C}=\overset{H}{C}-(CH_2)_3CO_2CH(CH_3)_2$	$C_{15}H_{28}O_2$ 240.39			Kovats indices[299]	299
51	Isopropyl (Z)-7-dodecenoate $CH_3(CH_2)_3-\overset{H}{C}=\overset{H}{C}-(CH_2)_5CO_2CH(CH_3)_2$	$C_{15}H_{28}O_2$ 240.39			Kovats indices[299]	299
52	Isopropyl (Z)-9-dodecenoate $CH_3CH_2-\overset{H}{C}=\overset{H}{C}-(CH_2)_7CO_2CH(CH_3)_2$	$C_{15}H_{28}O_2$ 240.39			GC data[299]	299
53	Methyl tetradecanoate Methyl myristate $CH_3(CH_2)_{12}CO_2CH_3$	$C_{15}H_{30}O_2$ 242.41	167—168/15			292,298
54	Isopropyl (Z)-9-tetradecenoate Isopropyl myristoleate $CH_3(CH_2)_3-\overset{H}{C}=\overset{H}{C}-(CH_2)_7CO_2CH(CH_3)_2CH(CH_3)_2$	$C_{17}H_{32}O_2$ 268.44			GC data[299]	299

Table 3 (continued)

PHYSICAL PROPERTIES AND CHROMATOGRAPHIC DATA ON COLEOPTERAN PHEROMONES

No.	Chemical name, common name, and synonym	Mol formula, mol wt	bp (°C/torr)	Refractive index (n_D) °C	$[\alpha]_D$ °C (solvent)	Additional analytical data[a]	Ref.
55	Ethyl hexadecanoate Ethyl palmitate $CH_3(CH_2)_{14}CO_2CH_2CH_3$	$C_{18}H_{36}O_2$ 284.49	156.5/2 192.6/10	1.4380^{25} 1.4363^{30}			292,298
56	Methyl (Z)-7-hexadecenoate $CH_3(CH_2)_7-C=C-(CH_2)_5CO_2CH_3$ (H H)	$C_{17}H_{32}O_2$ 268.44	150—155/6				292
57	Isopropyl (Z)-9-hexadecenoate Isopropyl palmitoleate $CH_3(CH_2)_5-C=C-(CH_2)_7CO_2CH(CH_3)_2$ (H H)	$C_{19}H_{36}O_2$ 296.50				GC data [299]	299
58	Ethyl octadecanoate Ethyl stearate $CH_3(CH_2)_{16}CO_2CH_2CH_3$	$C_{20}H_{40}O_2$ 312.54	181.1/2	1.4375^{35}			292
59	Methyl (Z)-9-octadecenoate Methyl oleate $CH_3(CH_2)_7-C=C-(CH_2)_7CO_2CH_3$ (H H)	$C_{19}H_{36}O_2$ 296.50	128/0.2 mp − 19.6	1.4466^{35}			292,298
60	Ethyl (Z)-9-octadecenoate Ethyl oleate $CH_3(CH_2)_7-C=C-(CH_2)_7CO_2CH_2CH_3$ (H H)	$C_{20}H_{38}O_2$ 310.52	143—145/0.01 207/13	1.4536^{16} 1.4515^{20}		UV (EtOH)[300]	292,294,298

No.	Compound / Structure	Molecular formula / MW	bp (°C/mmHg)	n_D	$[\alpha]$	Other data	Ref.
61	Isopropyl (Z)-9-octadecenoate Isopropyl oleate $CH_3(CH_2)_7$-C=C-$(CH_2)_7CO_2CH(CH_3)_2$	$C_{21}H_{40}O_2$ 324.55	218/16				299
62	Ethyl (9Z,12Z)-9,12-octadecadienoate Ethyl linoleate $CH_3(CH_2)_4C=CCH_2C=C(CH_2)_7CO_2CH_2CH_3$	$C_{20}H_{36}O_2$ 308.54	175/2.5 212/12	1.4585[25] 1.4573[27] 1.4489[48]		UV (EtOH)[301]	294,298
63	Ethanol CH_3CH_2OH	C_2H_6O 46.07	78/760	1.3618[19.5]			292,294,298
64	Myrtenol CH_2OH	$C_{10}H_{16}O$ 152.24	107—110/13 221—222/750	1.4924 1.4943[20]	+43.4[20] −46.3		292
65	8-Methyldecan-2-ol-propanoate	$C_{14}H_{28}O_2$ 228.36	60—65/0.005 (bath temp.)			GC data[271] LC data[271] CI mass spectrum[221]	271
66	10-Methyl-2-tridecanone	$C_{14}H_{28}O$ 212.36				GC and LC data[273] CI mass spectrum[273]	273
89	(E,E)-4,8-dimethyl-4,8-decadien-10-olide Ferrulactone I	$C_{12}H_{18}O$ 194.26				GLC and HPLC data[302]	302
90	(3Z,11S)-3-dodecen-11-olide Ferrulactone II	$C_{12}H_{20}O_2$ 196.28				GLC and HPLC data[302]	302

[a] Superscript numbers show additional references.

Table 4
NMR, IR, AND MASS SPECTRAL DATA ON COLEOPTERAN PHEROMONES

No.	Chemical name, common name and synonym	NMR spectra (δ-values) Nucleus (solvent): chemical shift (line shape, number of nuclei)[a]	IR spectrum (cm^{-1})[b]	Mass spectrum m/z (rel. intensity in %)	Ref.
1	(1R,5S,7R)(+)-exo-7-ethyl-5-methyl-6,8-dioxabicyclo[3.2.1]octane exo-Brevicomin	^{1}H(CCl$_4$): 0.86 (t, 3H), 1.30 (s, 3H), 3.79 (t, 1H), 3.99 (s, 1H) ^{13}C-NMR[129,289]	Film: 2940s, 2880m, 2850m, 1470m, 1390s, 1360w, 1340w, 1310w, 1280m, 1265w, 1250s, 1200m, 1190m, 1180s, 1140w, 1110m, 1085w, 1080w, 1040w, 1035s, 1020m, 1010s, 995m, 970m, 930m, 900w, 880m, 860s, 850s, 790w	156 (M$^+$, 4), 127 (6), 99 (8), 98 (16), 86 (16), 85 (44), 81 (10), 73 (8), 71 (10), 68 (14), 57 (8), 43 (100), 41 (16)	116, 120, 125, 126, 287
2	endo-7-Ethyl-5-methyl-6,8-dioxabicyclo[3.2.1]octane endo-Brevicomin	^{1}H (CCl$_4$): 0.97 (t, 3H), 1.30 (s, 3H), 3.84 (m, 1H), 4.03 (broad, 1H)	Film: 2950, 1560, 1460, 1380, 1240, 1175, 1110, 1035, 1005, 980, 905, 855	M-EtCO, M-EtCOH 156 (M$^+$), 140, 85, 68, 43 (100), 29	116, 120, 126, 131, 288
3	(1S,5R)(−)-1,5-dimethyl-6,8-dioxabicyclo[3.2.1]octane Frontalin	^{1}H (CDCl$_3$): 1.49 (s, 3H), 1.59 (3H), 1.80 (s, 6H), 3.58, 4.04 (AB-q, 2H) ^{1}H-NMR + Eu (facam)$_3$[137]	Film: 2960m, 2920m, 2860m, 2810m, 1445w, 1390m, 1380m, 1340w, 1325w, 1285w, 1270, 1240m, 1200m, 1170m, 1120s, 1060m, 1020s, 970w, 925m, 910w, 890m, 860m, 840s, 815m, 750w	142.1015 (M$^+$), 142 (M$^+$, 13), 112 (12), 100 (35), 72 (78), 71 (21), 67 (13), 54 (11), 43 (100), 41 (19), 39 (16), 27 (6)	118, 126, 132, 133, 137

#	Compound	¹H / ¹³C NMR	IR	MS	Ref.
4	2-Ethyl-1-6-dioxaspiro[4,4]nonane Chalcograne	^{1}H (C$_6$D$_6$): 0.86 and 0.92 (t, 3H), 3.73, 3.91, and 4.10 (m, 1H each) ^{13}C (C$_6$D$_6$): 114.6 (s), 81.4 (d), 79.3 (d), 66.8 (t), 66.7 (t), 36.2 (t), 34.5 (t), 35.3 (t), 39.9 (t), 30.9 (t), 30.7 (t), 30.3 (t), 29.0 (t), 25.0 (t), 24.9 (t), 10.5 (q), 10.2 (q)	Film: 2960s, 2920sh. 2870s, 1460m, 1440w, 1380w, 1345m, 1295w, 1245w, 1210w, 1175m, 1080m, 1060m, 1045s, 1015s, 1000sh. 950m, 925m, 915m, 850m, 830w, 780w	156 (M$^+$, 3), 127 (100), 126 (3), 115 (8), 114 (5), 98 (27), 97 (30), 87 (64), 86 (2), 85 (61), 69 (28), 56 (42), 55 (37)	141, 142, 144—147, 290, 291
5	(1S,2R,4S,5R)-2,4-dimethyl-5-ethyl-6,8-dioxabicyclo-[3.2.1]octane α-Multistriatin	^{1}H (CDCl$_3$): 0.81 (d, 6H), 0.92 (t, 3H), 1.4—2.2 (m, 6H), 1.68 (q, 2H), 3.68 (dd, 1H), 3.85 (d, 1H), 4.14 (m, 1H) ^{1}H-NMR + Eu (facam)$_3$[154]	Film: 2960s, 2930s, 2880s, 1485m, 1460s, 1440m, 1385m, 1365m, 1340w, 1320s, 1095m, 1055m, 1035s, 1000w, 990m, 955m, 920s, 895s, 820w, 790w, 760w	170.1326 (M$^+$) 170 (M$^+$, 4), 128 (27), 96 (42), 81 (22), 71 (34), 57 (100), 55 (42), 54 (14), 43 (14), 41 (26), 39 (14)	74, 148, 153—155, 210
6	(1S,2S,4S,5R)(−)-δ-multistriatin	^{1}H (CDCl$_3$): 0.78 (d, 3H), 0.90 (t, 3H), 1.13 (d, 3H), 1.2—2.2 (m, 6H), 3.82 (d, 2H), 4.22 (m, 1H) ^{1}H-NMR + Eu (facam)$_3$[154]	Film: 2970vs, 2880s, 1485m, 1460s, 1380m, 1365m, 1330w, 1310w, 1285w, 1250s, 1200s, 1170m, 1145w, 1130m, 1115w, 1050vs, 1020m, 990s, 980s, 955m, 910s, 895s, 860w, 765w	170.1312 (M$^+$, 11), 128 (22), 96 (36), 86 (16), 81 (38), 71 (24), 58 (12), 57 (100), 55 (56), 54 (26), 53 (12), 42 (12), 41 (44), 39 (30)	154
7	(S)(−)-2-methyl-6-methylene-7-octen-4-ol (S)(−)-ipsenol	^{1}H (CCl$_4$): 0.86 (d, 3H), 0.90 (d, 3H), 1.28 (q, 2H), 1.45—2.0 (m, 1H), 1.70	Film: 3350, 3080, 1600, 1387, 1370, 990, 900, 888	154 (M$^+$, 1), 68 (100)	78, 168, 170, 277

Table 4 (continued)
NMR, IR, AND MASS SPECTRAL DATA ON COLEOPTERAN PHEROMONES

No.	Chemical name, common name and synonym	NMR spectra (δ-values) Nucleus (solvent): chemical shift (line shape, number of nuclei)[a]	IR spectrum (cm⁻¹)[b]	Mass spectrum m/z (rel. intensity in %)	Ref.
		(s, 1H), 2.30 (m, 2H), 3.74 (sept, 1H), 4.9—5.4 (m, 3H), 6.10—6.70 (dd,1H)			
8	(S)(+)-2-methyl-6-methylene-2,7-octadien-4-ol (S)(—)-ipsdienol	^{1}H (CCl₄): 1.66 (d, 3H), 1.70 (d, 3H), 2.39 (d, 2H), 4.48 (m, 1H), 4.90—5.40 (5H), 6.38 (dd, 1H)	CCl₄: 3300 (OH), 1020 (C-OH), 992 and 900 (vinyl absorption)	152 (M⁺, 1), 134 (18), 119 (42), 105 (16), 93 (17), 92 (18), 91 (58), 85 (100), 81 (12), 79 (63), 77 (27), 67 (19), 66 (16), 65 (15), 55 (20), 53 (28), 51 (20), 43 (27), 41 (71), 39 (54)	170, 175, 176, 277
9	(1R,2S)(+)-1-methyl-1-(2-hydroxy)-ethyl-2-isopropenylcyclobutane (1R,2S)(+)-grandisol	^{1}H (CCl₄): 1.3—2.2 (m, 6H), 1.22 (s, 3H), 1.72 (s, 3H), 2.59 (s, OH), 2.60 (t, 1H), 3.63 (t, 2H), 4.88 and 4.71 (s, 2H)	CCl₄: 3630 (OH), 3250—3350 (OH), 1643 and 885 (=CH₂)	154 (M⁺, 2), 139 (6), 121 (10), 109 (27), 93 (15), 81 (17), 68 (100), 53 (23), 41 (42)	61, 92, 191
10	(1R,4S,5R,7R)(+)-3,3,7-trimethyl-2,9-dioxatricyclo-[3.3.1.0⁴,⁷]nonane Lineatin	^{1}H (CCl₄): 1.09 (s, 3H), 1.14 (s, 6H), 1.5—2.4 (m, 5H), 4.34 (t, 1H), 4.86 (d, 1H)	CCl₄: 2940s, 2900s, 2850m, 1465m, 1450m, 1380m, 1365m, 1340w, 1315m, 1240m,	168 (M⁺), 153 (M-15), 140, 125, 111, 109, 107, 96, 85, (100), 83, 55, 43	195, 196, 200

No.	Compound	NMR	IR	MS	Ref.
11	(Z)-3,3-dimethyl-Δ$^{1,\beta}$-cyclohexaneethanol	^{1}H (CCl$_4$): 0.95 (s, 6H), 1.12—1.83 (m, 4H), 2.0 (s, 2H), 2.09 (m, 2H), 4.05 (d, 2H)	1225m, 1205m, 1185m, 1170s, 1125s, 1100m, 1075m, 1015w, 995m, 965s, 920w, 900s, 830w; CCl$_4$: 3610 (OH)	154 (M$^+$, 7), 136 (40), 121 (48), 107 (25), 93 (63), 79 (53), 69 (100), 55 (35), 41 (88)	92
12	(Z)-3,3-dimethyl-Δ$^{1,\alpha}$-cyclohexaneacetaldehyde	^{1}H (CCl$_4$): 0.93 (s, 6H), 1.2—2.0 (m, 4H), 2.17 (t, 2H), 2.42 (s, 2H), 5.47 (d, 1H)		152 (M$^+$, 34), 137 (90), 109 (45), 95 (28), 81 (45), 69 (59), 55 (30), 53 (30), 41 (100)	92
13	(E)-3,3-dimethyl-Δ$^{1,\alpha}$-cyclohexaneacetaldehyde	^{1}H (CCl$_4$): 0.95 (s, 6H), 1.2—1.9 (m, 4H), 2.0 (s, 2H), 2.61 (t, 2H), 5.6 (d, 1H), 9.78 (d, (1H)		152 (M$^+$, 46), 137 (46), 119 (24), 109 (63), 93 (29), 81 (38), 69 (63), 55 (33), 41 (100)	92
14	(1S,4S,5S)(+)-2-pinen-4-ol (+)-*trans*-verbenol	^{1}H (CCl$_4$): 0.94 (s, 3H), 1.3 (s, 3H), 1.6 (s, 3H), 1.9—2.4 (m, 4H), 4.16 (s, 1H), 5.3 (s, 1H) NMR data of α-methoxy-α-(trifluoromethyl)-phenylacetate[72]		134.1096 (M$^+$·H$_2$O), 152 (M$^+$, 1), 134 (31), 119 (28), 109 (29), 94 (58), 92 (100), 91 (83), 83 (72), 79 (60), 70 (78), 59(55), 55 (80), 43 (52), 41 (83)	72, 210, 292
15	(1S,4S,5S)-2-Pinen-4-ol (S)-*cis*-Verbenol	^{1}H (CCl$_4$): 1.03 (s, 3H), 1.32 (s, 3H), 1.7 (s, 3H), 4.29 (1H), 5.27 (s, 1H)	CCl$_4$: 1670 (C=C)	152 (M$^+$), 137 (M − 15), 134 (M − 18), 119 (M − 33), 109 (M − 43), 43 (100)	73, 277
16	(−)-α-Cubebene	^{1}H (CDCl$_3$): 0.21 (cyclopropane H), 1.10	3020sh, 2950s, 2890s, 1600m,	204 (M$^+$, 20), 161 (98), 120 (28), 119 (100)	292, 293

Table 4 (continued)
NMR, IR, AND MASS SPECTRAL DATA ON COLEOPTERAN PHEROMONES

No.	Chemical name, common name and synonym	NMR spectra (δ-values) Nucleus (solvent): chemical shift (line shape, number of nuclei)[a]	IR spectrum (cm^{-1})[b]	Mass spectrum m/z (rel. intensity in %)	Ref.
		(cyclopropane H)			
17	Verbenone		1410s, 1395s, 1340sh, 1320s, 1275w, 1200w, 880m, 860m, 845m, 770m, 730s	150 (M$^+$, 60), 135 (80), 107 (100), 91 (61), 80 (60), 79 (39), 67 (22), 55 (27), 43 (41), 41 (42)	292
18	Methyl (*R,E*)(−)-2,4,5-tetradecatrienoate	^{1}H (CCl$_4$): 0.89 (t, 3H), 1.3 (m, 12H), 2.0 (m, 2H), 3.65 (s, 3H), 5.3 (m, 2H), 5.76 (d, 1H), 7.07 (dd, 1H) Racemic (*E*)-isomer[221] (CCl$_4$): 3.62, 5.30, 5.52, 6.39, 7.19	Film: 2800—3000, 1940m, 1721s, 1630s, 1462m, 1435s, 1305, 1262, 1240, 1177, 1140, 981s	236 (M$^+$), 205 (M-OCH$_3$), 177 (M-COOCH$_3$), 138 (C$_8$H$_{10}$O$_2^+$), 79 (100)	33, 214, 220, 221
19	(*R,Z*)(−)-14-methyl-8-hexadecen-1-ol	^{1}H (CCl$_4$): 0.88 (t, 3H), 0.88 (d, 3H), 1.35 (m, 19H), 2.0 (d, 4H), 2.67 (s, 1H), 3.52 (t, 2H), 5.28 (m, 2H)	Film: 3350s, 2940s, 1660w, 1470s, 1380, 1060m, 970w, 730w	80 eV: 254 (M$^+$, 0.5), 236 (1), 95 (47), 83 (54), 82 (62), 81 (62), 70 (80), 67 (69), 55 (100)	225—229
20	Methyl (*R,Z*)(−)-14-methyl-8-hexadecenoate	^{1}H (CCl$_4$): 0.88 (t, 3H), 0.88 (d, 3H),	Film: 3020m, 2930s, 2860s, 1750s,	282 (M$^+$), 253, 251, 250, 212, 85, 74,	229

No.	Compound	^{1}H NMR	IR	MS	Ref.
21	(R,Z)-14-methyl-8-hexadecenal	1.34 (m, 17H), 2.0 (d, 4H), 2.22 (t, 2H), 3.62 (s, 3H), 5.28 (m, 2H); ^{1}H (CCl_4): 0.87 (t, 3H), 0.87 (t, 3H), 1.33 (br, 17H), 2.0 (br, 4H), 2.37 (br, 2H), 5.3 (m, 2H), 9.71 (t, 1H) Additional NMR data[225]	1470m, 1440m, 1380m, 1250m, 1200m, 1175m, 1130m, 1085w, 1030w, 970w, 880w; Film: 2985, 2940, 2910, 2840, 2695, 1710, 1440, 1360, 723	70; 252 (M^+), 223 (M-C_2H_5 or -CHO), 208 (M − 44), 44, 49	225, 226, 228, 231; 225, 226, 228
22	(E)-14-methyl-8-hexadecen-1-ol	^{1}H (CCl_4): 0.88 (t, 3H), 0.88 (d, 3H), 1.33 (br, 19H), 1.97 (M, 4H), 3.25 (2, 1H), 3.5 (t, 2H), 5.16 (m, 2H)	Film: 3320, 2960sh, 2930, 2875sh, 2860, 1455, 1370, 1050, 960, 720	80 eV: 254 (M^+, 1), 236 (2), 225 (1), 95 (48), 83 (61), 82 (56), 81 (67), 70 (90), 67 (70), 55 (100)	225, 226, 228
23	(R,E)-14-methyl-8-hexadecenal	^{1}H (CCl_4): 0.87 (t, 3H), 0.87 (d, 3H), 1.33 (br, 17H), 2.0 (br, 4H), 2.37 (br, 2H), 5.3 (m, 2H), 9.7 (t, 1H)	Film: 2940, 2840, 2695, 1713, 1440, 1360, 965, 722	252 (M^+), 223 (M-29), 208 (M−CH_3CHO), 44 (CH_3CHO), 29	225, 228, 231 232
24	(Z)-3-decenoic acid			Methyl ester: 184 (M^+), 152 (M − 32), 123 (M − 61), 110 (M − 74), 71 (100)	233, 234, 294
25	(3E,5Z)-3,5-tetradecadienoic acid Megatomic acid	^{1}H of methyl ester (CCl_4): 0.89: (t, 3H) 1.28 (m, 12H), 2.12 (q, 2H), 3.02 (d, 2H), 3.59 (s, 3H), 5.1—6.45 (m, 4H)	Film: 3300-2350br, 1720s, 1620w, 1460, 1410, 1283, 1220, 980, 952, 935br, 827, 719 Methyl ester film: 3020, 1735s,	Methyl ester: 238 (M^+), 207 (M-OCH_3), 206 (M-CH_3OH), 179 (M-CO_2CH_3), 74 ($CH_3CO_2CH_3$)	236

Table 4 (continued)
NMR, IR, AND MASS SPECTRAL DATA ON COLEOPTERAN PHEROMONES

No.	Chemical name, common name and synonym	NMR spectra (δ-values) Nucleus (solvent): chemical shift (line shape, number of nuclei)[a]	IR spectrum (cm⁻¹)[b]	Mass spectrum *m/z* (rel. intensity in %)	Ref.
26	(3Z,5Z)-3,5-tetradecadienoic acid		1665w, 1620w, 1245, 1198, 1163, 1014w, 980, 945, 720		236
27	(S)(+)-1-methylbutyl (E)-2-methyl-2-pentenoate Dominicalure 1	¹H (CDCl₃): 0.9 (t, 3H), 1.03 (t, CDCl₃; 1705s, 1648s 3H), 1.23 (d, 3H), 1.82 (3H), 2.16 (pent, 2H), 4.98 (m, 1H), 6.73 (t, 1H) Shift reagent NMR data[68]	IR spectrum of meth- ylester essentially the same as for ester of **25** except for the absence of the 980 and 945 cm⁻¹ bands	184 (M⁺, 1), 155 (1), 141 (1), 137 (1), 131 (1), 119 (1), 115 (50), 97 (91), 87 (8), 70 (40), 69 (55), 55 (30), 43 (65), 41 (100)	68
28	(S)(+)-1-methylbutyl (E)-2,4-dimethyl-2-pentenoate Dominicalure 2	¹H (CDCl₃): 1.12 (d, 6H), 1.34 (d, CDCl₃; 1705s, 1648s, 3H), 1.93 (s, 3H), 2.70 (m, 1H), 5.05 (m, 1H), 6.64 (d, 2H) Shift reagent data[68]		198 (M⁺, 1), 128 (70), 111 (65), 83 (55), 70 (35), 59 (45), 55 (68), 43 (100), 41 (92)	68
29	2,3-Dihydro-2,3,5-trimethyl-6-(1-methyl-2-oxobutyl)-4H-pyran-4-one	¹H (CDCl₃): 1.07 (d, 3H), 1.09 (t, 3H),	Film: 2970m, 2930m, 2870m,	224 (M⁺), 168, 139, 124, 113, 112, 111,	241

	Compound	^{1}H NMR	IR	MS	Ref.
	Stegobinone	1.31 (d, 3H), 1.32 (d, 3H), 1.8 (s, 3H), 2.26—2.67 (m, 3H), 3.65 (q, 1H), 4.42 (m, 1H) for (2S,3R,7RS)	1725s, 1665s, 1610s, 1460sh, 1455m, 1415w, 1385s, 1365m, 1345m, 1305w, 1290w, 1265w, 1245w, 1215m, 1185w, 1170m, 1145s, 1120m, 1095m, 1070sh, 1050m, 1015m, 1000m, 966m, 910w, 860w, 835w, 805w, 760w, 710m	109, 97, 57	
30	3-Methyl-2-cyclohexen-1-ol Seudenol	^{1}H (CCl$_4$): 1.65 (6H), 1.85 (br s, 3H), 3.64 (s, 1H), 4.1 (br s, 1H), 5.48 (br s, 1H) MTPA ester + Eu(fod)$_3^{242}$	Film: 3350s, 3020w, 2960s, 2880s, 2840m, 1680w, 1460m, 1450m, 1390m, 1350w, 1290m, 1180m, 1160w, 1130w, 1120m, 1080m, 1070m, 1040s, 1020m, 1000m, 970s, 915m, 860w, 850w, 820w, 720w		242
31	(3S-4S)-threo-(−)-4-methyl-heptan-3-ol	^{1}H (CCl$_4$): 0.8—1.0 (m, 9H), 1.0—1.6 (m, 7H), 2.38 (s, 1H), 3.3 (dt, 1H)	Film: 3360s, 2970s, 2940s, 2280s, 1470s, 1385m, 1340w, 1310w, 1250w, 1150w, 1110m, 1075m, 1040w, 1020w, 970s, 950s, 910w, 880w, 860w, 820w, 780w, 750w	101 (M-29, 20), 83 (61), 71 (12), 59 (100), 57 (45), 55 (41), 43 (41), 41 (52)	244, 246, 292

Table 4 (continued)
NMR, IR, AND MASS SPECTRAL DATA ON COLEOPTERAN PHEROMONES

No.	Chemical name, common name and synonym	NMR spectra (δ-values) Nucleus (solvent): chemical shift (line shape, number of nuclei)[a]	IR spectrum (cm⁻¹)[b]	Mass spectrum m/z (rel. intensity in %)	Ref.
32	4-Hydroxydodecanoic acid lactone Dodecan-4-olide	¹H (CDCl₃): 0.9 (CH₃), 1.2, 2.3, and 4.4 (−CH₂−)		198 (M⁺), 180 (M − 18), 128, 85 (100)	248, 249, 294, 295
33	(R,Z)(−)-4-hydroxy-5-tetradecenoic acid lactone (Z)-5-tetradecen-4-olide (R,Z)(−)-5-(1-Decenyl)dihydro-2 (3H)-furanone	¹H (CDCl₃): 0.88 (t), 1.28 (m), 2.08 (m), 2.54 (m), 5.44 (m, 3H)	CCl₄: 3010 (=C−H), 1785 (C=O), 1160, 1010, 980, 900 (C−O)	CI-MS: 265 (M⁺ + 41), 253, (M⁺ + 29), 225 (M⁺ + 1), 224 (M⁺), 223 (M⁺ − 1), 207 (M⁺ + 1 − 18), 189 (M⁺ + 1 − 36), 165 (M⁺ + 1 − 60) M⁺ = 224.1776	251—254, 256
34	(R)(+)-γ-caprolactone			114 (M⁺, 1), 85 (100), 70 (10), 66 (3), 57 (16), 56 (18), 55 (14), 42 (18)	249, 257, 258, 292
35	(S)(−)-2,5-dimethyl-2-isopropyl-2,3-dihydrofuran	¹H (C₆D₆): 0.85—0.95 (2d, 6H), 1.2 (s, 3H), 1.67 (s, 3H), 1.83 (m, 1H), 2.03—2.13 (m, 1H), 2.43—2.53 (m, 1H), 4.35 (m, 1H)	Film: 1675s, 1185s, 950s	80 eV: 140 (M⁺, 32), 125 (18), 122 (3), 107 (10), 97 (100), 83 (14), 70 (42), 55 (52), 43 (93)	260, 261
36	(4S,6S,7S)-4,6-dimethyl-7-hydroxynonan-3-one Serricornin	¹H (C₆D₆): 0.77 (d, 3H), 0.83 (d, 3H), 0.86 (t, 3H), 0.95 (t, 3H), 1.13—1.32 (m,	(4RS,6R,7R): 3400m, 1710m (4RS,6R,7R)-acetate: 1735s, 1710s, 1240s	(4S,6R,7R)-acetate: 168 (M-HOAc, 9), 157 (18), 139 (18), 111 (21), 86 (61),	261, 262, 264—267, 269

No.	Name	NMR	IR	MS	Ref.
		5H), 1.84—2.18 (m, 3H), 2.3 (m, 1H), 3.09 (m, 1H) (4S,6R,7R)-acetate ^{1}H (CDCl$_3$): 0.8—1.1 (m, 12H), 1.3—1.8 (m, 5H), 2.06 (s, 3H), 2.2—2.8 (m, 3H), 4.74 (dt, 1H) (4S,6R,7R)-acetate ^{13}C (CDCl$_3$): 214.8 (C3), 78.04 (C7), 43.53 (C4), 35.98 (C8), 34.22 (C2), 16.67 (C10), 14.45 (C11). 10.18 (C9), 7.84 (C1)	(4S,6R,7R)-acetate: 2950s, 2925s, 2860m, 1730s, 1708s, 1455m, 1375m, 1240s, 1100m, 1015m, 955m	57 (85), 43 (100)	
37	2,6-Diethyl-3,5-dimethyl-3,4-dihydro-2H-pyran Anhydroserricornin	^{1}H (C$_6$D$_6$): 0.73 (d, 3H), 1.11 (t, 3H), 1.25 (t, 3H), 1.45 (m, 2H), 1.55 (s, 3H), 1.55—1.8 (m, 2H), 2.06 (m, 1H), 2.16 (q, 2H), 3.21 (m, 1H) Two diastereomers, ratio 2:3 ^{1}H (C$_6$D$_6$): 0.83 (d, 3H), 0.93 (t, 3H), 1.12 (t, 3H), 1.45 (m, 2H), 1.55 (s, 3H), 1.55—1.8 (m, 2H), 2.10 (m, 1H), 2.16 (q, 2H), 3.52 (m, 1H) NMR data of MTPA ester[69]	Film: 1685s, 1460s, 1180s	80 eV: 168 (M$^+$, 21), 153 (6), 139 (10), 125 (26), 111 (9), 99 (100), 85 (27), 83 (21), 71 (28), 70 (31), 69 (35), 57 (39), 55 (50), 43 (41), 41 (36)	261, 262, 266

Table 4 (continued)
NMR, IR, AND MASS SPECTRAL DATA ON COLEOPTERAN PHEROMONES

No.	Chemical name, common name and synonym	NMR spectra (δ-values) Nucleus (solvent): chemical shift (line shape, number of nuclei)[a]	IR spectrum (cm^{-1})[b]	Mass spectrum m/z (rel. intensity in %)	Ref.
38	(S)(+)-6-methyl-5-hepten-2-ol Sulcatol	¹H (CDCl₃): 1.20 (d, 3H), 1.25—1.60 (m, 3H), 1.64 (br s, 3H), 1.71 (br s, 3H), 1.92—2.22 (m, 2H), 3.60—3.95 (m, 1H), 4.97—5.24 (br t, 1H)	Film: 1670, 1645w, 1120, 990, 908, 832, 745	128 (M$^+$), 110 (M − 18)	69, 294
39	3-Hydroxy-3-methylbutan-2-one			102 (M$^+$, 1), 87 (4), 84 (1), 69 (2), 59 (100), 43 (23), 41 (14)	296
40	2,3-Dihydroxy-2-methylbutane			80 eV: 104 (M$^+$), 89 (M − 15), 86 (M-H₂O), 71 (M-CH₃-H₂O), 59 (M − 45), 43 (M − 71)	296
41	Phenol	¹H (CCl₄): 6.7 (2 arom. H), 6.81 (1 arom. H), 7.14 (2 arom. H)	KBr: 3500, 3360, 3040w, 1600m, 1590s, 1510s, 1490s, 755s, 690s	94 (M$^+$, 100), 66 (20), 65 (18), 40 (8), 39 (13)	34, 294
42	Hexan-2-ol			87 (M$^+$ − 15, 9), 84 (11), 69 (18), 56 (4), 55 (3) 45 (100), 41 (10)	292, 296
43	4,8-Dimethyldecanal	¹H (CDCl₃): 0.80—1.00 (9H), 1.26 (br s, 6H), 1.40—1.80	2960, 2920, 2700, 1725, 1460, 1370	140 (M − 44, 14), 57 (100)	275, 297

No.	Compound	¹H NMR	IR	MS	Ref.
44	1-Pentadecene	(m, 4H), 2.43 (dt, 2H), 9.77 (t, 1H)		210 (M⁺, 1), 111 (12), 97 (37), 83 (50), 69 (57), 57 (65), 55 (83), 41 (100), 39 (28)	292
45	Hexadecane		Film: 2930, 2865, 1467, 1379, 721	226 (M⁺, 3), 169 (1), 155 (1), 141 (1), 127 (2), 113 (5), 99 (8), 85 (32), 71 (50), 57 (95), 43 (100), 41 (52)	292, 298
46	1-Heptadecene			238 (M⁺, 7), 125 (11), 111 (27), 97 (62), 83 (65), 69 (60), 55 (88), 43 (95), 41 (100)	292
47	Pentanoic acid		Film: 30sh, 2950s, 2880m, 1722s	87 (M − 15, 1), 73 (32), 60 (100), 55 (7), 45 (16), 43 (18), 41 (20)	292, 298
48	Hexanoic acid		Film: 3040br, 2960s, 1721s	101 (M − 15, 2), 87 (9), 73 (40), 60 (100), 45 (19), 43 (30), 41 (35), 39 (18)	292, 298
49	Isopropyl dodecanoate Isopropyl laurate			242 (M⁺), 201 (M − 41, 200 (M − 42, 28), 115 (11), 102 (14)	299
50	Isopropyl (Z)-5-dodecenoate	¹H (CDCl₃): 0.9 (t, 3H), 1.23 (d, 6H), 1.25—1.4 (8H), 1.65 (m, 2H), 1.95—2.15 (m, 4H), 2.25 (t, 2H), 4.9 (m, 1H), 5.25 (m, 2H)		240 (M⁺, 22), 198 (15), 197 (7), 196 (3), 181 (34), 180 (39), 179 (16), 163 (25), 162 (14), 161 (13)	299

Table 4 (continued)
NMR, IR, AND MASS SPECTRAL DATA ON COLEOPTERAN PHEROMONES

No.	Chemical name, common name and synonym	NMR spectra (δ-values) Nucleus (solvent): chemical shift (line shape, number of nuclei)[a]	IR spectrum (cm^{-1})[b]	Mass spectrum m/z (rel. intensity in %)	Ref.
51	Isopropyl (Z)-7-dodecenoate	^{1}H-NMR similar to that of **50**		240 (M$^+$, 12), 198 (6), 197 (4), 196 (1), 181 (31), 180 (19), 179 (9), 163 (5), 162 (8), 161 (11)	299
52	Isopropyl (Z)-9-dodecenoate	^{1}H-NMR see **50**		240 (M$^+$, 12), 198 (6), 197 (4), 196 (1), 181 (22), 180 (19), 179 (16), 163 (8), 162 (7) 161 (6)	299
53	Methyl tetradecanoate Methyl myristate		Film: 2895s, 2825s, 1739s, 1464m, 1433m	242 (M$^+$, 27), 211 (10), 199 (19), 143 (16), 87 (60), 74 (100), 55 (27), 43 (30)	292, 298
54	Isopropyl (Z)-9-tetradecenoate Isopropyl myristoleate	^{1}H (CDCl$_3$): 0.83 (t, 3H), 1.1 (d, 6H), 1.15—1.25 (12H), 1.5 (m, 2H), 1.85 (m, 4H), 2.15 (t, 2H), 4.9 (m, 1H), 5.25 (m, 2H)		268 (M$^+$, 5), 226 (3), 225 (2), 224 (1), 209 (28), 208 (19), 207 (4), 191 (2), 190 (2), 189 (2)	299
55	Ethyl hexadecanoate Ethyl palmitate		Film: 2890s, 2835s, 1727s, 1456s, 1368m, 1171m, 1164m, 1032m, 717m	284 (M$^+$, 20), 241 (9), 239 (11), 101 (55), 88 (100), 73 (10), 55 (15), 43 (24)	292, 298

No.	Compound	IR	MS	Ref.
56	Methyl (Z)-7-hexadecenoate		268 (M$^+$, 1), 237 (15), 236 (21), 194 (18), 152 (35), 97 (75), 87 (76), 83 (75), 74 (80), 69 (90), 59 (29), 55 (100), 43 (52), 41 (56)	292
58	Ethyl octadecanoate / Ethyl stearate CH$_3$(CH$_2$)$_{16}$CO$_2$CH$_2$CH$_3$		312 (M$^+$, 60), 269 (20), 267 (19), 213 (15), 157 (30), 101 (62), 88 (100), 73 (10), 55 (12), 43 (11)	292
59	Methyl (Z)-9-octadecenoate / Methyl oleate	Film: 2925s, 1745s	296 (M$^+$, 10), 265 (24), 264 (49), 221 (43), 88 (50), 74 (62), 55 (100), 41 (70), 39 (75)	292, 298
60	Ethyl (Z)-9-octadecenoate / Ethyl oleate	Film: 2890s, 2835s, 1730s, 1468m, 1179m	310 (M$^+$, 20), 265 (38), 264 (50), 222 (25), 180 (19), 88 (60), 55 (100), 41 (70)	292, 294, 298
62	Ethyl (9Z,12Z)-9,12-octadecadienoate / Ethyl linoleate	Film: 3010s, 2955s, 2925s, 2855s, 1750s, 1467m, 1439m		294, 298
63	Ethanol	Film: 3300s, 2925s, 2890sh, 2840s, 1418w, 1381m, 1086s, 1047, 881s	46 (M$^+$, 15), 45 (39), 31 (100), 29 (20)	292, 294, 298
64	Myrtenol		152 (M$^+$, 3), 137 (1), 134 (2), 108 (30), 91 (45), 79 (100), 77 (20)	292

Table 4 (continued)
NMR, IR, AND MASS SPECTRAL DATA ON COLEOPTERAN PHEROMONES

No.	Chemical name, common name and synonym	NMR spectra (δ-values) Nucleus (solvent): chemical shift (line shape, number of nuclei)[a]	IR spectrum (cm⁻¹)[b]	Mass spectrum m/z (rel. intensity in %)	Ref.
65	8-Methyldecan-2-ol propanoate	^{1}H(d₆-acetone): 1.06 (t, 3H), 1.17 (d, 3H), 2.26 (q, 2H), 4.86 (m, 1H)	CCl₄: 1740s	177 (7), 156 (7), 154 (9, M − 74) 125 (38), 101 (26), 83 (25), 75 (20), 70 (59), 57 (100)	271
66	10-Methyl-2-tridecanone	^{1}H (CDCl₃): 2.13 (s, 3H), 2.42 (t, 2H)	CCl₄: 1720s	212 (M⁺, 0.2) 197 (M − 15, 0.2), 194 (M − 18, 0.6), 71 (40), 58 (92), 43 (100) CI(CH₄): 253 (M + 41, 5), 241 (M + 29, 20), 213 (M + 1, 100), 211 (M − 1, 25)	273
89	(E,E)-4,8-dimethyl-4,8-decadien-10-olide Ferrulactone I	^{1}H (CDCl₃): 1.57 (d, 3H), 1.64 (d, 3H), 4.56 (m, 2H), 4.82 (m, vinyl H), 5.54 (t, vinyl H) ^{13}C-NMR spectrum[302]	Neat: 1730s, 1132s	194.1291 (M⁺, 5), 127 (79), 99 (78), 68 (100), 67 (80) CI(NH₃)[616]	302
90	(3Z,11S)-3-dodecen-11-olide Ferrulactone II	^{1}H (CDCl₃): 1.26 (d, 3H), 1.3—1.7 (m, 10H), 5.50 and 5.62 (2 olef. H) ^{13}C-MR spectrum[302]	Neat: 970s	196.1449 (M⁺, 11), 136 (39), 82 (78), 81 (100), 67 (80), 54 (60) CI(NH₃)[302]	302

[a] For NMR spectra s = singlet, d = doublet, t = triplet, q = quartet, m = multiplet, br = broad, dd = doublet of doublet and dt = doublet of triplet.

[b] For IR data s = strong, m = medium, w = weak, sh = shoulder, and br = broad.

FUNCTION AND BIOLOGICAL ACTIVITY OF BEETLE PHEROMONES

Insect pheromones generally can be classified into three groups according to their function in (1) orientation of the insects towards or away from a location, (2) signaling danger, and (3) providing information for intraspecific social interactions. In the order Coleoptera only pheromones are known belonging to type (1), where in particular sex pheromones, aggregation pheromones and, in a functionally relevant sense, response inhibitors are found.

The attack of host trees by most bark beetles (Scolytidae) is mediated by population aggregation pheromones and happens in different phases. For polygamous species the host plants are selected mostly by the male insects, whereas for monogamous species the females are responsible for the selection of suitable host trees (primary attraction). The aggregating pheromone which can be emitted by one or both sexes and which now attracts both sexes to the source (secondary attraction) thus provides (1) the assembling of both sexes and (2) the recruitment of a large number of beetles to overwhelm the natural resistance of the host tree.

Male and female pheromones with sex-specific activities are found widely spread among the different Coleoptera families. The pheromone of the male boll weevil *Anthonomus grandis*, Curculionidae, attracts the females of that species from greater distances[303,304] and has been identified as a four-component pheromone complex,[60,92] the same as for the sugarcane weevil *Rhabdoscelus obscurus*.[305] In another curculionid species a male-produced aggregation pheromone was detected which attracts both sexes, the chemical composition of which is still unknown.[306] Alkanals with 15-, 16-, and 18-carbon chain lengths were identified from *Hylobius abietis* as potential pheromones.[307] In the sweet potato weevil evidence for a multiple component sex pheromone system was reported.[308] Many volatiles were found in the *Curculio caryae*.[309,310] From female cowpea weevils, *Callosobruchus maculatus*, calling behavior and pheromonal release was demonstrated.[311] The pheromone obtained by aeration was utilized for dose-response studies.[311] Female beetles of the Elateridae family produce sex pheromones in abdominal glands which cause orientation reactions and attraction of the males. Pentanoic and hexanoic acid could be isolated from the abdomina of unmated *Limonius* females.[312,313]

In the Bruchidae family up to now only one sex pheromone has been identified from male dried bean beetles, *Acanthoscelides obtectus*, and its structure elucidated as methyl tetradecatrienoate.[33]

Olfactory reactions of male khapra beetles *Trogoderma granarium* towards unmated females have been known for about 60 years.[314] In recent years, however, some 15 pheromones were detected in the genera *Trogoderma*, *Attagenus*, *Anthrenus*, and *Dermestes*, belonging to the Dermestidae family (Table 5).

In the Tenebrionidae for the flour beetles (*Tribolium* spp.) sex pheromones have been isolated and identified.

From female cigarette and drugstore beetles, Anobiidae, pyran derivatives have been isolated as sex stimulants (Table 5).

The female attractant of *Diabrotica balteata*, Chrysomelidae, seems to be an epoxyaldehyde or epoxyketone.[315] Only recently the pheromones of the western rootworm *Diabrotica virgifera virgifera*[271] and *D. undecimpunctata*[273] have been identified.

Sexual attraction by means of volatiles of male insects to unmated females has been proven for beetles of the genera *Phyllophaga*, *Lachnosterna*, *Popillia*, *Pachypus*, *Polyphylla*, *Plectis*, *Melolontha*, *Costelytra*, *Rhizotrogus*, and *Rhaepaea* belonging to the Scarabaeidae. Phenol and tetradecenolide were isolated from female *Costelytra zealandica*[34] and *Popillia japonica*,[251] respectively.

A typical example of the aphrodisiac activity of a beetle pheromone is the scent fanning of the male Meloidae beetle *Cerocoma schaefferi*. The odor secretion is produced in glands

Table 5

COLEOPTERA PHEROMONES: OCCURRENCE AND REFERENCES

Family, species (common name)	Pheromone or components identified (numbers refer to the note at the end of the table)	References for Identification	References for Laboratory and field bioassay
Anobiidae			
Lasioderma serricorne (cigarette beetle, Japan)	36	322	322, 323
L. serricorne (cigarette beetle, Germany)	37	324	324
Stegium paniceum (drugstore beetle)	29	325	
Bostrichidae			
Rhyzopertha dominica F. (lesser grain borer)	27, 28	68	326
Bruchidae			
Acanthoscelides obtectus (dried bean beetle)	18	33, 327	33
Cerambycidae			
Hylotrupes bajulus (house long-horn beetle, old house borer)	17, 64, 86	320	320
Chrysomelidae			
Diabrotica balteata (banded cucumber beetle)	Partial structure identified	315	
D. cristata	92		328
D. undecimpuncta (southern corn rootworm)	66	273	273
D. virgifera virgifera (western corn rootworm)	65	271	271
Cucujidae			
Cryptolestes ferrugineus (rusty grain beetle)	89, 90	302	302
Curculionidae			
Anthonomus grandis (boll weevil)	9, 11, 12, 13	52, 60, 92, 329	60, 329—335
Callosobruchus chinensis (azuki bean weevil)	Mixture of methyl-branched hydrocarbons and 85 mono- and dimethyl-branched hydrocarbons 11, many volatiles[a]	336	336
C. maculatus (cowpea weevil)		311, 337	311
Curculio caryae (pecan weevil)	Mixture of methyl-branched hydrocarbons and (85) mono- and dimethyl-branched hydrocarbons 11, many volatiles[a]	309, 310	309, 310
Hylobius abietis (large pine weevil)	67, 68, 69, and numerous terpenoids[a]	307	
Pissodes approximatus (white pine weevil)	9, 91	338	338
Pissodes strobi	9, 91	338	338
Rhabdoscelus obscurus (New Guinea sugarcane weevil)	9, 11, 12, 13	305, 306	305, 306
Sitophilus oryzae (rice weevil)	No structure	306, 339	339
Dermestidae			
Anthrenus flavipes (furniture carpet beetle)	24	41, 233	41
Attagenus elongatulus	25[b], 26	340	341

Table 5 (continued)
COLEOPTERA PHEROMONES: OCCURRENCE AND REFERENCES

Family, species (common name)	Pheromone or components identified (numbers refer to the note at the end of the table)	References for	
		Identification	Laboratory and field bioassay
A. megatoma (black carpet beetle)	25	235	
Dermestes ater (black larder beetle)	52[b], 54[b], 57[b]		342
D. lardarius (larder beetle)	52[b], 54[b], 57[b]		342
D. maculatus (hide beetle)	52, 54, 57, 61, various fatty acids from fecal pellets[a]	342—344	342, 344, 345
Trogoderma glabrum	19[b], 22, 23, 34, 48, 56	276, 346, 347	222, 276, 348
T. granarium (khapra beetle)	19[b], 20, 21, 23, 34, 48	276, 349, 350	276, 344, 348—355
T. grassmani	19[b]		222, 348
T. inclusum	19, 20, 21, 34	222, 276, 356	226, 276
T. simplex	19[b], 20[b], 21[b]		222, 348
T. sternale	19[b]		222, 348
T. variabile (warehouse beetle)	19, 20, 21, 34	276, 356, 357	276
Elateridae			
Limonius californicus (sugar beet wireworm)	47	312	
L. canus (Pacific coast wireworm)	48, and unsaturated branched C_{10}-acid	313	313
Lymexylidae			
Hylecoetus dermestoides	35	260, 261	
Platypodidae			
Platypus flavicornis (pinhole borer)	38, 63[b], 87, 88	358, 359	358
Scarabaeidae			
Costelytra zealandica (grass grub beetle)	41	34	34, 360
Popillia japonica (Japanese beetle)	33	251	251, 361, 362
Scolytidae			
Blastophagus piniperda	21 terpenes and 70[a]	363	
Dendroctonus adjunctus (round-headed pine beetle)	1, 3	364	364
D. brevicomis (western pine beetle)	1, 2, 3, 14, 15, 17, 64, 71, 72, 77	44, 95, 133, 365—370, 384, 399	368, 371—383
D. frontalis (southern pine beetle)	2, 3, 14, 15, 17, 64, 71, 72, 73, 74, 75, 76, 77, 78[a], 88[b]	43, 70, 86, 133, 365, 367—370, 384—387, 390, 391, 399	43, 133, 368, 379, 385—389, 392—398
D. jeffreyi (Jeffrey pine beetle)	1, 14, 15, 17, 73, 74	368, 400	368
D. ponderosae (mountain pine beetle)	1, 2, 3, 14, 15, 17, 79	43, 368, 385, 401, 402	43, 346, 368, 401, 403
D. pseudotsugae (Douglas fir beetle)	3, 30, 63[b], 79	42, 404—408	366, 404—407, 409—417
D. rufipennis (spruce bark beetle)	3[b], 30, 79[b]	404	416, 418, 419
D. simplex (eastern larch beetle)	30[b]		420
D. valens (red turpentine beetle)	8, 14, 15, 64, 71, 72	369	421
D. vitei	3, 14	387	387
Gnathotrichus retusus	38		422

Table 5 (continued)
COLEOPTERA PHEROMONES: OCCURRENCE AND REFERENCES

Family, species (common name)	Pheromone or components identified (numbers refer to the note at the end of the table)	References for Identification	Laboratory and field bioassay
G. sulcatus (ambrosia beetle)	38, 63[b]	69, 408	69, 408, 422, 423
Ips acuminatus	7, 8[a], 15	424, 425	
I. amitinus	7, 8, 14, 15, 64, 84	288, 426	
I. avulsus (small southern pine engraver)	8[b], 14, 15	424, 427, 428	428, 429
I. bonanseai	8, 14, 15[a]	424	
I. calligraphus (southern pine engraver)	7, 8, 14, 15	424, 428, 429	428, 430, 433
I. cembrae (larch bark beetle)	7, 8, 81, 88	431	431, 432
I. confusus (paraconfusus) (California five-spined ips)	7, 8, 14, 15, 17	158, 434	434, 435
I. cribricollis (six-spined engraver beetle)	7, 8[a]	424	
I. duplicatus	8	436	436
I. grandicollis (five-spined engraver beetle)	3, 7, 14, 15	424, 429, 437	169, 438
I. integer	8, 14, 15[a]	424	
I. interstitialis	8, 15	439	
I. knausi	8, 14, 15[a]	424	
I. (= Orthotomicus) latidens	7[b], 14[b], 15[b]	440	440, 441
I. paraconfusus (confusus) (California five-spined ips)	7, 8, 14, 15, 64, 75, 77	158, 277, 384, 408, 424, 429, 441, 443	169, 381, 383, 440, 442
I. pini (pine engraver)	8, 14, 15, 75	71, 408, 424, 441, 444	444—446
I. sexdentatus (six-toothed engraver beetle)	7, 8	424, 447	447, 448
I. typographus (spruce engraver, eight-toothed engraver beetle)	7, 8, 14, 15, 80	424, 426, 449	449—456
Leperisinus varius (ash bark beetle)	1, 82, 83	291	291
Orthotomicus caelatus	7[b]		441
O. erosus	14 or 15, 17[b]		457
Pityogenes chalcographus	4, 87	141	
Pityokteines curvidens (crooked-toothed fir bark beetle, European fir engraver)	7	458	458
P. spinidens	7	459	
P. vorontzovi	7, 8	459	
Scolytus multistriatus (smaller European elm bark beetle)	5, 16[b], 31	96, 370, 408, 462	96, 156, 460, 461, 463—468
S. pigmaeus	5[b], 16[b], 31[b]		464, 469
S. scolytus (large elm bark beetle)	5, 16[b], 31	470	470—473
Trypodendron (= Xyloterus) domesticum	10[b], 39, 40, 42, 53, 55, 63[b]	296, 474—476	474, 476—479
T. lineatum (striped ambrosia beetle)	10, 39, 63[b]	193, 296, 474, 476	193, 197, 474, 476, 478—484
Xyloterus signatus	10, 39	296, 474	474, 476
Staphylinidae			
Bledius mandibularis	32	295	
B. spectabilis	32	295	

Table 5 (continued)
COLEOPTERA PHEROMONES: OCCURRENCE AND REFERENCES

Family, species (common name)	Pheromone or components identified (numbers refer to the note at the end of the table)	References for	
		Identification	Laboratory and field bioassay
Tenebrionidae			
Tribolium castaneum (red flour beetle)	43	297, 485—487	275
T. confusum (confused flour beetle)	43, 44, 45, 46	297, 485, 487, 488	275, 488
Tenebrio molitor (mealworm beetle)	No structure		489

Note: 1 exo-Brevicomin, 2 endo-brevicomin, 3 frontalin, 4 chalcograne, 5 α-multistriatin, 6 δ-multistriatin, 7 ipsenol, 8 ipsdienol, 9 grandisol, 10 lineatin, 11 (Z)-3,3-dimethyl-$\Delta^{1,\beta}$-cyclohexane ethanol,12 (Z)-3,3-dimethyl-$\Delta^{1,\alpha}$-cyclohexane acetaldehyde, 13 (E)-3,3-dimethyl-$\Delta^{1,\alpha}$-cyclohexane acetaldehyde, 14 *trans*-ver-benol, 15 *cis*-verbenol, 16 α-cubebene, 17 verbenone, 18 methyl (R,E)(−)-2,4,5-tetradecatrienoate, 19 (Z)-14-methyl-8-hexadecen-1-ol, 20 methyl (Z)-14-methyl-8-hexadecenoate, 21 (Z)-14-methyl-8-hexade-cenal, 22 (E)-14-methyl-8-hexadecen-1-ol, 23 (E)-14-methyl-8-hexadecenal, 24 (Z)-3-decenoic acid, 25 (3E,5Z)-3,5-tetradecadienoic acid, 26 (3Z,5Z)-tetradecadienoic acid, 27 dominicalure 1, 28 dominicalure 2, 29 stegobinone, 30 seudenol, 31 4-methylheptan-3-ol, 32 dodecan-4-olide, 33 (Z)-5-tetradecen-4-olide, 34 γ-caprolactone, 35 2,5-dimethyl-2-isopropyl-2,3-dihydrofuran, 36 serricornin, 37 anhydroserricornin, 38 sulcatol, 39 3-hydroxy-3-methylbutan-2-one, 40 2,3-dihydroxy-2-methylbutane, 41 phenol, 42 hexan-2-ol, 43 4,8-dimethyldecanal, 44 1-pentadecene, 45 hexadecane, 46 1-heptadecene, 47 pentanoic acid, 48 hexanoic acid, 49 isopropyl laurate, 50 isopropyl (Z)-5-dodecenoate, 51 isopropyl (Z)-7-dodecenoate, 52 isopropyl (Z)-9-dodecenoate, 53 methyl myristate, 54 isopropyl myristoleate, 55 ethyl palmitate, 56 methyl (Z)-7-hexadecenoate, 57 isopropyl palmitoleate, 58 ethyl stearate, 59 methyl oleate, 60 ethyl oleate, 61 isopropyl oleate, 62 ethyl linoleate, 63 ethanol, 64 myrtenol, 65 8-methyldecan-2-ol propanoate, 66 10-methyl-2-tridecanone, 67 pentadecanal, 68 hexadecanal, 69 octadecanal, 70 naphthalene, 71 pinocarvone, 72 *trans*-pinocarveol, 73 heptan-1-ol, 74 heptan-2-ol, 75 linalool, 76 myrtenal, 77 phenylethanol, 78 4-methylpentan-2-ol, 79 3-methyl-2-cyclohexenone, 80 2-methyl-3-buten-2-ol, 81 3-methyl-3-buten-1-ol, 82 7-methyl-1,6-dioxaspiro[4.5]decane, 83 2-nonanone, 84 (E)-2-methyl-6-methylene-3,7-octadien-2-ol, 85 (E)-3,7-dimethyl-2-octene-1,8-dioic acid, 86 *p*-cymen-8-ol, 87 hexan-1-ol, 88 3-methylbutan-1-ol, 89 fer-rulactone I, 90 ferrulactone II, 91 grandisal, 92 8-methyl-2-decyl acetate.

[a] Compounds have been identified from the insects; however, their functions as pheromones have not been established.

[b] Although attractive or synergistic, the compounds have not been identified from the species.

of the mandible and the third antennal link and is fanned with the hairy front legs towards the female during mating.[316,317]

In the Staphylinidae a species-specific sex pheromone has been found on the cuticle of the female *Eusphalerum minutum*[318] and *Aleochara curtula*.[319]

Two esters have been identified from the lesser grain borer, *Rhyzopertha dominica*, a Bostrichidae species.[68]

The ovipositional behavior of the timber pest *Hylotrupes bajalus*, Cerambycidae, is mediated by pheromones produced in the frass of the wood-boring larvae of this species.[320] Investigating the mating behavior of *Xylotrechus pyrrhoderus*, a male-produced sex pheromone could be proven.[321]

Frass volatiles of *Cryptolestes ferrugineus*, belonging to the Cucujidae family, acting as aggregation pheromones were very recently identified as macrocyclic lactones.[302]

In the family Platypodidae, males of *Platypus flavicornis* produce volatiles during their attack on pine trees. Furthermore, sex-specific pheromones have been identified in *Trypodendron* and *Xylosterus* species (see Table 5).

One of the major difficulties in dealing with beetle pheromones is the difficulty of clear differentiation between actual pheromones and other biologically active compounds like host scents, synergists, and substances of unknown function, occurring in the insects. Thus, Table 5 can give an overview of chemicals only, that have been isolated from the insect or have an activity on another species than the one from which they have been isolated. Furthermore, the table presents substances the biological functions of which are still unknown, but it strictly excludes chemicals like warning substances and defensive secretions.

It can be clearly seen from Table 5 that most of the beetles use more than one compound for intraspecific and/or sexual communication. This agrees with a suggestion of Wright[490] in 1964, who supposed that multicomponent pheromones would be widely used because of their greater content of information than that of a single compound pheromone. This conforms to many other communication systems like the Morse code, computer languages etc., which always use more than one "letter" (point and dash, 0 and 1, respectively) for specific signaling and correct recognition.[29,491,492] Thus, chemical communication using binary, ternary, or multicomponent mixtures of semiochemicals offers the advantage of distinguishing not only certain compounds of a complex mixture, but also recognizing the defined ratio of the single components, thus improving the content of information of the chemical message.

The findings of multicomponent pheromones in different orders of insects as well as repeated failures of field tests based on a single compound, led to the conclusion that insects generally use pheromone complexes for intraspecific communication.[29,491-495] Beetle pheromones, from the beginning, presented a complex picture. Sex pheromones and aggregation pheromones can be emitted by one sex or both sexes, and "female" as well as "male" pheromones are found. In the Scolytidae, in general, the pheromone is released by the female of monogamous species, and by the male of the polygamous species.[365] In the genus *Dendroctonus*, brevicomin has been found in the males in one case only (*D. adjunctus*) attracting females,[364] whereas it has been identified in the female and its frass, respectively, of *D. brevicomis*, attracting the corresponding males.[95] Frontalin, another bicyclic ketal, is the active principle of the female pheromone of some *Dendroctonus* species (*D. adjunctus*, *D. frontalis*, *D. ponderosae* and *D. pseudotsugae*, c.f. Table 5), but was also identified in the male *D. brevicomis*, attracting the females.[366] Most *Dendroctonus* species furthermore produce oxidation products of pinene as their second pheromone main component (see the chapter on biosynthesis of beetle pheromones). The terpenoid alcohols ipsenol and ipsdienol are produced as the main components of the male aggregants in the genus *Ips*, attracting both sexes.

The pheromone-regulated colonization mechanism of host trees by bark beetles can be divided into three phases, i.e., the initial attack, the mass attack, and the shifting of the attack. During the initial attack, the "pioneer" beetles will recognize a suitable host tree and signal this suitability to recruit large numbers of its species colleagues to overwhelm the natural resistance of healthy trees. This mass attack is mostly initiated by one sex only. As the responding beetles enter the bark, pheromones are released, usually by defecation, and/or host volatiles are generated. During this stage of attack (*secondary attraction*) the focus of attraction may extend from one part of the tree to the whole plant, as well as on several trees. To prevent overcrowding and subsequent mortality, now the attack can be stopped and shifted to another host plant, also by pheromone regulation.[365, 496]

In contrast to the female sex pheromones of butterflies and moths (Lepidoptera), many bark beetles appear to aggregate in response to a single compound only. For example, *Ips sexdentatus*[447] and *I. duplicatus*[436] are attracted by ipsdienol alone, and the southern pine beetle *D. frontalis* is attracted by frontalin[133] in North and Central America. On the other hand it can be clearly seen from Table 4 that multicomponent pheromone systems exist in the Scolytidae, and as in the case of *I. paraconfusus*,[158] the three terpenoid alcohols ipsenol, ipsdienol, and *cis*-verbenol are required for attraction.

In the Curculionidae family the male boll weevil *Anthonomus grandis* attracts females of the same species over distances more than 80 m and to a lower extent also males.[304] With the four compounds, isolated from the male insects and identified, in laboratory bioassay[60] a complete behavioral sequence could be achieved using only a combination of all four substances. When one component of this blend was missing, only weak or almost no response was found in the females. Furthermore, pheromones were proven in the *Curculio, Rhabdoscelus,* and *Sitophilus* genus (c.f. Table 5). For *Hylobius abietis* a type of sensory cell was detected specifically responding to the bark beetle pheromone frontalin.[497]

In 1924 olfactory responses of the male khapra beetle *Trogoderma granarium,* Dermestidae, towards unmated females were reported. The pheromone responsible like all the other *Trogoderma* pheromones, is produced by the females only and attracts the males even in very low concentrations.[498,499] Olfactometer tests of six different *Trogoderma* species resulted in males of *T. granarium, T. inclusum, T. simplex,* and *T. variabile* responding mutually to the foreign species female odor, whereas the pheromones of *T. glabrum* and *T. sternale* showed no interspecific activity.[499-501] Since the pheromones of the *Trogoderma* genus are chiral compounds, different activities are found for each of the enantiomers (c.f. stereobiology). Intraspecific attractants, consisting of various isopropyl esters of fatty acids, were found in a gland secretion of male hide beetles *Dermestes maculatus,* attracting adult *D. maculatus* males and females and promoting the recognition of sexually mature males. On the other hand, males and females of *D. maculatus* utilize excreted fatty acids as feeding aggregants and nutrients.[402]

Female beetles of the Elateridae genus *Hypolithus, Hemicrepidius, Limonius, Ctenicera,* and *Agriotes* produce sex pheromones eliciting orientation reactions and the attraction of the males.[18,502-508] While the range of attractiveness of *Limonius* females is restricted to some 12 m, an unmated *Agriotes litigiosus* female attracted 2000 males from a distance up to 800 m within 4 hr.[504]

SPECIFICITY OF BEETLE PHEROMONES — STEREOBIOLOGY

Dealing with the specificity of beetle pheromones, one should distinguish between sex pheromones, the primary function of which is to bring males and females of a species together for the purpose of mating, and aggregation pheromones, often eliciting responses in both sexes. The role of sex pheromones as an isolating mechanism, as it is found with lepidopterous sex attractants, may be misleading, especially in the bark beetles. In the Coleoptera family it can be seen, that the specificity of chemical communication by volatiles may enter at all levels of the communication system, i.e., either the pheromone production, the transmission, the pheromone perception, and the behavioral response.

It is now already well established, that insects mostly use pheromone blends. But, again, in the Coleoptera a lot of examples are known, where only one single compound has been identified as the pheromone or gives a sufficient and complete behavior response.

Variation of the position or geometry of double bonds is one of the common ways of guaranteeing isolation mechanisms in the pheromones of Lepidoptera (e.g., defined mixtures of (Z)- and (E)-11-tetradecenyl acetate with the European corn borer *Ostrinia nubilalis* and the red-banded leaf roller *Argyrotaenia velutinana* and (Z)-9- and (Z)-11-tetradecenyl acetate for the tortricid moth *Adoxophyes* spp., etc.). Similarly, this can be found in a few Coleoptera species also, using olefinic pheromones. Four species of *Trogoderma* beetles use methylbranched alkenals as their major sex pheromone components.[276] Aeration of *T. granarium,* trapping of the volatiles on Porapak Q (polystyrene), and analysis of the pheromone gave a 92:8 ratio of (Z)- to (E)-14-methyl-8-hexadecenal (trogodermal). The (Z)-isomer furthermore was the main component in female *T. inclusum* and *variabile* and the (E)-isomer was obtained from *T. glabrum.*[276] In laboratory bioassays male beetles were able to discriminate between the geometric isomers. *T. angustum* is exceptional by neither releasing trogodermal nor responding to it.[509]

The receptor potentials recorded from the antennae of male *Trogoderma* species by graded stimulation with trogodermal mostly coincided with the isomer production of this attractant. Olfactory sensillae of species exclusively emitting (Z)-trogodermal were approximately 10^3 times more responsive to the (Z)-isomer than to the (E)-aldehyde. The response of *T. granarium* (releasing both geometric isomers) to (Z)-trogodermal was tenfold higher than to the (E)-isomer.[351] Surprisingly, the olfactory sensilla of *T. glabrum* react to the (Z)-configurated aldehyde some 10^3 times better than to the (E)-isomer, despite the exclusive production of the latter in the conspecific females.[509]

Aspects of the value of pheromones in speciation of *Coleoptera* have been discussed by Lanier and Burkholder,[510] as have consequences of interspecific attraction of *Trogoderma* species to extracts of female beetles by Vick et al.[501] and by Levinson and Bar Ilan.[498] The most active compound of this pheromone complex is the 14-methyl-8-hexadecenal, the isomeric composition of which spread among the different species as mentioned above. However, this aldehyde is not found in macerated females, the corresponding alcohol was isolated instead (Table 5). Interspecific responses between *T. inclusum*, *T. variabile*, and *T. granarium* certainly come from the common (Z)-aldehyde. Neither *T. sternale* nor *T. grassmani* respond to the (Z)- or (E)-aldehyde.[348] *T. simplex* males strongly respond to the females of *T. variabile*, *T. granarium*, and *T. inclusum*, indicating that *T. simplex* probably emits the (Z)-aldehyde.

Females of the black carpet beetle *Attagenus megatoma* produce megatomic acid, (3E,5Z)-3,5-tetradecadienoic acid, as their sex pheromone.[235] A geometric isomer of this chemical, (3Z,5Z)-3,5-tetradecadienoic acid, has been identified in the related species *A. elongatus*.[340] Here again the geometry of the chemical structure of the pheromone could represent an isolation factor.

Another factor of odor discrimination and thus of species or sex specificity is chirality often found in beetle semiochemicals. In Table 5 many compounds are found existing in two enantiomeric forms (i.e., like-image and mirror-image). Since chirality is often found associated with compounds of biological origin, insects often biosynthesize and utilize as their pheromone one enantiomer only or a specific ratio of the antipods. Only in recent years have stereospecific syntheses of sufficient quantities of pure enantiomers of beetle pheromones been possible to show that beetles can often distinguish between enantiomers in laboratory and field bioassays (the subject of chirality in insect semiochemicals has been reviewed quite thoroughly by Silverstein,[511] Brand et al.,[512] and Mori[513]).

The ambrosia beetle *Gnathotrichus sulcatus* produces 6-methyl-5-hepten-2-ol (sulcatol), its aggregant, as a 65:35 mixture of (S)(+)- and (R)(−)-enantiomers[69] but responds significantly better to racemic 50:50 mixture.[422] The aggregation pheromone of *G. retusus* was isolated and identified as (S)(+)-sulcatol. In field and laboratory experiments, *G. retusus* responded to (S)(+)-sulcatol, but not to (±)-sulcatol[422,514] which was attractive to a sympatrically occurring species *G. sulcatus*. The latter did not respond to optically pure (S)(+)-sulcatol, but began to respond when ≥1% (R)(−)-sulcatol was present in the enantiomeric mixture.[514] Thus, the species isolation may be achieved by the different enantiomeric composition of the pheromone.

The smaller European elm bark beetle *Scolytus multistriatus* responds to a pheromone blend[461] consisting of the three synergistically acting compounds (−)-4-methyl-3-heptanol, (−)-α-multistriatin, and (−)-α-cubebene (for formulas, see Table 2). The absolute configuration of the heptanol was determined as (3S,4S)[244] and that of multistriatin with (1S,2R,4S,5R).[74] The multistriatin in *S. multistriatus* was determined as exclusively the (−)-α-enantiomer by NMR spectra in the presence of a chiral shift reagent.[70]

In field tests in England, *S. multistriatus* was attracted by (±)-α-multistriatin and (−)-

α-multistriatin.[460] The attraction of *S. scolytus* to the four enantiomers of 4-methylheptan-3-ol in the field was studied by Blight et al.[472] Only the (−)-threo-isomer attracted significantly more beetles than the blank control. There was no evidence for inhibition of attraction by the other three isomers. Commercially available racemic 4-methylheptan-3-ol could therefore be used, in combination with the other pheromone components, for detection and control of *S. scolytus*.[472] Electroantennogram and single-cell recording techniques demonstrated the presence of separate receptors for (−)-threo- and (−)-erythro-4-methylheptan-3-ol on the *S. scolytus* antenna.[472]

The flight responses of both sexes of *Dendroctonus brevicomis* to a combination of exo-brevicomin, frontalin, and myrcene were studied.[382] Stewart et al.[70] established that the natural enantiomers were $(1S,5R)$(−)-frontalin in males and $(1R,5S,7R)$(+)-exo-brevicomin in females using NMR spectral analysis.

Frontalin has also been identified as a main pheromone constituent produced by the female southern pine beetle *D. frontalis*.[133] However, the full activity of frontalin is dependent on the presence of α-pinene.[515,516] Males of *D. frontalis* are more responsive to (−)-frontalin than to the (+)-frontalin, while females are not responsive to either isomer.[517] The enantiomeric composition of natural frontalin in *D. frontalis* females is 85:15 [(−):(+)].[70] Payne et al.[395] studied the olfactory receptor and behavior discrimination of enantiomers of the attractant with the southern pine beetle. The combination of frontalin and α-pinene attracts predominantly males; the addition of verbenone seems to equalize the sex ratio of attacking beetles; high doses of verbenone inhibit the attraction for both sexes.[393,515,518] endo-Brevicomin, produced by male beetles, together with verbenone has a strong inhibitory effect on the response to the attractant,[379] stopping the mass attack and shifting the attack to a fresh tree. Data on field tests using optically pure isomers of frontalin, brevicomin, and ipsenol resulted in a preferred activity of only one enantiomer of each of the test compounds.[519]

Male spruce engraver beetle (*Ips typographus*) utilizes a synergistic two-component pheromone, 2-methyl-3-buten-2-ol and (*S*)-*cis*-verbenol.[424,450-452] Dickens found a tenfold higher electrophysiological response to the (*S*)-*cis*-isomer than to that of its antipode.[520] The response to the pheromone complex is partially inhibited by the male-specific ipsenol. *I. calligraphus* responds to ipsdienol and *cis*-verbenol,[428] (*S*)-*cis*-verbenol being the active enantiomer. The response is inhibited by the (*R*)-*cis*-verbenol when it is present in a tenfold higher concentration than the (S)-antipode.[430]

In laboratory bioassays Birch and co-workers proved that *I. pini* females from the western U.S. respond well to (−)-ipsdienol, but neither to the (+)-isomer nor a racemic mixture.[445,446] This is in contrast to the response of insects from the eastern U.S., which responded to the corresponding (−)-enantiomer and the racemic mixture, but not to the (+)-ipsdienol. Furthermore, the response to the attractant blend produced by the conspecific males is inhibited by (*S*)(−)-ipsenol.[521] The natural ipsdienol composition for *I. pini* from Idaho (U.S.) was established with 100% (−)-form.[71]

(+)-Ipsdienol is a component of the pheromone of the competing species Ips paraconfusus, and interrupted the response of Californian *I. pini* to a natural source of attraction in field tests.[445] Ponderosa pine logs infested with *I. paraconfusus* males inhibited the attraction of *Dendroctonus brevicomis* to either pine tree logs or to the synthetic pheromones of the latter, exo-brevicomin, frontalin, and myrcene. Logs cut from trees under attack by *D. brevicomis* inhibited the response of *I. paraconfusus* to logs infested with male *I. paraconfusus*.[383] exo-Brevicomin, frontalin, and myrcene did not inhibit their response, but verbenone did.[383] *trans*-Verbenol and exo-brevicomin were found in female *D. brevicomis* while verbenone, *trans*-verbenol and frontalin were found in male *D. brevicomis* near the beginning of the aggregation phase of host colonization.

Dendroctonus brevicomis was attracted to a mixture of the *paraconfusus* pheromones, ipsenol, ipsdienol, and *cis*-verbenol, at 10^{-9} g each per μℓ, but not at higher or lower release rates. Increased amounts of a mixture of the three pheromones of *I. paraconfusus* inhibited the attraction of *D. brevicomis* to its synthetic pheromones.[381] The pheromones of *D. brevicomis* did not inhibit the attraction of *I. paraconfusus* to its pheromone; however, verbenone was a potent inhibitor.[381]

Of the 12 oxygenated compounds identified in *I. amitinus* pheromone bouquet, 9 were not sex-specific, but differences in the quantities of the verbenols, verbenone, and myrtenol were observed between the sexes. The absolute configuration of ipsdienol produced in male *I. amitinus* was determined with 95% (R)(−)-configuration.[288]

The stimulation effect of the two enantiomers of the male produced aggregation pheromone, ipsdienol, was tested by electrophysiological recordings from single olfactory cells in females of *I. pini* and *I. paraconfusus*.[522,523] Two types of receptor cells were found, each specialized to one of the optical configurations. The western *I. pini* had more cells tuned to (−)-ipsdienol than to (+)-ipsdienol; *I. paraconfusus* apparently had the majority of its ipsdienol cells keyed to the (+)-enantiomer.[522,523] This difference is consistent with behavioral responses where these species are sympatric in California. More (−)- than (+)-ipsdienol cells were also obtained in the eastern *I. pini*[522,523] even though this population produces more (+)- than (−)-ipsdienol (65:35) and requires both enantiomers for aggregation behavior.

The ambrosia beetle *Trypodendron* (− *Xyloterus*) *lineatum* females produce lineatin (3,3,7 trimethyl-2,9-dioxatricyclo[3.3.1.0^{4,7}]nonane, see Table 2). Primary and secondary host odors[481,483] "synergize" the response of *T. lineatum* to its aggregation pheromone. Each compound is responsible for a specific step in the colonization sequence, the mechanism[483] of which is thought to be initiated with lineatin, emitted from the female entering a new host tree and attracting in-flight beetles. α-Pinene, a common monoterpene hydrocarbon in coniferous host plants, affords the recognition of the host; at the same time α-pinene repels *T. domesticum*, a sympatrically occurring hardwood pest in Europe, also responding to lineatin.[524] The actual invasion then occurs in response to ethanol[483,525] which is formed by anaerobic fermentation in tree trunks. In field trials with lineatin isomers, *T. lineatum* was attracted to the 4,6,6-lineatin isomer and gave a lesser response to the structural analog 4,5,6-lineatin. The responses to (+)- and (±)-lineatin were essentially equal.[480] Electrophysiological and field responses of *Trypodendron* spp. to enantiomers of lineatin have been reported recently by Payne et al.[484]

(Z)-5-(1-decenyl)dihydro-2(3H)-furanone is secreted by the female Japanese beetle *Popillia japonica*.[251] Only the pure (R,Z)-isomer is responsible for the attraction of conspecific males, the response is strongly inhibited by the presence of a small amount of the (S,Z)-compound. Also, both the (E)-isomers and the saturated analog are produced in low yields by the female, the function of these compounds has not yet been clarified.

Chirality seems to be a factor of pheromone specificity also in the Dermestidae. Female *Trogoderma granarium* use 14-methyl-(Z)-8-hexadecenal and 14-methyl-(E)-8-hexadecenal as their sex attractant[276] (chiral at C-14). Whereas Rossi and Niccoli[355] tested the (S)(Z)- and (S)(E)- form only, Levinson et al.[353,354] were able to compare the activity of the (R)- as well as the (S)-isomers of (Z)- and (E)-trogodermal (the aldehyde mentioned) in electrophysiological and behavioral tests. In both cases the male beetles responded to the (R)-configurated compounds some 10^2 to 10^3 times better than to the corresponding (S)-aldehydes. Remarkably, a change of double bond geometry [(S,Z) ⇒ (S,E); (R,Z) ⇒ (R,E)], however, caused a loss of activity of a factor of 10 only.[354] Since *T. granarium* and *T. inclusum* are cross attracted to the females of either species,[498,501] probably due to the emission of (R,Z)-trogodermal by both, the release of different enantiomers of trogodermal cannot be the isolation factor between the two species, as it was speculated by Rossi et al.[225,352]

Finally it should be mentioned (or emphasized) that additional sensory stimuli like the odor of host plants can act synergistically and become an essential factor of behavioral

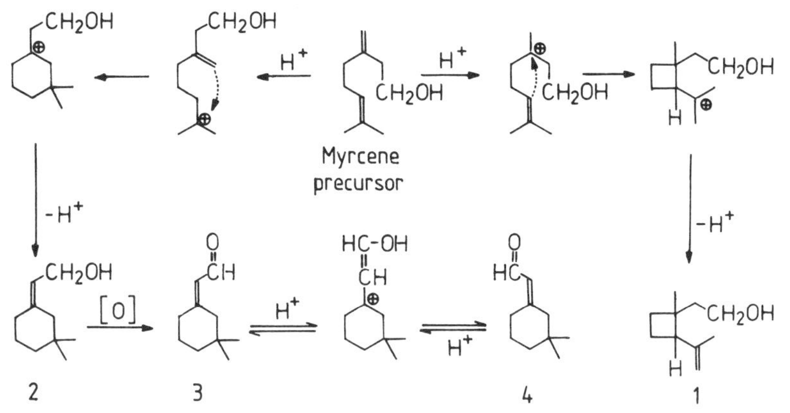

Scheme 8.

release and the sequence of pheromone regulated behavior, and thus of pheromone specificity. (The biology of pheromones has been reviewed in Reference 496.)

BIOSYNTHESIS

The isolation of the boll weevil pheromone led to the identification of four terpenoid compounds (1, 2, 3, and 4 in Scheme 8)[60] for which a hypothetical biosynthesis scheme (Scheme 8) was suggested by Tumlinson et al.[526] With this formula scheme it could be shown that all the four pheromone components can be derived from a myrcene precursor such as geraniol. Myrcene and β-ocimene are the major constituents of the essential oil of cotton buds,[527] furthermore, ten other monoterpene hydrocarbons,[527] several monoterpene alcohols,[528] and myrtenal[529] were found in cotton buds. Hardee[530] reported that male weevils required feeding for the synthesis of the attractants. Nevertheless, Mitlin and Hedin[531] demonstrated the *de novo* biosynthesis of the pheromone from ^{14}C-labeled acetate, mevalonic acid, and glucose. From later experiments it could be concluded that male abdomens possess three major enzymatic capabilities: (1) oxidation of alcohols to the corresponding aldehydes by a dehydrogenase, (2) dehydration of the alcohols to hydrocarbons by a hydrase, and (3) conversion to other alcohols by means of an isomerase.[532]

The pheromones of the bark beetles can be classified into three categories: (1) the bicyclic ketals, (2) terpene alcohols and a corresponding ketone, and (3) a group of simple cyclic and acyclic alcohols. The main compounds identified from the second category are *trans*-verbenol,[367] *cis*-verbenol, ipsenol, and ipsdienol[158] (see Table 5). The reaction for the biogenesis of all these compounds will certainly be an allylic hydroxylation of α-pinene and myrcene, both of which occur in considerable amounts in host pine trees. Since beetles of *Ips* and *Dendroctonus* species either ingest, or are in intimate contact with, numerous host plant terpenes, these terpene hydrocarbons are also found in the insects and show synergistic effects to the aggregating pheromone complex.

Analyzing the hindguts of 12 *Ips* species, Vité et al. found that the verbenols were biosynthesized even on exposure of the insects to oleoresin, whereas ipsdienol production required feeding.[424] Exposing several *Dendroctonus* species to an atmosphere saturated with α-pinene also resulted in an increase in *trans*-verbenol content[369,402] in their hindguts. In a more thorough study in hindguts of male *D. frontalis*, *cis*- and *trans*-verbenol, 4-methyl-2-pentanol, pinocarvone, and *trans*-pinocarveol were detected on exposure to α- and β-pinene.[391] In the guts of female beetles only 4-methyl-2-pentanol and *trans*-pinocarveol were found.[391]

Interesting results on the biosynthesis of the geometric and enantiomeric isomers of verbenol were obtained exposing beetles to different isomers of α-pinene. Adults of both sexes produced *trans*-verbenol and myrtenol after treatment with (+)-α-pinene, whereas after exposure to (−)-α-pinene predominantly *cis*-verbenol and myrtenol were found in the hindguts extract.[443] Hughes and Renwick found no appreciable *cis*-verbenol synthesis in the absence of exogenous (−)-α-pinene with *I. paraconfusus* males.[533]

Various *Ips* species produced ipsenol and/or ipsdienol when exposed to myrcene. But while *I. grandicollis* and *I. calligraphus* required feeding of myrcene before metabolization, *I. avulsus* and *I. paraconfusus* produced some products without prior feeding.[429] Ipsenol seemed to derive by reduction of ipsdienol, obtained by hydroxylation of myrcene.[429] Exposing *I. paraconfusus* to deuterium-labeled myrcene-^{2}H, Hendry et al.[534] observed deuterated ipsenol and ipsdienol in the abdomens of male beetles, thus proving the in vivo conversion of host plant chemicals to pheromones.

Similar oxidation experiments were carried out with *D. brevicomis* and *D. frontalis*. Both species were able to hydroxylate camphene to camphenol, both sexes of *D. brevicomis* oxidized myrcene into myrcenol, whereas ipsdienol was a major hydroxy product in males only.[535,536] When *D. frontalis* adults were exposed to the nonterpenoid hydrocarbon 1-methylcyclohexene, 3-methyl-2-cyclohexen-1-ol (seudenol) and 3-methyl-2-cyclohexen-1-one (MCH) were detected afterwards[537] with seudenol[404] and MCH[405] being the identified pheromones of the Douglas fir beetle *D. pseudotsugae*.

Since bark beetles externally contact and/or ingest large quantities of monoterpene hydrocarbons known to be toxic to these insects, pheromone production (oxidation of the hydrocarbons) also represents an important detoxification mechanism. Possibly, pheromones of that kind are "waste products from the detoxification of host terpenes that, as a consequence of timing and conditions of their production and release, have secondarily been utilized as chemical messengers".[538]

For an allylic hydroxylation of α-pinene and myrcene microsomal mixed-function oxidases, a group of enzymes responsible for the degradation of insecticides and drugs could be of importance. This enzyme complex has been found in several insect gut tissue; in the midgut tissue of larvae of a Lepidoptera species these oxidases could have been induced best by α-pinene and myrcene from a selection of secondary plant substances.[539]

The involvement of microorganisms has also been taken into consideration in pheromone synthesis, since many insect pheromones are known also to be produced by microorganisms. The lower alimentary tract of insects usually contains a large number of these organisms, the enzyme system of which might produce the gut-associated pheromone compounds. Brand et al.[540] isolated several microorganisms from the gut of male and female *I. paraconfusus* which were able to ferment α-pinene into *cis*- and *trans*-verbenol. One of the isolated organisms was *Bacillus cereus* which produced the verbenols together with *trans*-pinocarveol and myrtenol.[540] However, no isolation of a microorganism, able to convert myrcene into ipsdienol, has been reported. Furthermore, it is unlikely, that the microflora of different sexes and species could vary to such an extent to explain the specificity of pheromone production. A bacterium of the genus *Serratia* was isolated from the posterior part of the digestive tract of the scolytid beetle *Phloeosinus armatus*, which produced a bioconversion of sabinene into terpinen-4-ol and α-terpineol. Tested separately or in synergy, these two compounds attract both females and males at low concentrations, whereas sabinene had a repulsive effect.[541]

Two mycangial fungi and two yeasts have been found in female *Dendroctonus frontalis*[542,543] usually introduced into the phloem of host pines upon attack. One of the fungi in chemical transformation experiments could quantitatively convert either *cis*- or *trans*-verbenol into verbenone.[544] Consequently, it was suggested, that microorganisms growing in the phloem, *i.e.*, external to the beetle, also could be responsible for a production of pheromonal and behavioral chemicals.[544]

Since an increase of verbenone concentration around a colonized tree results in an inhibition of further attack by *D. frontalis* and *D. brevicomis*,[518] microorganisms like the fungi mentioned above, could also play a role in regulation of host plant colonization. More recently, endo-brevicomin has been implicated in the shifting of attack from one tree to another.[365]

Three yeasts have been obtained from *D. frontalis*, and the main volatile substances produced identified as isoamyl alcohol, 2-phenylethanol, isoamyl acetate, and 2-phenylethyl acetate.[545] 2-Phenylethanol was identified only recently in the hindguts of emergent *D. brevicomis* males and feeding *Ips paraconfusus* males.[384]

The involvement of juvenile hormone (JH) in pheromone biosynthesis was proven for *I. paraconfusus*[533,546] and *I. typographus*.[547] Thus, a topical treatment of male *I. paraconfusus* with 10,11-epoxyfarnesenic acid methylester (JHIII) improved the response to the hindgut region.[546] The feeding requirement of this beetle for pheromone production can be bypassed by distension of the gut through air injection and by topical JH application before exposure to myrcene.[533] Transplantation of glands and decapitation experiments suggested that the stimulation of pheromone production occurs indirectly. First, the distension of the gut by feeding (or air injection) is believed to induce neural stimulation of the corpora allata. The resulting JH acts through the brain neurosecretory cells and/or the corpora cardiaca to stimulate brain hormone (BH) release. The latter appears to control myrcene oxidation to ipsdienol and ipsenol, certainly by controlling the synthesis of specific enzymes.[533]

The same principle appears in species of *Pityokteines*, demonstrated by Harring.[459] Treatment with JH and exposure to myrcene led to the biosynthesis of ipsdienol and ipsenol in both sexes. However, female beetles, which are not responsible for pheromone release in nature, were not affected by gut distension, whereas males were stimulated to produce pheromone.[459] In other beetles like *Tenebrio molitor*[548] and in certain bark beetle species a direct introduction of pheromone biosynthesis by JH treatment is found. For example, *D. brevicomis* produces large quantities of exo-brevicomin after JH application, but the presence of some exogenous source of precursor seems to be necessary.[549] The use of JH to induce pheromone biosynthesis has proved to be extremely valuable in the isolation and structure elucidation of the pheromone of the *Pitygenes chalcographus*. Treatment of several thousands of these tiny beetles with JH resulted in the isolation of sufficient material for the identification of the pheromone, chalcograne.[141,550]

No details are known about the biosynthesis of the four bicyclic ketals: frontalin, exo-brevicomin, endo-brevicomin, and multistriatin (formulas given in Table 2). All four compounds theoretically can be derived from the corresponding dihydroxy- or epoxyketones, and the latter can be originated from the olefinic ketones 6-methyl-6-hepten-2-one, 6-nonen-2-one, and 4,6-dimethyl-7-octen-3-one, respectively. But unfortunately, none of these compounds has been detected in bark beetles up to now. Only isomeric compounds like 6-methyl-5-hepten-2-one are produced by one of the mycangial fungi of female *D. frontalis*,[551] by certain other microorganisms, and by ants. Similar compounds like the acyclic multistriatin precursor, 4,6-dimethyl-4-octen-3-one and 4,6-dimethyl-6-octen-3-one, were identified as alarm and defensive secretions from ants and from the daddy long legs (*Opiliones*).

The production of phenol, the sex attractant of the New Zealand grass grub beetle *Costelytra zealandica*, is claimed to be due to a bacterium.[552]

BEETLE PHEROMONES IN PLANT PROTECTION: A MEANS OF BIOLOGICAL INSECT CONTROL

Human communities as well as accidental transport of insect pests gave rise to an ecological imbalance in nature. Furthermore, extensive use of pesticides in the past 40 years has led to a critical loading of the biosphere with toxic chemicals. Admittedly, insects that destroyed crops and were vectors of disease, were killed, the crop yields increased, pest populations declined, and millions of lives saved throughout the world; but the long-term effects of extensive, indiscriminate insecticide application were neglected. Broad spectrum pesticides also wiped out the natural enemies of the target pest and gave rise to a rebound of the pest. New major pests were created, freed of their natural enemies, and the development of resistance to insecticides arose. Because of resistance and environmental problems new insecticides were produced, many of which were biodegradable but much more toxic to nontarget organisms. Beneficial insects, birds, other animals, and humans continued to be at risk, and newly arisen resistance effects led to the application of increased insecticide dosage. For all these reasons, the need for biological control of pest insects seems to be self-evident.

Biological control can be defined as a pest control utilizing either living organisms or their natural (and nontoxic) products. At the present time a number of such alternatives are known, represented by e.g., mass release and manipulation of predators or parasitoids, use of insect pathogens, hormonal control or use of hormone analogs, reproductive suppression by radiative or chemical sterilization, genetic selection of resistant plants, introduction of reproductively incompatible strains of insects, use of behavior modifying chemicals like pheromones or pheromone analogs, release of sterile insects, and any combination of these methods.

Integrated pest management is the concept of an integration of all factors "impinging on the pest control decision so as to determine when control should be practiced and what would comprise the best method or combination of methods to employ".[553] This does not mean a total ban of synthetic pesticides but leads to their relegation into an appropriate role within a holistic program that considers entire ecosystems. Sex and aggregation pheromones (a clear distinction between the two groups is not always possible), stimulate or disrupt certain behavioral modes after passing a multistep path; i.e., release onto a substrate or into the surrounding medium, perception with the corresponding sensory organ, transduction to the nervous system, and finally the release of a specific behavioral reaction. But contrary to conventional pest management techniques, a wide variety of exogenous and endogenous factors which interfere with these processes must be taken into consideration: the vapor pressure and volatility of pheromones, chemical stability of the compounds against light, humidity and air oxygen, wind or other climate conditions, season, daily rhythm of the insects, certainty and rapidity of response, running or flight activity, and other competing factors. The high degree of specificity of pheromones on one hand allows a specific interference but on the other hand makes more difficult the contemporary combating of sympatrically occurring species.

Practical applications of pheromones can be categorized as follows:

1. Trapping to collect species that are otherwise difficult to obtain.
2. Trapping insects for monitoring and survey; insect populations can thus be estimated and new areas of infestation detected. Timing and dosage of pesticide treatment can be based on the results.
3. Attracting to areas treated with insecticides.
4. Luring to areas treated with pathogens, which can then be spread by the infected individuals onto the rest of the population.
5. Mass trapping (= trap-out technique) for population suppression.
6. Disruption of communication by permeation of areas of insect population with pheromone. In the case of mating disruption population suppression will be the result.
7. Parapheromones, which are pheromone-mimic chemicals, and antipheromones (inhibitors), which block the behavioral release, are not part of the natural communication system, but can also be used in some of these categories.

(Reviews on the application of insect pheromones and pheromone analogs in the management of insect pests are given in References 25 and 554 to 566.)

Monitoring with Pheromones

The suitability of attractive pheromones for the indication of the occurrence of pest insects has been recognized for a long time. For decades tethered or caged insects as well as pheromone gland extracts have been used for specific luring. This has given indications of the occurrence and extension of a pest epidemic and can be used in quarantine precautions or to help decide on preventative precautions in integrated plant protection. In this case pheromones are superior to conventional techniques like the use of light traps, control of hatching, or visual control. But it must be mentioned that results from pheromone monitoring often are not strictly comparable to those of one of the other methods.

A characteristic property of this technique is an underproportional increase of trap catches compared to the population density because of the competition with conspecific insects, which also release pheromone. This conclusively can cause a shift of the maximum of the catches.[567] Another factor, considerably affecting the number of insects trapped, can be an additional luring of insects from surrounding areas, which is not the target of the control precaution. Thus, a strict numerical correlation between the number of trap catches and population density is found in very few cases only.

A decisive condition for reproducibility of trapping results is a permanent, long lasting, and constant release of the pheromone chemical from the bait. Most of the semiochemicals are volatile, and a pheromone bait remains attractive for a few days only when no evaporation inhibitors are used. In some cases a satisfactory reproducibility of the evaporation rate could be achieved using trioctanoin[568] or oils from plants[569] as diluents, absorbing the pheromones on sand, silica gel, paper, cellulose or similar substrates, or diluting the pheromones in solvents like hexane, benzene, and xylene.

The insects attracted from the pheromone in most cases remain in the vicinity of the odor source for a short time only. Thus, a mechanism must be applied to trap them or to prevent escape. Generally, sticky glue foils or surfaces are used. Polybutene, beetle lime, or similar substances preserve their capturability for many weeks as long as their capacity is not exhausted by contamination of the surface or by heavy occupation by the insects. Less volatile, conserving liquids can be used in the case of a great accumulation of insects.[569,570] The combination of pheromones with electrical trap grids[571] is rather effective but for obvious reasons can be applied only near buildings and the availability of electricity. In a very few cases satisfactory results were achieved using traps impregnated with contact insecticides. Respiration poisons like dichlorvos could only be employed when the exchange of air was restricted and the entering animals were prevented from escaping from the insecticide;[572,573] furthermore, repelling effects may arise.[574]

A huge number of publications deals with the construction, design, and positioning of pheromone traps. Thus, different trap design often gives deviating results and these results mostly cannot be generalized or transferred to different insect species. Visual observation can be carried out with day-active insects only, and is almost impossible with species predominantly or exclusively night-active. While the color of the traps sometimes seems to be an essential factor of capture efficiency, a vertical, trunklike structure often becomes important for the design of traps used for catches of insects living on trees. In many cases the height of the trap position also is of importance.

Probably the greatest number of different trap systems and design has been developed for the bark beetles, ranging from simple sticky traps to elaborate fan-driven, tree-stimulating olfactometers.[575] Fan-driven funnel olfactometers are convenient for the flight control of *Ips* species,[576] whereas *Dendroctonus* species require a tree-simulating sleeve olfactometer, representing an additional visual stimulus.[372] Window traps are widely used to test and evaluate bark beetle pheromones[577] as well as hardware cloth sticky screens, perforated cylinders, rotary nets, bucket traps, nonsticky traps, and transparent sticky traps.[575]

Monitoring bark beetles with pheromone traps is problematic because it accumulates the insects where they are least desired: in the forest stand to be protected.[365] Monitoring may eliminate predators of the insects which respond to the pheromone as to a kairomone[377] and furthermore result in additional infestations of the trees surrounding the traps.

A great advantage of pheromone traps compared to light- or bait-traps is the extraordinary specificity, but difficulties can arise with sympatric insect species, using the same attractant blend. Because of the permanent offer of pheromone, species with different daily activity rhythms are also attracted, and thus additional isolation mechanisms are sometimes overruled. Occasionally, two different attractive compounds can be combined in one trap to lure two species at the same time. In these cases it must be assured that no synergistic or antagonistic effects of the co-attractant arise thus falsifying the actual rate of captures. Combined baiting with pheromones and feeding attractants[578] as well as the addition of pheromone baits to light traps[579] can also be of advantage.

Direct Combating of Pest Insects: Mass Trapping and Population Suppression

Pest insects can also be decimated with attractive pheromones. With this technique the same peripheral conditions and factors are required as with monitoring. But contrary to the latter technique, now the calculation of costs and the problems of a rational and practiceable realization of mass trapping are brought to the foreground. Since trapping with pheromones is restricted to adult animals only, whereas the damage of plants mostly is caused from immature stages of the animals (i.e., the next generation), pheromone pest management of this kind has to be carried out preventively. Therefore, it is the rate of suppression of the next generation only, which is decisive, whereas large numbers of captured insects or a high percentage of their population do not guarantee the success of this kind of plant protection. Factors like lifetime, natural mortality, maturity period, number of descendants, copulatory activity, competition with the population present, and the sequence of generations might become of crucial importance. A mathematical treatment of population models led to the conclusion that for a gradual reduction of the daughter population in some cases, catch contingents of 90 to 95% of the parent generation would be necessary. This difficult to achieve requirement means that a pheromone-protected area has to remain almost completely and without pause within a range of pheromone concentration, able to compete with the attractiveness of the "conpheromonal" insects.

Successful mass trapping is evidently dependent on the size and compactness of the pheromone-treated area. With increasing size of the test area the relative amount of the border zones becomes smaller; it is that area which is not entirely covered by pheromone because of different wind directions. When the original infestation is too great, a preceding pesticide treatment or other precautions are recommended to reduce the population density to a level for a satisfactory efficiency of the subsequent pheromone trapping. An uninfested area, which is in danger of being invaded from the neighborhood, can be protected by a ring of traps, preventing or at least reducing infestation.[580]

When pheromones in combination with attractants from host plants act as aggregants (e.g., the terpenoids of host trees with bark beetles), suitable catch-trees or a single stand of trees can replace a trap. These trees can be treated with appropriate insecticides, or the infested trees liquidated together with its pest and the corresponding brood. But this technique of biological control can also be improved considerably by additional use of pheromone traps.[581-583]

Quite often a significant reduction of the pest population was achieved in combination with direct combat methods. But in most cases the efficiency of other precaution techniques could not be attained and the expenses and waste of material and time meant that use of pheromones could not compete with conventional techniques.

Disrupting Chemical Communication

In 1960 Beroza proposed for the first time to use pheromones for disturbance and interruption of chemical communication by a permanent saturation of infested areas with sex pheromones.[584] This "confusion technique" or "disorientation technique" prevents the target insect from locating its opposite sex and thus hinders mating. Other consequences of the method might be the appearance of fatigue for the case of temporally limited response readiness, as well as the raising of the habituation threshold or the direction to a wrong target.

In numerous field trials it could be demonstrated that this concept could be realized for some insect species, whereas it failed for others. The reasons for the differences are not normally obvious, but among others it certainly can be optical searching behavior predominating over odorous communication. However, all the available data show that specific sex attractants or pheromones do not only give the highest degree of attraction but also the most intensive disorientation. But in contrast to attractants which have to be chemicals or blends of ultimate purity and defined composition, for disruptants this requirement should be not necessary. Even for the disruption of a multicomponent pheromone it is conceivable that the predominant presence of a major component could satisfy this condition. Sympatric species, using pheromone blends with one common component are also likely to be disturbed in their chemical communication by applying this single common compound.[585,586]

Pheromone inhibitors or antipheromones are chemicals which reduce or stop the rate of captures of a pheromone trap when they are exposed in combination with the corresponding pheromone or in its neighborhood. Since the mechanism of the inhibition by these substances is not clear as yet, it is often difficult to correctly interpret the results of bioassays and field trials. Experiments using pheromone inhibitors for disruption in many cases failed totally. In some cases the response readiness towards conspecific calling insects or towards synthetic sex pheromones was even enhanced by the inhibitors.[587,588] Positive results were obtained for pheromone inhibitors of bark beetles which could reduce the colonization of host trees or sawn logs, e.g., methylcyclohexanone with *Dendroctonus pseudotsugae*[589,590] and brevicominisomers and verbenone with *D. frontalis*.[394]

The concentration of these chemicals to obtain the desired disorientation effect in each case had to be evaluated empirically, literature data varying from a few grams per hectare to hundredfold higher doses, especially with pheromone-mimics (showing similar behavioral reactions to pheromones, but with much higher response thresholds). Moreover, as in the mass trapping technique, a permanent, long-lasting and constant release of the semiochemical from its source is required. This can be achieved by spraying microcapsules treated with pheromone and covered with material of suitable permeability, exposure of capillaries and hollow fibers filled with pheromone, or by the distribution of a relatively small number of pheromone dispensers with a very high (and therefore long-lasting) pheromone content. Sometimes an accumulation of the volatiles near the ground, caused by convection, wind, weather, or microclimate factors was observed. Since this is not desired for the combat of insects living in tree tops, auxiliary addition of adhesives, fibrous or ribbon-like carrier materials, was used to keep the chemicals in the foliage or to prevent loss of activity through washing into the soil.

Host Baiting

In some instances pheromones are preferred to attract insects to the host plant rather than to traps. It could be demonstrated in southern Idaho, that traps for the Douglas fir beetle, *Dendroctonus pseudotsugae*, baited with a blend of the pheromone frontalin, α-pinene, and camphene (4:1:1), could not compete with trees under natural attack.[591] This led to studies on baiting selected trees with attractive pheromones to concentrate the pest insects on those trees or at least near them.[592] The trees infested with beetles were removed in standard clear-cutting operations after the infestation. A different strategy for host baiting is to attract insects to selected host plants treated with insecticide. This technique, used with the cotton boll weevil, *Anthonomus grandis*, played an important role in the Pilot Boll Weevil Eradication Experiment in southern Mississippi and parts of Louisiana and Alabama.[19,593]

Pheromones as an Auxiliary Means to other Insect Control Procedures

All nonpheromonal procedures for direct combat or indirect manipulation of pest insects require a large amount of the population being brought into contact with the particular agents. This makes the use of the strong and specific attraction of pheromones evident as an auxiliary technique for these methods. The aggregation pheromones, e.g., are utilized as additives to some insecticides to reduce the repellancy of the latter.[594] They could also promote the uptake of insect pathogens[595,596] and could make possible the application of chemosterilants in free-living populations.

In the following sections the use of beetle pheromones in biological or integrated pest management will be presented, listed according to target pests occurring in forest and shade trees, in field crops, in greenland and pasture, and as stored products in insect pests. (Reviews on the practical application of pheromones are given in References 25 and 554 to 563.)

Forest and Shade-Tree Insect Pests

The Western Pine Beetle Dendroctonus brevicomis

The aggregation pheromone system of the western pine beetle, *Dendroctonus brevicomis*, which attacks ponderosa pines, consists of one male-produced, one female-produced, and one component, produced by the host tree (c.f. Table 5, References 399, 597, and references therein). The results of large-scale attempts at population suppression based on mass trapping have been reported[598,599] and strategies for a biological and integrated pest management for this beetle developed.[600,601]

Large-baited sticky traps, positioned in a grid pattern within a 65-km² area at Bass Lake, California, captured approximately 1 million beetles (essentially the total population estimated for the area), the tree mortality dropped from 283 ± 89 before the experiment to 91 ± 28 afterwards, and remained low for the next 4 years. Other experiments of a much larger scale in a more heavily infested California district (125 km² at McCloud Flats) resulted in the trapping of almost 7 million pine beetles. But in this case, even with this high number of trap catches, the total beetle population was not decreased enough to affect tree mortality.[600] Data from these experiments were analyzed by a computer system to determine the impact of the trap catches on population location and trend and on the structure of the insect community associated with the western pine beetle.

Permeation with the three-component pheromone, racemic exo-brevicomin, racemic frontalin, and myrcene over a 0.81-ha plot prevented the trapping of beetles in traps baited with the same pheromone blend and situated in the center of the permeated area.[380,602] Formulations with verbenone, a pheromone inhibitor of the western pine beetle, attached to pheromone-impregnated ponderosa pine trees, prevented a mass attack of the beetles, as was observed with trees simply baited with the pheromone only.

The Ambrosia Beetle Gnathotrichus sulcatus

The male ambrosia beetle, *Gnathotrichus sulcatus*, utilizes a 65:35 mixture of the enantiomers of the chiral alcohol sulcatol (Table 5). Ethanol and α-pinene act as synergists for the aggregation as was determined for two different *Gnathotrichus* species.[603] In field tests, traps baited with racemic sulcatol caught large numbers of both sexes of *G. sulcatus* in competition with natural host and beetle odors. Since the infestation of freshly sawed lumber by ambrosia beetles is a continuing concern of wood industries, programs were run for the control and suppression of populations in sawmills.[604-606] As a result it was recommended to place piles of freshly sawed sapwood slabs around the mill site and to position sulcatol-baited traps alongside each slab pile. These traps will capture most of the beetles, the remainder attack the slabbing, thus preventing damage of the higher valued logs. Possibly the first commercial application of pheromone traps for scolytid control was initiated in 1975 by the Chemainus sawmill, British Columbia, Canada, to protect its timber and sawed lumber from the ambrosia beetle.

The Smaller European Elm Bark Beetle Scolytus multistriatus

The European elm bark beetle, *Scolytus multistriatus*, was accidently introduced half a century ago from Europe to the U.S. It attacks weakened or dying elm trees which emit attractive volatiles, and the consequent addition of attractants by boring females greatly increases the rate of attack of both sexes. Then the tree is rapidly overwhelmed and hence infected by the pathotoxic fungus which is the origin of Dutch elm disease and associated with the beetle. This tree disease has virtually eliminated the elm as a shade-tree throughout most of the U.S. By aerating the boring females three compounds were identified as the aggregation pheromone, multilure,[462] two produced by the insect, and one by the stressed tree presenting a powerful aggregating agent.[96]

Since that time, baited traps have been used for detection, survey, and suppression of populations.[607] Mass trapping of some million beetles in 1974 and 1975 in Michigan showed no significant impact on tree mortality and moreover in some areas an actual increase in the number of attacked trees in the treatment area. Apparently the grid deployment of the sticky traps within the treatment area gave rise to a significant beetle immigration from adjacent areas.

A pheromone-barrier system has been tested in the U.S. for protection of American and Asiatic elms (*Ulmus americana* and *U. pumila*). Multilure-baited traps were deployed in a ring around an area containing high-value elms. This trap barrier should lure beetles out of the area within the barrier, and on the other hand, prevent beetles from flying in from outside the barrier. With these precautions a large drop in beetle population occurred in the areas examined, but no significant effect on tree mortality because of DED (Dutch elm disease) was observed after the first trapping season. Nevertheless, the initiators of this project proposed a continuation of these experiments for a total of 5 years to eliminate annual variations and the lag between the inception of the method and expression of the symptom.

In eight eastern states of the U.S. Lanier et al.[444] encircled 12 small areas containing elms with trapping barriers.[444] In over 3 years the rate of elm disease was down in every area. Birch initiated a study for survey and suppression of elm bark beetles at different areas,[608] each containing some hundreds of elm trees and being separated by at least 24 km of open, high elevation desert. With these studies some capture of elm bark beetles was observed by traps which were 8 km away from any known elm. Apparently, the beetles may disperse over greater distances than previously suspected. From two field experiments the rate of recapture of released beetles was determined and this rate related to the attraction of beetles to naturally occurring brood sources vs. pheromone-baited traps.[463] The effectiveness of pheromone mass trapping of the European elm bark beetle is discussed by the same author.[609]

To check the flight activity of *Scolytus multistriatus* and *Scolytus scolytus* in Germany and the Netherlands and to estimate the effect of sanitary measures such as the removal of infested trees, sex attractant traps are used.[464,465]

Arciero studied the seasonal variation in populations of *S. multistriatus* by multilure-trapping in order to estimate the percentage of beetles carrying the Dutch elm disease fungus.[466]

Multilure-baited trap trees were mass attacked by beetles and few of the resulting brood survived through the adult stage. Application of cacodylic acid or monosodium methano-monate (MSMA) caused rapid death of the trees.[467]

Douglas Fir Beetle Dendroctonus pseudotsugae

Douglas fir beetles periodically build up epidemic levels mainly in windthrown Douglas fir trees. The female beetles produce two attractive compounds, both synergized by host volatiles. However, 3-methyl-2-cyclohexen-1-one, produced by the males and apparently functioning to shut down the aggregation, was used by Furniss et al. as an "anti-aggregative pheromone" to reduce the level of attack on felled trees.[415,589] *D. pseudotsugae* has also been attracted with pheromones to "bait trees" that were scheduled for logging.[411]

The Southern Pine Beetle Dendroctonus frontalis

Early studies involved the use of the attractant pheromone frontalin and the host terpene α-pinene to bait trees and to attract beetles. The infested trees were killed with cacodylic acid to reduce brood survival.[412,610] endo-Brevicomin is also produced by the males of *D. frontalis* and has a strong inhibitory effect on the response to the attractant.[379] Payne et al.[397] used 36 dispensers releasing ca. 300 mg/ha/day of a mixture of endo- and exo-brevicomin from a controlled-release formulation, and achieved reduced landings of beetles on host trees.[397] The addition of verbenone to the brevicomin mixture gave a reduction of 74% in beetles landing on treated trees and resulted in a significant decrease in the number of galleries constructed by boring beetles.[398]

Tests with mixtures containing frontalin have also demonstrated a strong response of *Thanasimus dubius*, a clerid predator.[611] The use of frontalin as attractant for the southern pine beetle hence must take into account this kairomonal effect.

Trypodendron (= Xyloterus) lineatum

The ambrosia beetle *Trypodendron lineatum* is considered to be the most serious economic pest among the bark beetles in Europe. Present conditions in the lumber market and forestry practices actually favor increases in pest populations.[612,613]

The female-produced aggregation pheromone lineatin is synergized by primary and secondary host odors,[481,482] each component responsible for a specific step of colonization.[483] α-Pinene affords host recognition, and the actual invasion occurs in response to ethanol, formed by anaerobic fermentation in the trunks.[479,483,525] Catches of *Trypodendron lineatum* doubled when ethanol and α-pinene were added to the lineatin. The trap design influenced the sex ratio of beetles corresponding to lineatin per se.[479] Perforated cylinders, which simulate a tree trunk, baited with lineatin, α-pinene, and ethanol, were efficient trapping devices for *T. lineatum*.[483]

The Spruce Engraver Ips typographus

The spruce engraver males excrete synergistically acting attractants, 2-methyl-3-buten-2-ol, ipsenol, ipsdienol, and the verbenols, by defecation.[424,426,449-452] Pheromone trapping with these chemicals has now replaced and improved the former applied "trap tree" technique. The response to the pheromone mixture is somewhat enhanced by ipsdienol[454,455] and partially inhibited by ipsenol. Research projects sponsored by the Norwegian, Swedish, and West German governments have led to marketable pheromone formulations and trapping devices for a combat of *I. typographus*.[614-617] During the past 10 years *I. typographus* has become a serious threat to the entire spruce forests in southeast Norway and Sweden. Slow release pheromone dispensers were tested from 1976 to 1979 in cooperation with a commercial company in Norway, and the action was highly successful, with actually 1 million traps operating in the season 1979.[616,618] The total trap catch for 1979 was some 2.9 billion beetles, and the estimated direct costs, not including manpower, was $23 million.[619] In 1980, the program was repeated with some modifications, and the estimated number of beetles trapped in Norway alone was estimated at 4.5 billion.[493] This project was probably the most ambitious undertaken anywhere, aimed at controlling tree-infesting bark beetles on the basis of synthetic pheromones. It entailed production of more than 100 kg of the attractants, a massive quantity considering that insects normally respond sensorically to odorous chemicals at a near-molecular level.

Heavy fractions of essential oils of pine and fir, enriched with *trans*-verbenol, *cis*-verbenol, ipsenol, ipsdienol, and methylbutenol, and applied to firs killed with 2,4-D amine salt, increased settling on the "trap trees" by *I. typographus*.[620]

Methylbutenol in combination with (*S*)-*cis*-verbenol and ipsdienol affected response in the predator *Thanasimus formicarius*.[621] The sympatric species *T. femoralis* responds to (*S*)-*cis*-verbenol, while ipsenol and ipsdienol synergize the response.[621]

Miscellaneous Bark Beetles

Furniss reported tests with the Douglas fir beetle pheromone components against the Eastern larch beetle (*Dendroctonus simplex*), the use of methylcyclohexenone against the spruce beetle *D. rufipennis* in Alaska, the anti-aggregative effect of ipsenol against the pine engraver, *Ips pini* in ponderosa pine in Idaho, and similar experiments with ipsenol and ipsdienol.[622] *D. rufipennis* were attracted to lindane-treated trees baited with frontalin[623] but the attractancy of these trees could not compete with wind-thrown trees.[624] The antiaggregative 3-methyl-2-cyclohexen-1-one, released by the female *Dendroctonus pseudotsugae*[405] was found to reduce attack by Douglas fir beetles.[416] The granular corn cob formulation gave 93% reduction of attack of the spruce beetle, *D. rufipennis*, a cross-attractive insect to the Douglas fir beetle.[416]

Field Crop Insect Pests

The Cotton Boll Weevil Anthonomus grandis

The cotton boll weevil, *Anthonomus grandis*, is responsible for almost 30% of cotton damage in the U.S. Approximately one third of the insecticides produced worldwide are applied to cotton crops. The cost of spraying against the weevil in the U.S. alone was estimated at $50 million in 1974. The male beetle produces a four-component sex pheromone which attracts females and serves as aggregant in the springtime and the fall. The synthetic mixture (grandlure) is available in several different formulations and traps of different design. Mitchell et al.[334] reported capture rates, varying from 76 to 96%, of overwintered beetles, using grandlure-baited traps and catches, and killing up to 100% from in-field traps in combination with insecticides. In contrast to this report, Huddleston et al.[332] published disappointing results in an attempt to disrupt communication with grandlure. Further disruption experiments with the boll weevil in the laboratory and small field plots were reported by Villavaso and McGovern.[333] In 1978 the Trial Boll Weevil Eradication Program was started in North Carolina and Virginia, in the course of which grandlure represents an important technical component in the program. (Results of boll weevil mass trapping are found in References 625 and 626.)

Stored Products Insect Pests

Practical application of stored-products pests are advanced for several *Trogoderma* and *Attagenus* species. Numerous traps baited with pheromones and insecticides are used in storage facilities, warehouses, and on ships for quarantine purposes in the detection of the insects.

Mating disruption has been demonstrated for the black carpet beetle, *Attagenus megatoma*. But it is difficult to apply in storage facilities, since it is to be expected, that the stored products and the containers, permeated with aggregants or absorbed residues, may still attract insects when the material has left the warehouses.

For population suppression, pheromone-insecticide baited traps are used, as well as pheromone traps treated with spores of a pathogenic protozoon.[596] The attracted beetles were contaminated with the spores, thus transmitting them to the rest of the population.

The entry of the khapra beetle, *Trogoderma granarium*, a notorious pest in warmer regions throughout Europe, Asia, and Africa, into the U.S. is only prevented by tight quarantine precautions in ships and port facilities. Small detection traps containing the pheromone and an insecticide are now in use in a number of storage facilities and on ships. Since many *Trogoderma* species share some of the same pheromone components, the two trapping systems mentioned above can be used for a combat of mixed populations.

The combined use of sex and food attractants is proposed by Levinson and Levinson as a concept for an integrated manipulation of storage pests.[509,627] With an increasing number of insects an increasing number of traps should be used for mass trapping in attempts to keep the population at a level that does not cause economic damage ("insectistasis"). Since 1975 a number of large and small food stores have been treated in this way in West Germany. The main coleopteran species involved are *Attagenus megatoma, A. elongatulus, Acanthoscelides obtectus, Oryzaeplutus surinamensis, Sitophilus granarius, S. oryzae, Tenebrio molitor, Tribolium confusum, T. castaneum, Trogoderma granarium,* and *T. inclusum.*

Grassland and Pasture

To the group of pest insects causing damage on pasture belong numerous lepidopteran species. From the order Coleoptera only the New Zealand grass grub beetle, *Costelytra zealandica*, should be mentioned here; the males are attracted by the sex attractant phenol (see Table 5), and this gives a base for a pheromone defense against the beetles in grassland.[360]

Until the discovery of the female sex pheromone of the Japanese beetle *Popillia japonica*, phenethyl propionate and eugenol were used as baits in current standard survey traps.[628,629] With the availability of the synthetic pheromone chemical japonilure, blends of these three attractants showed improved attractancy compared to each of the single lures.[361] Exposure of a mixture of phenethyl propionate, eugenol, geraniol, and japonilure resulted in a lure of outstanding attractancy. The combination was more than 3.6 times as attractive as phenethyl propionate and eugenol.[362] (A survey on trapping Japanese beetles with synthetic female sex pheromone and food-type lures is given in Reference 630).

Pheromone Formulations

To gain satisfactory and reproducible trapping results a pheromone formulation with a constant release rate lasting over longer periods, is required.[631] Furthermore, pheromone baits should be easy to store, easy to handle, and simple to produce. Early workers used filter papers, dental rolls, and cotton wicks, but the release from these dispensers was nonuniform. The release characteristics were often improved by the addition of relatively nonvolatile liquids, such as high-boiling esters, trioctanoin, corn oil and silicone oil, and similarly, pheromones adsorbed on sand, silica gel, etc. were used. But for the use of attractants on a more commercial basis new formulation technologies had to be developed.

Formulations generally can be classified into three main groups. One group of formulations consists of almost homogeneous matter. To the second group belong closed hollow pheromone reservoirs, and the third is represented by open hollow reservoir formulations.[632] The homogeneous matter baits may consist of rubber or plastic material which is used like a "solvent" for the pheromone. They do not show an absolutely constant release rate over longer periods. But in choosing a suitable material, a release lasting over several weeks (i.e. one season) can be achieved, and exchanging the bait is not necessary. The second group, the closed hollow reservoir type, represents small pheromone-containers with a permeable wall. After a short time of permeation of the volatile through the wall and equilibration, a constant release rate is given. When the chemical from the reservoir is exhausted (normally after a few months) and all of it dissolved in the wall, the former hollow reservoir becomes a homogenous matter bait. The third class of lures is the type of open hollow reservoirs. The odor chemical vaporizes inside and is released by diffusion through the air. The latter formulation type shows a good constancy with respect to the rate of release. A disadvantage of this type of pheromone formulation is the free access of air oxygen to the (often chemically unstable) compounds and the occurrence of condensation of pheromone on the outside of the bait body. This can lead to an accumulation of dust and particles and hinder the free vaporization of the pheromone.[632]

To the group of closed hollow reservoir formulations belong the multilayer laminated dispenser sheets. They consist of a bottom protective layer, a pheromone reservoir layer in the middle, and a top permeable layer throughout which the chemical evaporates.[633,634] The sheets can be cut into ribbons and applied over an area. (Controlled release of pheromones from different formulations is reviewed in References 635 to 640, mostly dealing with lepidoptera pheromone examples.)

Industrially produced hollow fibers belong to the last type of formulation. These hollow fibers are microcapillary reservoirs, sealed at one end, and the wall almost impermeable. The volatiles, retained by capillary forces, are released by diffusion from the surface of the liquid-vapor interface into the atmosphere through the open end.[641] These fibers may be supplied in tape form, which is a parallel array of fibers on an adhesive tape, and in chopped form, suitable for dissemination with a ground rig or from the air. Some companies use microencapsulated formulations. Two main types exist, gelatin capsules produced by co-acervation, and polyurea/polyamide capsules produced by interfacial polymerization.

Several companies are now producing beetle pheromone formulations and trapping devices based on these semiochemicals. Albany International produces pheromones controlling the European elm bark beetle. The Hercon Division of Health-Chem. Corp., New York, provides pheromone formulations in laminated dispensers that can be used for trapping of the Japanese beetle and also manufactures pheromones for mass trapping of the European elm bark beetle. With the introduction of baited traps for the Japanese beetle the J.T. Baker Chemical Co. started its pheromone marketing. Borregard Industries Ltd. (Norway) furnishes traps and pheromone components for trapping the spruce bark beetle and an ambrosia beetle. Celamerck (West Germany) provides pheromone baits for different *Ips* bark beetle species.

Other companies and research institutions applying pheromones also for other insect orders include Zoecon Corp., Bend Research Inc., U.S.; Takeda Chemical Comp. Ltd., Shin-Etsu Chemical Co., Japan; Plant Protection Division of ICI, Wolfson Unit of Chemical Ecology, Oecos Ltd. Monitoring and Control Development, Murphy Chemical, and International Pheromones Ltd., Great Britain; Hoechst AG, Wacker-Chemie, Spieβ-Urania, and Sigma Chemie GmbH, West Germany; Eidgenössische Forschungsanstalt Wädenswil, Switzerland; INRA, France; Institute for Pesticide Research, Netherlands; Montedison, Italy; Chemada Chemical Co., Israel; and others.

CONCLUSIONS

Several general conclusions may be drawn from the results of the use of beetle pheromones as a component of pest control. Monitoring and survey traps are becoming widely used and are very successful even though correlations between the rate of trap catches and the population density are not readily determined. Reductions in the use of conventional insecticides have been achieved by using the monitoring information to mediate the spray schedule. Population suppression by mass trapping or disruption has been demonstrated in a few cases of low-density populations only. Attempts to control high-density populations have often failed. Failures to suppress populations can be deduced from an inadequate understanding of insect biology and behavior, high insect density, improper pheromone formulations, improper distribution of traps and dispensers, invasion from outside the treated area, lack of chemical definition of the natural communication system, or poor timing. Nevertheless, pheromones are becoming an important component of integrated pest management techniques.

REFERENCES

1. **Butenandt, A., Hecker, E., Hopp, M., and Koch, W.,** Über den Sexuallockstoff des Seidenspinners. IV. Die Synthese des Bombykols und der *cis-trans*-isomeren Hexadecadien-(10,12)-ole-(1), *Liebigs Ann. Chem.,* 658, 39, 1962.
2. **Nordlund, D. A. and Lewis, W. J.,** Terminology of chemical releasing stimuli in intraspecific interactions, *J. Chem. Ecol.,* 2, 211, 1976.
3. **Law, J. H. and Regnier, F. H.,** Pheromones, *Annu. Rev. Biochem.,* 40, 533, 1971.
4. **Whittaker, R. H. and Feeny, P. P.,** Allelochemics: chemical interactions between species, *Science,* 171, 757, 1971.
5. **Brown, W. L. Jr., Eisner, T., and Whittaker, R. H.,** Allomones and kairomones, transspecific chemical messengers, *BioScience,* 20, 21, 1970.
6. **Karlson, P. and Butenandt, A.,** Pheromones (ectohormones) in insects, *Annu. Rev. Entomol.,* 4, 39, 1959.
7. **Karlson, P. and Lüscher, M.,** Pheromone. Ein Nomenklaturvorschlag für eine Wirkstoffklasse, *Naturwissenschaften,* 46, 63, 1959.
8. **Rutowski, R. L.,** The function of pheromones, *J. Chem. Ecol.,* 7, 481, 1981.
9. **Wilson, E.,** Chemical communication in the social insects, *Science,* 149, 1064, 1965.
10. **Wilson, E. O. and Bossert, W. H.,** Chemical communication among animals, *Rec. Prog. Hormone Res.,* 19, 673, 1963.
11. **Eiter, K.,** Insektenlockstoffe, *Fortschr. Chem. Org. Naturst.,* 28, 204, 1970.
12. **Wood, D. L., Silverstein, R. M., and Nakajima, M., Eds.,** *Control of Insect Behavior by Natural Products,* Academic Press, New York, 1970.
13. **Beroza, M., Ed.,** *Chemicals Controlling Insect Behavior,* Academic Press, New York, 1970.
14. **Jacobson, M., Ed.,** *Insect Sex Pheromones,* Academic Press, New York, 1972.
15. **MacConnell, J. G. and Silverstein, R. M.,** Recent results in insect pheromone chemistry, *Angew. Chem.,* 85, 647, 1973; *Angew. Chem. Int. Ed. Engl.,* 12, 644, 1973.
16. **Evans, D. A. and Green, C. L.,** Insect attractants of natural origin, *Chem. Soc. Rev.,* 2, 75, 1973.
17. **Shorey, H. H.,** Behavioral responses to insect pheromones, *Annu. Rev. Entomol.,* 18, 349, 1973.
18. **Priesner, E.,** Artspezifität und Funktion einiger Insektenpheromone, *Fortschr. Zool.,* 22, 49, 1973.
19. **Birch, M. C., Ed.,** *Pheromones,* American Elsevier, New York, 1974.
20. **Jacobson, M., Ed.,** *Insecticides of the Future,* Marcel Dekker, New York, 1974.
21. **Baker, R. and Evans, D. A.,** Biological chemistry. I. Insect chemistry, *Ann. Rep. Prog. Chem. Sect. B,* 72, 347, 1975.
22. **Beroza, M., Ed.,** *Pest Management with Insect Sex Attractants,* ACS Symp. Series No. 23, Am. Chem. Soc., Washington, D.C., 1976.
23. **Shorey, H. H. and McKelvey, J. J., Eds.,** *Chemical Control of Insect Behavior: Theory and Application,* John Wiley & Sons, New York, 1977.
24. **Brown, K.,** Chemical communication between animals, in *Chemical Influence on Behaviour,* Brown, K. and Copper, S. J., Eds., Academic Press, New York, 1979, 599.
25. **Brand, J. M., Young, J. C., and Silverstein, R. M.,** Insect pheromones: a critical review of recent advances in their chemistry, biology, and application, *Prog. Chem. Org. Nat. Prod.,* 37, 1, 1979.
26. **Masson, C. and Brossut, R.,** Chemical communication among insects, La communication chimique chez les insectes, *Recherche,* 12, 406, 1981.
27. **Baker, R. and Bradshaw, J. W. S.,** Insect pheromones and related behavior-modifying chemicals, *Aliphatic Relat. Nat. Prod. Chem.,* 2, 46, 1981.
28. **Wegler, R., Ed.,** *Chemie der Pflanzenschutz- und Schädlingsbekämpfungsmittel,* Vol. 6, Springer-Verlag, Berlin, 1981.
29. **Bestmann, H. J. and Vostrowsky, O.,** Insektenpheromone. Struktur und Vorkommen, Isolierung und Strukturaufklärung, Synthese, Biologische Aktivität und Verhaltensauslösung, Anwendung im Pflanzenschutz, *Naturwissenschaften,* 69, 457, 1982.
30. **Inscoe, M. N.,** Insect sex attractants, pheromones and related compounds, in *Insect Suppression with Controlled Release Pheromone Systems,* Vol. 2, Kydonieus, A. F. and Beroza, M., Eds., CRC Press, Boca Raton, Fla., 1982, 201.
31. **Birch, M. C. and Hayes, K. F.,** *Insect Pheromones,* Frontiers in Biology, No. 147, Edward Arnold Ltd., London, 1982.
32. **Hummel, H. E. and Miller, T. E., Eds.,** *Techniques in Pheromone Research,* Springer-Verlag, Berlin, 1984.
33. **Horler, D. F.,** (−)-Methyl *n*-tetradeca-*trans*-2,4,5-trienoate, an allenic ester produced by the male dried bean beetle, *Acanthoscelides obtectus* (Say), *J. Chem. Soc. C,* 1970, 859.
34. **Henzell, R. F. and Lowe, M. D.,** Sex attractant of the grass grub beetle, *Science,* 168, 1005, 1970.

35. **Morgan, E. D. and Wadhams, L. J.**, Chemical constituents of Dufour's glands in the ant, *Myrmica rubra, J. Insect Physiol.*, 18, 1125, 1972.

36. **Bergstrøm, G. and Løfquist, J.**, *Camponotus ligniperda*. Model for the composite volatile secretions of Dufour's gland in formicine ant, in *Pesticide Chemistry*, Vol. 3, Tahori, A. S., Ed., Gordon & Breach, London, 1971, 195.

37. **Bergstrøm, G., Kullenberg, B., and Ställberg-Stenhagen, S.**, Natural odoriferous compounds. VII. Recognition of two forms of *Bombus lucorum* (Hymenoptera, Apidae) by analysis of the volatile marking secretion from individual males, *Chem. Scripta*, 4, 174, 1973.

38. **Buser, H. R. and Arn, H.**, Analysis of insect pheromones by quadrupole mass fragmentography and high-resolution gas chromatography, *J. Chromatogr.*, 106, 83, 1975.

39. **Browne, L. E., Birch, M. C., and Wood, D. L.**, Novel trapping and delivery systems for airborne insect pheromones, *J. Insect Physiol.*, 20, 183, 1974.

40. **Byrne, K. J., Gore, W. E., Pearce, G. T., and Silverstein, R. M.**, Porapak-Q collection of airborne insect pheromones, *J. Chem. Ecol.*, 1, 1, 1975.

41. **Ma, M., Hummel, H. E., and Burkholder, W. E.**, Estimation of single furniture carpet beetle (*Anthrenus flavipes* LeConte) sex pheromone release by dose-response curve and chromatographic analysis of penta-fluorobenzyl derivative of (Z)-3-decenoic acid, *J. Chem. Ecol.*, 6, 597, 1980.

42. **Rudinsky, J., Morgan, M. E., Libbey, L., and Michael, R. R.**, Sound production in Scolytidae. 3-Methyl-2-cyclohexen-1-one released by female Douglas fir beetle in response to male sonic signal, *Environ. Entomol.*, 2, 505, 1973.

43. **Rudinsky, J., Morgan, M. E., Libbey, L. M., and Putnam, T. B.**, Antiaggregative-rivalry pheromone of the mountain pine beetle, and a new arrestant of the southern pine beetle, *Environ. Entomol.*, 3, 90, 1974.

44. **Libbey, L. M., Morgan, M. E., Putnam, T. B., and Rudinsky, J. A.**, Pheromones released during inter- and intra-sex response of the scolytid beetle, *Dendroctonus brevicomis, J. Insect. Physiol.*, 20, 1667, 1974.

45. **Inscoe, M. N. and Beroza, M.**, Analysis of pheromones and other compounds controlling insect behavior, in *Analytical Methods for Pesticides and Plant Growth Regulators*, Vol. 8, Zweig, G. and Sherma, J., Eds., Academic Press, New York, 1976, 31.

46. **Young, J. C. and Silverstein, R. M.**, Biological and chemical methodology in the study of insect communication, in *Methods in Olfactory Research*, Moulton, D. G., Türk, A., and Johnson, W. Jr., Eds., Academic Press, New York, 1975, 75.

47. **Roelofs, W.**, Chemical control of insects by pheromones, in *Biochemistry of Insects*, Rockstein, M., Ed., Academic Press, New York, 1979, 419.

48. **Bestmann, H. J. and Vostrowsky, O.**, Chemistry of insect pheromones, in *Chemie der Pflanzenschutz-und Schädlingsbekämpfungsmittel*, Wegler, R., Ed., Vol. 6, Springer-Verlag, Berlin, 1981, 29.

49. **Ritter, F. J. and Persoons, C. J.**, Isolation and identification of pheromones, in *Integrated Control of Insect Pests in the Netherlands*, Pudoc, Wageningen, The Netherlands, 1980, 203.

50. **Golub, M. A. and Weatherston, I.**, Techniques for extracting and collecting sex pheromones from live insects and from artificial sources, in *Techniques in Pheromone Research*, Hummel, H. E. and Miller, T. E., Eds., Springer-Verlag, Berlin, 1984, 223.

51. **Engelhardt, H.**, *Hochdruck-Flüssigkeits-Chromatographie*, 2nd ed., Springer-Verlag, Berlin, 1977.

52. **Wright, J. E. and Thomas, B. R.**, Determination of the components of the boll weevil pheromone with a high pressure liquid chromatographic method, *J. Liquid Chromatogr.*, 4, 1409, 1981.

53. **Phelan, P. L. and Miller, J. R.**, Separation of isomeric insect pheromonal compounds using reversed-phase HPLC with silver nitrate in the mobile phase, *J. Chromatogr. Sci.*, 19, 13, 1981.

54. **Zlatkis, A. and Kaiser, R. E.**, *HPTLC — High Performance Thin Layer Chromatography, J. Chromatography Library*, Vol. 9, Elsevier, Amsterdam, 1977.

55. **Koppenhoefer, B., Hintzer, K., Weber, R., and Schurig, V.**, Quantitative Trennung der Enantiomerenpaare des Pheromons 2-Ethyl-1,6-dioxaspiro[4.4]nonan durch Komplexierungsgaschromatographie an einem optisch aktiven Metallkomplex, *Angew. Chem.*, 92, 473, 1980; *Angew. Chem. Int. Ed. Engl.*, 19, 471, 1980.

56. **Houx, N. W. H., Voerman, S., and Jongen, W. M. F.**, Purification and analysis of synthetic sex attractants by liquid chromatography on a silver-loaded resin, *J. Chromatogr.*, 96, 25, 1974.

57. **Heath, R. R., Tumlinson, J. H., Doolittle, R. E., and Proveaux, A. T.**, Silver nitrate high pressure liquid-chromatography of geometrical isomers, *J. Chromatogr. Sci.*, 13, 380, 1975.

58. **Houx, N. W. H. and Voerman, S.**, High-performance liquid-chromatography of potential insect sex attractants and other geometrical isomers on a silver-loaded ion-exchanger, *J. Chromatogr.*, 129, 456, 1976.

59. **Vostrowsky, O. and Bestmann, H. J.**, Isolierung und Strukturaufklärung von Pheromonen, *Mitt. Dtsch. Ges. Allg. Angew. Entomol.*, 1, 152, 1978.

60. **Tumlinson, J. H., Hardee, D. D., Gueldner, R. C., Thompson, A. C., and Minyard, Y. P.,** Sex pheromones produced by male boll weevil: isolation, identification, and synthesis, *Science,* 166, 1010, 1969.

61. **Tumlinson, J. H. and Heath, R. R.,** Structure elucidation of insect pheromones by microanalytical methods, *J. Chem. Ecol.,* 2, 87, 1976.

62. **Heath, R. R. and Tumlinson, J. H.,** Techniques for purifying, analyzing, and identifying pheromones, in *Techniques in Pheromone Research,* Hummel, H. E. and Miller, T. E., Eds., Springer-Verlag, Berlin, 1984, 287.

63. **Price, G. D., Sunas, E. C., and Williams, J. F.,** Microcell for obtaining normal contrast ir solution spectra of the five-γ level, *Anal. Chem.,* 39, 138, 1967.

64. **Cournoyer, R., Shearer, J. C., and Anderson, D. H.,** Fourier transform infrared analysis below the one-nanogram level, *Anal. Chem.,* 49, 2275, 1977.

65. **Riley, R. G., Silverstein, R. M., Carroll, B., and Carroll, R.,** Methyl 4-methylpyrrole-2-carboxylate: a volatile trail pheromone from the leaf-cutting ant *Atta cephalotes, J. Insect Physiol.,* 20, 651, 1974.

66. **Persoons, C. J.,** Structure Elucidation of some Insect Pheromones. A Contribution to the Development of Selective Pest Control Agents, Ph.D. thesis, University of Wageningen, The Netherlands, 1977.

67. **Tumlinson, J. H., Yonce, C. E., Doolittle, R. E., Heath, R. R., Gentry, C. R., and Mitchell, E. R.,** Sex pheromones and reproductive isolation of the lesser peachtree borer and the peachtree borer, *Science,* 185, 614, 1974.

68. **Williams, J. H., Silverstein, R. M., Burkholder, W. E., and Khorramshahi, A.,** Dominicalure 1 and 2: components of aggregation pheromone from male lesser grain borer *Rhyzopertha dominica* (F.) (Coleoptera: Bostrichidae), *J. Chem. Ecol.,* 7, 759, 1981.

69. **Byrne, K. J., Swigar, A. A., Silverstein, R. M., Borden, J. H., and Stokkink, E.,** Sulcatol: population aggregation pheromone in *Gnathotrichus sulcatus* (Coleoptera: Scolytidae), *J. Insect Physiol.,* 20, 1895, 1974.

70. **Stewart, T. E., Plummer, E. L., McCandless, L. L., West, J. R., and Silverstein, R. M.,** Determination of enantiomer composition of several bicyclic ketal insect pheromone components, *J. Chem. Ecol.,* 3, 27, 1977.

71. **Plummer, E. L., Stewart, T. E., Byrne, K., Pearce, G. T., and Silverstein, R. M.,** Determination of the enantiomorphic composition of several insect pheromone alcohols, *J. Chem. Ecol.,* 2, 307, 1976.

72. **Mori, K.,** Synthesis of optically pure (+)-*trans*-verbenol and its antipode, the pheromone of *Dendroctonus* bark beetles, *Agric. Biol. Chem.,* 40, 415, 1976.

73. **Mori, K., Mizumachi, N., and Matsui, M.,** Pheromone synthesis. 12. Synthesis of optically pure (1*S*,4*S*,5*S*)-2-pinen-4-ol (*cis*-verbenol), the pheromone of *Ips* bark beetles, *Agric. Biol. Chem.,* 40, 1611, 1975.

74. **Pearce, G. T., Gore, W. E., and Silverstein, R. M.,** Synthesis and absolute configuration of multistriatin, *J. Org. Chem.,* 41, 2797, 1976.

75. **Pearce, G. T., Gore, W. E., and Silverstein, R. M.,** Carbon-13 spectra of some insect pheromones and related compounds of the 6,8-dioxabicyclo[3.2.1]octane system, *J. Magn. Reson.,* 27, 497, 1977.

76. **Nishida, R., Fukami, H., and Ishii, S.,** Sex pheromone of the German cockroach (*Blatella germanica* L.) responsible for male wing raising: 3,11-dimethyl-2-nonacosanone, *Experientia,* 30, 978, 1974.

77. **Cooper, M. A., Salmon, J. R., Whittacker, D., and Scheidegger, V.,** Stereochemistry of the verbenols, *J. Chem. Soc. B,* p.1259, 1967.

78. **Reece, C. A., Rodin, J. O., Brownlee, R. G., Duncan, W. G., and Silverstein, R. M.,** Synthesis of principal components of the sex attractant from male *Ips confusus* [now *I. paraconfusus*] frass, 2-methyl-6-methylene-7-octen-4-ol, 2-methyl-6-methylene-2,7-octadien-4-ol, and (+)-*cis*-verbenol, *Tetrahedron,* 24, 4249, 1968.

79. **Beroza, M., Bierl, B. A., and Moffit, H. R.,** Pheromones: (*E,E*)-8,10-dodecadien-1-ol in the codling moth, *Science,* 183, 89, 1974.

80. **Bestmann, H. J., Vostrowsky, O., Koschatzky, K. H., Platz, H., Brosche, T., Kantardjiew, I., Rheinwald, M., and Knauf, W.,** (Z)-5-Decenylacetat, ein Sexuallockstoff für Männchen der Saateule *Agrotis segetum* (Lepidoptera), *Angew. Chem.,* 90, 815, 1978; *Angew. Chem. Int. Ed. Engl.,* 17, 768, 1978.

81. **Vostrowsky, O. and Bestmann, H. J.,** Der Sexualpheromon-Komplex der Noctuinae (Agrotinae), *Mitt. Dtsch. Ges. Allg. Angew. Entomol.,* 2, 252, 1981.

82. **Hendry, L. B., Anderson, M. E., Jugovich, J., Mumma, R. O., Robacker, D., and Kosarych, Z.,** Sex pheromone of the oak leaf roller: a complex chemical messenger system identified by mass fragmentography, *Science,* 187, 355, 1975.

83. **Kovats, E.,** Gas chromatographic characterization of organic substances in the retention index system, in *Advances in Chromatography,* Giddings, J. C. and Keller, R. A., Eds., Marcel Dekker, New York, 1966, 229.

84. **Ettre, L. S.,** The interpretation of analytical results: qualitative and quantitative analysis, in *The Practice of Gas Chromatography,* Ettre, L. S. and Zlatkis, A., Eds., Wiley-Interscience, New York, 1967, 381.

85. **McDonald, L. M. and Weatherstone, J.,** Gas chromatography and structural elucidation of lepidopteran pheromones, *J. Chromatogr.,* 118, 195, 1976.

86. **Bridges, J. R. and Guinn, F. H.,** A solid injection technique for studying bark beetle pheromones, *Z. Angew. Entomol.,* 89, 54, 1980.

87. **Beroza, M. and Coad, R. A.,** Reaction gas chromatography, *J. Gas Chromatogr.,* 4, 199, 1966.

88. **Beroza, M. and Coad, R. A.,** Reaction gas chromatography, in *The Practice of Gas Chromatography,* Ettre, L. S. and Zlatkis, A., Eds., Wiley-Interscience, New York, 1967, 461.

89. **Beroza, M. and Inscoe, M. N.,** Precolumn reactions for structure determinations, in *Ancillary Techniques of Gas Chromatography,* Ettre, L. S. and Fadden, W. H., Eds., Wiley-Interscience, New York, 1969, 89.

90. **Siggia, S., Ed.,** in *Instrumental Methods of Organic Functional Group Analysis,* Wiley-Interscience, New York, 1972.

91. **Talma, E., Verwiel, P. E. J., and Lakwijk, A. C.,** Reaction GC-MS in Biochemistry, *Adv. Mass Spectrom. Biochem. Med.,* 2, 215, 1977.

92. **Tumlinson, J. H., Gueldner, R. D., Hardee, D. D., Thompson, A. C., Hedin, P. A., and Minyard, J. P.,** Identification and synthesis of the four compounds comprising the boll weevil sex attractant, *J. Org. Chem.,* 36, 2616, 1971.

93. **Beroza, M. and Sarmiento, R.,** Determination of the carbon skeleton and other structural features of organic compounds by gas chromatography, *Anal. Chem.,* 35, 1353, 1963.

94. **Beroza, M. and Sarmiento, R.,** Carbon-skeleton chromatography by using hot-wire thermal-conductivity detector, *Anal. Chem.,* 36, 1744, 1964.

95. **Silverstein, R. M., Brownlee, R. G., Bellas, T. E., Wood, D. L., and Browne, L. E.,** Brevicomin. Principal sex attractant in the frass of the female western pine beetle, *Science,* 159, 889, 1968.

96. **Pearce, G. T., Gore, W. E., Silverstein, R. M., Peacock, J. W., Cuthbert, R. A., Lanier, G. N., and Simeone, J. B.,** Chemical attractants for the smaller European elm bark beetle, *Scolytus multistriatus* (Coleoptera: Scolytidae), *J. Chem. Ecol.,* 1, 115, 1975.

97. **Bierl, B. A., Beroza, M., and Ashton, W. T.,** Reaction loops for reaction gas chromatography. Subtraction of alcohols, aldehydes, ketones, epoxides, and acids and carbon-skeleton chromatography of polar compounds, *Mikrochim. Acta,* p. 637, 1969.

98. **Regnier, F. E. and Huang, J. C.,** Identification of some oxygen-containing functional groups by reaction gas chromatography, *J. Chromatogr. Sci.,* 8, 267, 1970.

99. **Bierl, B. A., Beroza, M., and Collier, C. W.,** Potent sex attractant of the gypsy moth: its isolation, identification, and synthesis, *Science,* 170, 87, 1970.

100. **Beroza, M. and Bierl, B. A.,** Apparatus for ozonolysis of microgram amounts of compounds, *Anal. Chem.,* 38, 1976, 1966.

101. **Moore, B. P. and Brown, W. V.,** Gas-liquid chromatographic identification of ozonolysis fragments as a basis for micro-scale structure determination, *J. Chromatogr.,* 60, 157, 1971.

102. **Tumlinson, J. H., Heath, R. R., and Doolittle, D. A.,** Application of chemical ionization mass spectrometry of epoxides to the determination of olefin position in aliphatic chains, *Anal. Chem.,* 46, 1309, 1974.

103. **Bierl, B. A. and Beroza, M.,** Electron impact mass spectrometry for location of epoxide position in long chain vic-dialkyl and trialkyl epoxides, *J. Am. Oil Chem. Soc.,* 51, 466, 1974.

104. **Bestmann, H. J., Brosche, T., Koschatzky, K. H., Michaelis, K., Platz, H., Vostrowsky, O., and Knauf, W.,** Identifizierung eines neuartigen Pheromonkomplexes aus der Graseule *Scotia exclamationis, Tetrahedron Lett.,* p. 747, 1980.

105. **Blum, W. and Richter, W. J.,** Einsatz von Hochleistungs-Trennkapillaren in der GC.-EIMS./GC.-CIMS.-Analyse: Eine Möglichkeit zur massenspektrometrischen Doppelbindungscharakterisierung in komplexen Monoalkengemischen, *Helv. Chim. Acta,* 57, 1744, 1974.

106. **Blum, W. and Richter, W. J.,** Analysis of alkene mixtures by combined capillary gas chromatography in chemical ionization mass spectrometry, *Tetrahedron Lett.,* p. 835, 1973.

107. **Vostrowsky, O. and Michaelis, K.,** Methoxymerkurierung-Demerkurierung zur Bestimmung der Doppelbindungsposition von Pheromonen, *Z. Naturforsch.,* 36b, 402, 1981.

108. **Vostrowsky, O., Michaelis, K., and Bestmann, H. J.,** Methoxymerkurierung-Demerkurierung zur Bestimmung der Doppelbindungspositionen doppelt-ungesättigter Lepidopteren-Pheromone, *Liebigs Ann. Chem.,* p. 1721, 1981.

109. **Baker, R., Bradshaw, J. W. S., and Speed, W.,** Methoxymercuration-demercuration and mass spectrometry in the identification of the sex pheromones of *Panolis flammea,* the pine beauty moth, *Experientia,* 38, 233, 1982.

110. **Tomida, I. and Ishii, S.,** Sex pheromone of the Eri-silkworm moth, *Philosamia cynthia ricini* Donovan, Lepidoptera: Saturniidae, *Appl. Entomol. Zool.,* 3, 103, 1968.

111. **Schneider, D.,** Elektrophysiologische Untersuchungen von Chemo- und Mechanorezeptoren der Antenne des Seidenspinners *Bombyx mori* L., *Z. Vgl. Physiol.,* 40, 8, 1957.

112. **Roelofs, W. L.,** Electroantennograms, *Chem. Technol.,* 9, 222, 1979.

113. **Roelofs, W. L., Kochansky, J., Cardé, R. T., Arn, H., and Rauscher, S.,** Sex attractant of the grape vine moth, *Lobesia botrana, Mitt. Schweiz. Entomol. Ges.,* 43, 71, 1973.

114. **Priesner, E., Jacobson, M., and Bestmann, H. J.,** Structure-response relationships in Noctuid sex pheromone reception, *Z. Naturforsch.,* 30c, 283, 1975.

115. **Moulton, D. G., Turk, A., and Johnston, J. W. Jr., Eds.,** in *Methods in Olfactory Research,* Academic Press, New York, 1975.

116. **Bellas, T. E., Brownlee, R. G., and Silverstein, R. M.,** Synthesis of brevicomin, principal sex attractant in the frass of the female western pine beetle, *Tetrahedron,* 25, 5149, 1969.

117. **Wasserman, H. H. and Barber, E. H.,** Carbonyl epoxide rearrangements. Synthesis of brevicomin and related [3.2.1]bicyclic systems, *J. Am. Chem. Soc.,* 91, 3674, 1969.

118. **Mundy, B. P., Otzenberger, R. D., and DeBernardis, A. R.,** A synthesis of frontalin and brevicomin, *J. Org. Chem.,* 36, 2390, 1971.

119. **Knolle, J. and Schäfer, H. J.,** Anodic oxidation of organic compounds. 15. Synthesis of brevicomin by Kolbe electrolysis, *Angew. Chem.,* 87, 777, 1975; *Angew. Chem. Int. Ed. Engl.,* 14, 758, 1975.

120. **Kociensky, P. J. and Ostrow, R. W.,** Stereoselective total synthesis of *exo-* and *endo*-brevicomin, *J. Org. Chem.,* 41, 398, 1976.

121. **Chaquin, P., Morizur, J. P., and Kossanyi, J.,** An easy access to *exo*-brevicomin, *J. Am. Chem. Soc.,* 99, 903, 1977.

122. **Coke, J. L., Williams, H. J., and Natarajan, S.,** A new preparation of acetylenic ketones and application to the synthesis of *exo*-brevicomin, the pheromone from *Dendroctonus brevicomis, J. Org. Chem.,* 42, 2380, 1977.

123. **Rodin, J. O., Reece, C. A., Silverstein, R. M., Brown, V. H., and DeGraw, J. I.,** Synthesis of brevicomin, principal sex attractant of western pine beetle, *J. Chem. Eng. Data,* 16, 381, 1971.

124. **Lipkowitz, K. B., Mundy, B. P., and Geeseman, D.,** Studies directed towards a practical synthesis of brevicomins. II. A novel synthesis of 1.5-dimethyl-8-oxabicyclo[3.2.1]octane-6-one, *Synth. Commun.,* 3, 453, 1973.

125. **Mori, K.,** Synthesis of *exo*-brevicomin, pheromone of western pine beetle, to obtain optically active forms of known configuration, *Tetrahedron,* 30, 4223, 1974.

126. **Joshi, N. N., Mamdapur, V. R., and Chadha, M. S.,** Convenient synthesis to 6,8-dioxabicyclo[3.2.1]octanes, the aggregation pheromone components of bark beetles, *J. Chem. Soc. Perkin Trans. 1,* p. 2963, 1983.

127. **Masaki, Y., Nagata, K., Serizawa, Y., and Kaji, K.,** Short step synthesis of optically and biologically active *exo*-brevicomin, *Tetrahedron Lett.,* 23, 5553, 1982.

128. **Ferrier, R. J. and Prasit, P.,** Unsaturated carbohydrates. 25. Abbreviated synthesis of the insect pheromone (+)-*exo*-brevicomin from a nona-3,8-dienulose derivative, *J. Chem. Soc. Perkin Trans. 1,* p. 1645, 1983.

129. **Mori, K.,** Stereoselective synthesis of (±)-*endo*-brevicomin, a pheromone inhibitor produced by *Dendroctonus* bark beetles, *Agric. Biol. Chem.,* 40, 2499, 1976.

130. **Look, M.,** Improved synthesis of *endo*-brevicomin for the control of bark beetles (Coleoptera: Scolytidae), *J. Chem. Ecol.,* 2, 83, 1976.

131. **Byrom, N. T., Grigg, R., and Kongkathip, B. J.,** Catalytic synthesis of *endo*-brevicomin and related di- and tri-oxabicyclo[x.2.1]systems, *J. Chem. Soc. Chem. Commun.,* p. 216, 1976.

132. **D'Silva, T. D. J. and Peck, D. W.,** Convenient synthesis of frontalin-1,5-dimethyl-6,8-dioxabicyclo[3.2.1]octane, *J. Org. Chem.,* 37, 1828, 1972.

133. **Kinzer, G. W., Fentiman, A. F. Jr., Page, T. F. Jr., Foltz, R. L., Vité, J. P., and Pitman, G. B.,** Bark beetle attractants: identification, synthesis, and field bioassay of a new compound isolated from *Dendroctonus, Nature (London),* 221, 477, 1969.

134. **Mori, K., Kobayashi, S., and Matsui, M.,** Pheromone synthesis. 7. Synthesis of (±)-frontalin, pheromone of *Dendroctonus* bark beetles, *Agric. Biol. Chem.,* 39, 1889, 1975.

135. **Sato, T., Yamaguchi, S., and Kaneko, H.,** Metal-catalyzed organic photoreactions. One-step synthesis of (±)-frontalin by titanium(IV) chloride-catalyzed photoreaction of heptane-2,6-dione, *Tetrahedron Lett.,* p. 1863, 1979.

136. **Wilson, R. M. and Rekers, J. W.,** Decomposition of bicyclic endoperoxides: an isomorphous synthesis of frontalin *via* 1.5-dimethyl-6,7-dioxabicyclo[3.2.1]octane, *J. Am. Chem. Soc.,* 103, 206, 1981.

137. **Mori, K.,** Synthesis of optically active forms of frontalin, the pheromone of *Dendroctonus* bark beetles, *Tetrahedron,* 31, 1381, 1975.

138. **Ohrui, H. and Emoto, S.,** A synthesis of (*S*)-(−)-frontalin from D-glucose, *Agric. Biol. Chem.,* 40, 2267, 1976.

139. **Hicks, D. R. and Fraser-Reid, B.,** Synthesis of one enantiomer, the other enantiomer and a mixture of both enantiomers of frontalin from a derivative of methyl-α-D-glucopyranoside, *J. Chem. Soc. Chem. Commun.,* p. 869, 1976.

140. **Magnus, P. D. and Roy, G.,** A short synthesis of (*R*)-(+)-frontalin and *Latia* luciferin using new organosilicon reagents, *J. Chem. Soc. Chem. Commun.*, p. 297, 1978.

141. **Francke, W., Heemann, V., Gerken, B., Renwick, J. A. A., and Vité, J. P.,** 2-Ethyl-1,6-dioxaspiro[4.4]nonane, principal aggregation pheromone of *Pityogenes chalcographus* (L), *Naturwissenschaften*, 64, 590, 1977.

142. **Francke, W. and Reith, W.,** Alkyl-1,6-dioxaspiro[4.4]nonane; eine neue Klasse von Pheromonen, *Liebigs Ann. Chem.*, p. 1, 1979.

143. **Phillips, C., Jacobson, R., Abrahams, B., Williams, H. J., and Smith, L. R.,** Useful route to 1,6-dioxaspiro[4.4]nonane and 1,6-dioxaspiro[4.5]decane derivatives, *J. Org. Chem.*, 45, 1920, 1980.

144. **Smith, L. R., Williams, H. J., and Silverstein, R. M.,** Facile synthesis of optically active 2-ethyl-1,6-dioxaspiro[4.4]nonane, component of the aggregation pheromone of the beetle *Pityogenes chalcographus*, *Tetrahedron Lett.*, p. 3231, 1978.

145. **Redlich, H. and Francke, W.,** Optisch aktives Chalcogran (2-Ethyl-1,6-dioxaspiro[4.4]nonan), *Angew. Chem.*, 92, 640, 1980; *Angew. Chem. Int. Ed. Engl.*, 19, 630, 1980.

146. **Mori, K., Sasaki, M., Tamada, S., Suguro, T., and Masuda, S.,** Synthesis of optically active 2-ethyl-1,6-dioxaspiro[4.4]nonane (Chalcograne), the principal aggregation pheromone of *Pityogenes chalcographus* (L.), *Tetrahedron*, 35, 1601, 1980.

147. **Hungerbühler, E., Naef, R., Wasmuth, D., Seebach, D., Loosli, H. R., and Wehrli, A.,** Synthese optisch aktiver 2-Methyl- und 2-Äthyl-1,6-dioxaspiro[4.4]nonan- und -[4.5]decan-Pheromone aus einem gemeinsamen chiralen Vorläufer, *Helv. Chim. Acta*, 63, 1960, 1980.

148. **Gore, W. E., Pearce, G. T., and Silverstein, R. M.,** Relative stereochemistry of multistriatin (2,4-dimethyl-5-ethyl-6,8-dioxabicyclo(3.2.1)octane), *J. Org. Chem.*, 40, 1705, 1975.

149. **Elliot, W. J. and Fried, J.,** Maytansinoids. Synthesis of a fragment of known absolute configuration involving chiral centers C-6 and C-7, *J. Org. Chem.*, 41, 2469, 1976; Stereocontrolled synthesis of α-multistriatin, an essential component for the European elm bark beetle, *J. Org. Chem.*, 41, 2475, 1976.

150. **Marino, J. P. and Abe, H.,** Stereospecific 1,4-additions of methyl cyanocuprate to enol phosphates of α,β-epoxycyclohexanones: application to the total synthesis of (±)-α-multistriatin, *J. Org. Chem.*, 46, 5379, 1981.

151. **Beck, K.,** Pheromone chemistry of the smaller European elm bark beetle, *J. Chem. Educ.*, 55, 567, 1978.

152. **Walbe, D. M. and Wand, M. D.,** Stereocontrolled synthesis of substituted 2-alkoxytetrahydropyrans from *meso*-2,4-dimethylglutaric anhydride, *Tetrahedron Lett.*, 23, 4995, 1982.

153. **Cernigliaro, G. J. and Kocienski, P. J.,** A synthesis of (−)-α-multistriatin, *J. Org. Chem.*, 42, 3622, 1977.

154. **Mori, K.,** Synthesis of (1*S*,2*R*,4*S*,5*R*)-(−)-α-multistriatin, the pheromone in the smaller European elm bark beetle, *Scolytus multistriatus*, *Tetrahedron*, 32, 1979, 1976.

155. **Phaik-Eng Sum and Weiler, L.,** Stereospecific synthesis of (−)-α-multistriatin from (D)-glucose, *Can. J. Chem.*, 56, 2700, 1978.

156. **Elliot, W. J., Hromnak, G., Fried, J., and Lanier, G. W.,** Synthesis of multistriatin enantiomers and their action on *Scolytus multistriatus*, *J. Chem. Ecol.*, 5, 279, 1979.

157. **Mori, K. and Iwasawa, H.,** Stereoselective synthesis of optically active forms of δ-multistriatin, the attractant for European populations of the smaller European elm bark beetle, *Tetrahedron*, 36, 87, 1980.

158. **Silverstein, R. M., Rodin, J. O., and Wood, D. L.,** Sex attractants in frass produced by male *Ips confusus* in ponderosa pine, *Science*, 154, 509, 1966.

159. **Vig, O. P., Anand, R. C., Kad, G. L., and Sehgal, J. M.,** Terpenoids. LVII. Syntheses of β-farnesene and methylene isomer of tagetol, *J. Ind. Chem. Soc.*, 47, 999, 1970.

160. **Katzenellenbogen, J. A. and Lenox, R. S.,** The generation of allyllithium reagents by lithium-tetrahydrofuran reduction of allylic mesitoates. A new procedure for selective allylic cross coupling and allylcarbinol synthesis, *J. Org. Chem.*, 38, 326, 1973.

161. **Wilson, S. R. and Phillips, L. R.,** Cyclobutene derivatives as isoprene equivalents in terpene synthesis. 1-Cyclobutenylmethyllithium, *Tetrahedron Lett.*, p. 3047, 1975.

162. **Karlsen, S., Froyen, P., and Skattebøl, L.,** New synthesis of the bark beetle pheromones 2-methyl-6-methylene-7-octen-4-ol (ipsenol) and 2-methyl-6-methylene-2,7-octadien-4-ol (ipsdienol), *Acta Chem. Scand.*, B30, 664, 1976.

163. **Kondo, S., Dobashi, S., and Matsumoto, M.,** The reaction of 2-(1,3-butadienyl)magnesium chloride, *Chem. Lett.*, p. 1077, 1976.

164. **Haslouin, J. and Rouessac, F.,** Coupures thermiques du type Rétro-Diels-Alder. V. Synthese du (±)-ipsenol, *Bull. Soc. Chim. Fr.*, p. 1242, 1977.

165. **Bertrand, M., and Viala, J.,** Nouvelle approche des dérivés du myrcene par transposition thermique des orthoesters alléniques mixtes, *Tetrahedron Lett.*, p. 2575, 1978.

166. **Clinet, J. C. and Linstrummele, G.,** An efficient method for preparation of conjugated allenic carbonyl compounds. The synthesis of two bark beetle pheromones, *Noveau J. Chim.*, 1, 373, 1977.

167. **Hosomi, A., Saito, M., and Sakurai, H.,** 2-Trimethylsilylmethyl-1,3-butadiene as a novel reagent for isoprenylation. New access to ipsenol and ipsdienol, pheromones of *Ips paraconfusus, Tetrahedron Lett.,* p. 429, 1979.

168. **Mori, K.,** Synthesis of optically active forms of ipsenol, the pheromone of *Ips* bark beetles, *Tetrahedron,* 32, 1101, 1976.

169. **Mori, K.,** Synthesis and absolute configuration of (−)-ipsenol (2-methyl-6-methylene-7-octen-4-ol), the pheromone of *Ips paraconfusus* Lanier, *Tetrahedron Lett.,* p. 2187, 1975.

170. **Mori, K., Takigawa, T., and Matsuo, T.,** Synthesis of optically active forms of ipsdienol and ipsenol, *Tetrahedron,* 35, 933, 1979.

171. **Riley, R. G., Silverstein, R. M., Katzenellenbogen, J. A., and Lenox, R. S.,** Improved synthesis of 2-methyl-6-methylene-2,7-octadien-4-ol, a pheromone of *Ips paraconfusus*, and an alternative synthesis of the intermediate, 2-bromomethyl-1,3-butadiene, *J. Org. Chem.,* 39, 1957, 1974.

172. **Mori, K.,** A new synthesis of 2-methyl-6-methylene-octa-2,7-dien-4-ol, a component of the pheromone of California five-spined *Ips, Agric. Biol. Chem.,* 38, 2045, 1974.

173. **Garbers, C. F. and Scott, F.,** Terpenoid synthesis. V. Electrophilic addition reactions in the synthesis of the ocimenones, the rose oxides, and a pheromone of *Ips paraconfusus, Tetrahedron Lett.,* p. 1625, 1976.

174. **Cheskis, B. A., Lebedeva, K. V., Kovaleva, T. I., and Kondrat'ev Yu, A.,** Synthesis of ipsdienol, *Khemoretseptsiya Nasekomykh,* 4, 129, 1979.

175. **Ohloff, G. and Giersch, W.,** Access to optically active ipsdienol from verbenone, *Helv. Chim. Acta,* 60, 1496, 1977.

176. **Mori, K.,** Absolute configuration of (+)-ipsdienol, the pheromone of *Ips paraconfusus* Lanier, as determined by the synthesis of its (*R*)(−)-isomer, *Tetrahedron Lett.,* p. 1609, 1976.

177. **Zurflüh, R., Dunham, L. L., Spain, V. L., and Siddall, J. B.,** Synthetic studies on insect hormones. IX. Stereoselective total synthesis of a racemic boll weevil pheromone, *J. Am. Chem. Soc.,* 92, 425, 1970.

178. **Gueldner, R. C., Thompson, A. C., and Hedin, P. A.,** Stereoselective synthesis of racemic grandisol, *J. Org. Chem.,* 37, 1854, 1972.

179. **Kosugi, H., Sekiguchi, S., Sekita, R., and Uda, H.,** Photochemical cycloaddition reactions of α,β-unsaturated lactones with olefins, and application to synthesis of natural products, *Bull. Chem. Soc. Jpn.,* 49, 520, 1976.

180. **Cargill, R. L. and Wright, B. W.,** A new fragmentation reaction and its application to the synthesis of (±)-grandisol, *J. Org. Chem.,* 40, 120, 1975.

181. **Billups, W. E., Cross, J. H., and Smith, C. V.,** A synthesis of (±)-grandisol, *J. Am. Chem. Soc.,* 95, 3438, 1973.

182. **Ayer, W. A. and Brown, L. M.,** Transformation of carvone into racemic grandisol, *Can. J. Chem.,* 52, 1352, 1974.

183. **Golob, N. F.,** α-Oxycyclopropylcarbinyl Rearrangement and its Application to the Total Synthesis of Grandisol, Ph.D. thesis, Indiana University, Bloomington, Ind., 1974; *Diss. Abstr. Int.,* B35, 4835, 1975.

184. **Wenkert, E., Berges, D. A., and Golob, N. F.,** Oxacyclopropanes in organochemical synthesis. Total synthesis of (−)-valeranone and (±)-grandisol, *J. Am. Chem. Soc.,* 100, 1263, 1978.

185. **Stork, G. and Cohen, J. F.,** Ring size in epoxynitrile cyclization. A general synthesis of functionally substituted cyclobutanes. Application to (±)-grandisol, *J. Am. Chem. Soc.,* 96, 5270, 1974.

186. **Babler, J. H.,** Base-promoted cyclization of a δ-chloroester: Application to the total synthesis of (±)-grandisol, *Tetrahedron Lett.,* p. 2045, 1975.

187. **Trost, B. M. and Keeley, D. E.,** New synthetic methods. Secoalkylative approach to grandisol, *J. Org. Chem.,* 40, 2013, 1975.

188. **Clark, R. D.,** Addition of organocuprates to cyclobutenyl esters. Synthesis of (±)-grandisol, *Synth. Commun.,* 9, 325, 1979.

189. **Hobbs, P. D. and Magnus, P. D.,** Synthesis of optically active grandisol, *J. Chem. Soc. Chem. Commun.,* p. 856, 1974.

190. **Hobbs, P. D. and Magnus, P. D.,** Studies on terpenes. 4. Synthesis of optically active grandisol, the boll weevil pheromone, *J. Am. Chem. Soc.,* 98, 4594, 1976.

191. **Mori, K.,** Synthesis of both the enantiomers of grandisol, the boll weevil pheromone, *Tetrahedron,* 34, 915, 1978.

192. **Mori, K., Tamada, S., and Hedin, P. A.,** (−)-Grandisol, the antipode of the boll weevil pheromone, is biologically active, *Naturwissenschaften,* 65, 653, 1978.

193. **MacConnell, J. G., Borden, H. J., Silverstein, R. M., and Stokkink, E.,** Isolation and tentative identification of lineatin, a pheromone from the frass of *Trypodendron lineatum* (Coleoptera: Scolytidae), *J. Chem. Ecol.,* 3, 549, 1977.

194. **Mori, K. and Sasaki, M.,** Synthesis of (±)-lineatin, the unique tricyclic pheromone of *Trypodendron lineatum* (Olivier), *Tetrahedron Lett.,* p. 1329, 1979.

195. **Mori, K. and Sasaki, M.,** Synthesis of racemic and optically active forms of lineatin, the unique tricyclic pheromone of *Trypodendron lineatum* (Olivier), *Tetrahedron,* 36, 2197, 1980.

196. **Mori, K., Uematsu, T., Minobe, M., and Yanagi, K.,** Synthesis and absolute configuration of (+)-lineatin, the pheromone of *Trypodendron lineatum, Tetrahedron Lett.,* 23, 1921, 1982.

197. **Borden, J. H., Handley, J. R., Johnston, B. D., MacConnell, J. G., Silverstein, R. M., Slessor, K. N., Swigar, A. A., and Wong, D. T. W.,** Synthesis and field testing of 4,6,6-lineatin, the aggregation pheromone of *Trypodendron lineatum* (Coleoptera: Scolytidae), *J. Chem. Ecol.,* 5, 681, 1979.

198. **Slessor, K. N., Oehlschlager, A. C., Johnston, B. D., Pierce, H. D. Jr., Grewal, S. K., and Wickremesinghe, L. K. G.,** Lineatin: regioselective synthesis and resolution leading to the chiral pheromones of *Trypodendron lineatum, J. Org. Chem.,* 45, 2290, 1980.

199. **Skattebøl, L. and Stenstrøm, V.,** Synthesis of (±)-lineatin, an aggregation pheromone component of *Trypodendron lineatum, Tetrahedron Lett.,* 24, 3021, 1983.

200. **Mori, K. and Uematsu, K.,** Synthesis and absolute configuration of both the enantiomers of lineatin, *Tetrahedron,* 39, 1735, 1983.

201. **Babler, J. H. and Mortell, T. R.,** Facile route to three of four terpenoid components of the boll weevil sex attractant, *Tetrahedron Lett.,* p. 669, 1972.

202. **Pelletier, S. W. and Mody, N. V.,** Facile synthesis of cyclohexyl constituents of the boll weevil sex pheromone, *J. Org. Chem.,* 41, 1069, 1976.

203. **Babler, J. H. and Coghlan, M. J.,** Facile method for bishomologation of ketones to α,β-unsaturated aldehydes. Application to synthesis of cyclohexanoid components of the boll weevil sex pheromone, *Synth. Commun.,* 6, 469, 1976.

204. **DeSouza, J. P. and Goncalves, A. M. R.,** Alternative route to three of the four terpenoid components of the boll weevil sex pheromone, *J. Org. Chem.,* 43, 2068, 1978.

205. **Bedoukian, R. H. and Wolinsky, J.,** Biogenetic type synthesis of cyclohexyl constituents of the boll weevil pheromone, *J. Org. Chem.,* 40, 2154, 1975.

206. **Traas, P. C., Boelens, H. and Takken, H. J.,** Two step synthesis of a sex attractant of male boll weevil from isophorone, *Rec. Trav. Chim.,* 95, 308, 1976.

207. **Traas, P. C., Boelens, H., and Takken, H. J.,** A convenient synthesis of 3,3-dimethylcyclohexylidene-acetaldehyde, as sex attractant of the male boll weevil, *Synth. Commun.,* 6, 489, 1976.

208. **Nakai, T., Mimura, T., and Ari-izumi, A.,** A new method for the bishomologation of carbonyl compounds to α,β-unsaturated aldehydes *via* the [3,3]sigmatropic rearrangement of thionocarbamates, *Tetrahedron Lett.,* p. 2425, 1977.

209. **Wollenberg, R. H. and Peries, R.,** Efficient synthesis of insect sex pheromones emitted by the boll weevil and the red bollworm moth, *Tetrahedron Lett.,* p. 297, 1979.

210. **Rossi, R.,** Insect pheromones. II. Synthesis of chiral components of insect pheromones, *Synthesis,* p. 413, 1978.

211. **Tanaka, A., Uda, H., and Yoshikoshi, A.,** Total synthesis of α-cubebene, β-cubebene, and cubenol, *J. Chem. Soc. Chem. Commun.,* p. 308, 1969.

212. **Whitham, C. H.,** The reaction of α-pinene with lead tetra-acetate, *J. Chem. Soc.,* p. 2232, 1961.

213. **Landor, P. D., Landor, S. R., and Mukasa, S.,** Synthesis of (±)-methyl tetradeca-*trans*-2,4,5-trienoate, the allenic sex pheromone produced by the male dried bean beetle, *J. Chem. Soc. Chem. Commun.,* p. 1638, 1971.

214. **Descoins, C., Henrick, C. A., and Siddall, J. B.,** Synthesis of a presumed sex attractant of dried bean beetle, *Tetrahedron Lett.,* p. 3777, 1972.

215. **Baudouy, R. and Gore, J.,** New synthesis of a pheromone with a vinyl allenic linkage, *Synthesis,* p. 573, 1974.

216. **Michelot, D. and Linstrumelle, G.,** Cuprates alléniques. I. Préparations et reactions. Synthèse stéréo-sélective de la phèromone de la bruche parasite du haricot, *Tetrahedron Lett.,* p. 275, 1976.

217. **Koziensky, P. J., Cerniglario, G., and Feldstein, G.,** A synthesis of (±)-methyl *n*-tetradeca-*trans*-2,4,5-trienoate, an allenic ester produced by the male dried bean beetle *Acanthoscelides obtectus* (Say.), *J. Org. Chem.,* 42, 353, 1977.

218. **Khimyan, A. P. and Badanyan, S. O.,** Reactions of unsaturated compounds. LXXI. New approach to the synthesis of *trans*-2,4,5-tetradecatrien-1-ol, the precursor of the pheromone of the parasitic bean beetle, *Arm. Khim. Zh.,* 34, 254, 1981.

219. **Pirkle, W. H. and Boeder, C. W.,** Synthesis and absolute configuration of (−)-methyl (*E*)-2,4,5-tetra-decatrienoate, the sex attractant of the male dried bean weevil, *J. Org. Chem.,* 43, 2091, 1978.

220. **Mori, K., Nukada, T., and Ebata, T.,** Pheromone synthesis. XLII. Synthesis of optically active forms of methyl (*E*)-2,4,5-tetradecatrienoate, the pheromone of the male dried bean beetle, *Tetrahedron,* 37, 1343, 1981.

221. **Westmijze, H., Ruitenberg, K., Meijer, J., and Vermeer, P.,** Stereospecific synthesis of vinylstannanes using triphenylstannylcopper (I) species. Preparation of the *cis*-isomer of the presumed sex attractant of *Acanthoscelides obtectus, Tetrahedron Lett.,* 23, 2797, 1982.

222. **Rodin, J. O., Silverstein, R. M., Burkholder, W. E., and Gorman, J. E.,** Sex attractant of female Dermestid beetle *Trogoderma inclusum* LeConte, *Science,* 165, 904, 1969.

223. **DeGraw, J. I. and Rodin, J. O.**, Synthesis of methyl 14-methyl-*cis*-8-hexadecenoate and 14-methyl-*cis*-hexadecen-1-ol. Sex attractant of *Trogoderma inclusum* LeConte, *J. Org. Chem.*, 36, 2902, 1971.

224. **Ronmestant, M. L., Place, P., and Gore, J.**, Vinylallenes. IV. Preparation et transposition de Cope de divers ethinyl-4-hexadiene-1.5-ols-3, *Tetrahedron Lett.*, p. 677, 1976.

225. **Rossi, R., Salvadori, P. A., Carpita, A., and Niccoli, A.**, Synthesis of the (R)(−)enantiomers of the pheromone components of several species of *Trogoderma* [Coleoptera: Dermestidae], *Tetrahedron*, 35, 2039, 1979.

226. **Mori, K., Suguro, T., and Uchida, M.**, Synthesis of optically active forms of (Z)-14-methylhexadec-8-enal, *Tetrahedron*, 34, 3119, 1978.

227. **Jensen, U. and Schäfer, H. J.**, Pheromones. 6. Synthesis of optically active *Trogoderma* pheromones by Kolbe electrolysis, *Chem. Ber.*, 114, 292, 1981.

228. **Rossi, R. and Carpita, A.**, Insect pheromones — synthesis of chiral sex pheromone components of several species of *Trogoderma* (Coleoptera: Dermestidae), *Tetrahedron*, 33, 2447, 1977.

229. **Mori, K.**, Absolute configuration of (−)-14-methyl-*cis*-8-hexadecen-1-ol and methyl (−)-14-methyl-*cis*-8-hexadecenoate, the sex attractant of female dermestid beetle, *Trogoderma inclusum* LeConte, *Tetrahedron Lett.*, p. 3869, 1973.

230. **Mori, K.**, Absolute configurations of (−)-14-methylhexadec-8-*cis*-en-1-ol and methyl (−)-14-methylhexadec-8-*cis*-enoate, the sex pheromone of female Dermestid beetle, *Tetrahedron*, 30, 3817, 1974.

231. **Sato, T., Naruse, K., and Fujisawa, T.**, A stereocontrolled synthesis of (R,Z)- and (R,E)-14-methyl-8-hexadecenals, (trogodermal), the sex pheromone of *Trogoderma* species, *Tetrahedron Lett.*, 23, 3587, 1982.

232. **Suguro, T. and Mori, K.**, Pheromones Synthesis. XXVII. Synthesis of optically active forms of (E)-14-methyl-8-hexadecenal, *Agric. Biol. Chem.*, 43, 409, 1979.

233. **Fukui, H., Matsumura, F., Ma, M. C., and Burkholder, W.**, Identification of the sex pheromone of the furniture carpet beetle, *Anthrenus flavipes*, *Tetrahedron Lett.*, p. 3563, 1974.

234. **Hofmann, K., O'Leary, W. M., Yoho, C. W., and Liu, T. Y.**, Lipid stimulation of bacterial growth, *J. Biol. Chem.*, 234, 1672, 1959.

235. **Silverstein, R. M., Rodin, J. O., Burkholder, W. E., and Gorman, J. E.**, Sex attractant of the black carpet beetle, *Science*, 157, 85, 1967.

236. **Rodin, J. O., Laeffer, M. A., and Silverstein, R. M.**, Synthesis of *trans*-3,*cis*-5-tetradecadienoic acid (megatomic acid), the sex attractant of the black carpet beetle, and its geometric isomers, *J. Org. Chem.*, 35, 3152, 1970.

237. **Yokoi, K. and Matsubara, Y.**, Synthesis of physiologically active substances. III. Synthesis of black carpet beetle sex pheromone, *Kinki Daigaku Rikogakubu Kenkyu Hokoku*, p. 65, 1979.

238. **Sakakibara, M. and Mori, K.**, Synthesis of a stereoisomeric mixture of 2,3-dihydro-2,3,5-trimethyl-6-(1-methyl-2-oxobutyl)-4H-pyran-4-one, the pheromone of the drugstore beetle, *Tetrahedron Lett.*, p. 2401, 1979.

239. **Hoffmann, R. W. and Ladner, W.**, On the absolute stereochemistry of C-2 and C-3 in Stegobinone, *Tetrahedron Lett.*, p. 4653, 1979.

240. **Hoffmann, R. W. and Ladner, W.**, (2S)-2,3-dihydro-2,3,5-trimethyl-6-(1-methyl-2-oxobutyl)-4H-pyran-4-one and its use as an insect lure and its intermediate products, *Ger. Offen. Patent*, 2, 947, 422, June 11, 1981.

241. **Mori, K., Ebata, T., and Sakakibara, M.**, Pheromone synthesis. XXXVIII. Synthesis of (2S,3R,7RS)-stegobinone [2,3-dihydro-2,3,5-trimethyl-6-(1-methyl-2-oxobutyl)-4H-pyran-4-one] and its (2R,3S,7RS)-isomer. The pheromone of the drugstore beetle, *Tetrahedron*, 37, 709, 1981.

242. **Mori, K., Tamada, S., Uchida, M., Mizumachi, N., Tachibana, Y., and Matsui, M.**, Pheromone synthesis. 20. Synthesis of optically active forms of seudenol, the pheromone of the Douglas fir beetle, *Tetrahedron*, 34, 1901, 1978.

243. **Einterz, R. M., Ponder, J. W., and Lenox, R. S.**, The synthesis of 4-methyl-3-heptanol and 4-methyl-3-heptanone. Two easily synthesized insect pheromones, *J. Chem. Educ.*, 54, 382, 1977.

244. **Mori, K.**, Absolute configuration of (−)-4-methylheptan-3-ol, a pheromone of the smaller European elm bark beetle, as determined by the synthesis of its 3R,4R- and 3S,4R-isomers, *Tetrahedron*, 33, 289, 1977.

245. **Vigneron, J. P., Meric, R., and Dhaenens, M.**, Preparation of optically pure *erythro*- and *threo*-4-methylheptan-3-ols, *Tetrahedron Lett.*, 21, 2057, 1980.

246. **Mori, K. and Iwasawa, H.**, Preparation of the both enantiomers of *threo*-2-amino-3-methylhexanoic acid by enzymatic resolution and their conversion to optically active forms of *threo*-4-methylheptan-3-ol, a pheromone component of the smaller European elm bark beetle, *Tetrahedron*, 36, 2209, 1980.

247. **Rossi, R. and Marasco, M.**, Insect pheromones by asymmetric synthesis. Asymmetric synthesis of (S)-4-methyl-3-heptanone, the alarm pheromone of *Atta texana*, of (S)-4-methyl-3-heptanol, the major aggregation pheromone component of *Scolytus scolytus*, and of (S)-3-methyl-2-heneicosanone, a structural analog of a sex pheromone component of *Blatella germanica*, *Chim. Ind. (Milan)*, 62, 314, 1980.

248. **Nishizawa, M., Yamada, M., and Noyori, R.,** Highly enantioselective reduction of alkynyl ketones by a binaphthol-modified aluminum hydride reagent. Asymmetric synthesis of some insect pheromones, *Tetrahedron Lett.,* 22, 247, 1981.

249. **Vigneron, J. P. and Bloy, V.,** Preparation d'alkyl-4-γ-lactones optiquement actives, *Tetrahedron Lett.,* 21, 1735, 1980.

250. **Doolittle, R. E., Tumlinson, J. H., Proveaux, A. T., and Heath, R. R.,** Synthesis of the sex pheromone of the Japanese beetle, *J. Chem. Ecol.,* 6, 473, 1980.

251. **Tumlinson, J. H., Klein, M. G., Doolittle, R. E., Ladd, T. L., and Proveaux, A. T.,** Identification of the female Japanese beetle sex pheromone: inhibition of male response by an enantiomer, *Science,* 197, 789, 1977.

252. **Sato, K., Nakayama, T., and Mori, K.,** Pheromone synthesis. XXXII. New synthesis of the both enantiomers of (Z)-5-(1-decenyl)oxacyclopentan-2-one, the pheromone of the Japanese beetle, *Agric. Biol. Chem.,* 43, 57, 1979.

253. Commercially available chemical.

254. **Midland, M. M. and Tramontano, A.,** The synthesis of naturally occurring 4-alkyl- and 4-alkenyl-γ-lactones using the asymmetric reducing reagent B-3-pinanyl-9-borabicyclo[3.3.1]nonane, *Tetrahedron Lett.,* 21, 3549, 1981.

255. **Midland, M. M. and Nguyen, N. H.,** Asymmetric synthesis of γ-lactones. A facile synthesis of the sex pheromone of the Japanese beetle, *J. Org. Chem.,* 46, 4107, 1981.

256. **Baker, R. and Rao, V. B.,** Synthesis of optically pure (*R,Z*)-5-dec-1-enyloxacyclopentan-2-one, the sex pheromone of the Japanese beetle, *J. Chem. Soc. Perkin Trans. 1,* p. 69, 1982.

257. **Ravid, U. and Silverstein, R. M.,** General synthesis of optically active 4-alkyl (or alkenyl)-γ-lactones from glutamic acid enantiomers, *Tetrahedron Lett.,* p. 423, 1977.

258. **Ravid, U., Silverstein, R. M., and Smith, L. R.,** Synthesis of enantiomers of 4-substituted γ-lactones with known absolute configuration, *Tetrahedron,* 34, 1449, 1978.

259. **Pirkle, W. and Adams, P. E.,** Broadspectrum synthesis of enantiomerically pure lactones. Synthesis of sex pheromones of the carpenter bee, love beetle, etc., *J. Org. Chem.,* 44, 2169, 1979.

260. **Redlich, H., Xiang-jun, J., Paulsen, H., and Francke, W.,** Darstellung von (*S*)-2,5-dimethyl-2-isopropyl-2,3-dihydrofuran, einem der beiden Enantiomeren des Sexuallockstoffs des Werftkäfers, *Hylecoetus dermestoides* L., *Tetrahedron Lett.,* 22, 5043, 1981.

261. **Francke, W., Mackenroth, W., Schröder, W., and Levinson, A. R.,** Cyclic enolethers as insect pheromones, in *Les Médiateurs Chimiques,* Les Colloques de l'INRA 7, 85, l'INRA, Paris, 1982.

262. **Francke, W. and Mackenroth, W.,** Alkylsubstituierte 3,4-dihydro-2*H*-pyrane: Massenspektrometrie, Synthese und Identifizierung als Insekteninhaltsstoffe, *Angew. Chem.,* 94, 704, 1982; *Angew. Chem. Int. Ed. Engl.,* 21, 698, 1982.

263. **Chuman, T., Kato, K., and Noguchi, M.,** Synthesis of (±)-serricornin, 4,6-dimethyl-7-hydroxynonan-3-one, a sex pheromone of cigarette beetle (*Lasioderma serricorne* F.), *Agric. Biol. Chem.,* 43, 2005, 1979.

264. **Ono, M., Onishi, I., Chuman, T., Kono, M., and Kato, K.,** A novel synthesis of (±)-serricornin, 4,6-dimethyl-7-hydroxynonan-3-one, the sex pheromone of cigarette beetle (*Lasioderma serricorne* F.), *Agric. Biol. Chem.,* 44, 2259, 1980.

265. **Mori, K., Nomi, H., Chuman, T., Kohno, M., Kato, K., and Noguchi, M.,** Determination of the absolute configuration at C-6 and C-7 of serricornin (4,6-dimethyl-7-hydroxy-3-nonanone), the sex pheromone of the cigarette beetle, *Tetrahedron Lett.,* 22, 1127, 1981.

266. **Hoffmann, R. W., Helbig, W., and Ladner, W.,** Synthesis of 6*S*,7*S*-anhydro-serricornin, *Tetrahedron Lett.,* 23, 3479, 1982.

267. **Mori, M., Chuman, T., Kohno, M., Kato, K., Noguchi, M., Nomi, H., and Mori, K.,** Absolute stereochemistry of serricornin, the sex pheromone of cigarette beetle, as determined by synthesis of its (4*S*,6*R*,7*R*)-isomer, *Tetrahedron Lett.,* 23, 667, 1982.

268. **Chuman, T., Kohno, M., Kato, K., Noguchi, M., Nomi, H., and Mori, K.,** Stereoselective synthesis of *erythro*-serricornin, (4*R*,6*R*,7*S*)- and (4*S*,6*R*,7*S*)-4,6-dimethyl-7-hydroxynonan-3-one, stereoisomers of the sex pheromone of the cigarette beetle, *Agric. Biol. Chem.,* 45, 2019, 1981.

269. **Mori, M., Chuman, T., Kato, K., and Mori, K.,** A stereoselective synthesis of "natural" (4*S*,6*S*,7*S*)-serricornin, the sex pheromone of cigarette beetle, from levoglucosenone, *Tetrahedron Lett.,* 23, 4593, 1982.

270. **Mori, K.,** Pheromone synthesis. XLI. A simple synthesis of (*S*)-(+)-sulcatol, the pheromone of *Gnathotrichus retusus,* employing baker's yeast for asymmetric reduction, *Tetrahedron,* 37, 1341, 1981.

271. **Guss, P. L., Tumlinson, H. J., Sonnet, P. E., and Proveaux, A. T.,** Identification of a female-produced sex pheromone of the western corn rootworm, *J. Chem. Ecol.,* 8, 545, 1982.

272. **Sonnet, P. E.,** Synthesis of the stereoisomers of the sex pheromone of the southern corn rootworm and lesser tea tortrix, *J. Org. Chem.,* 47, 3793, 1982.

273. **Guss, P. L., Tumlinson, J. H., Sonnet, P. E., and McLaughlin, J. R.,** Identification of female produced sex pheromone from the southern corn rootworm, *Diabrotica undecimpuncta howardi* Barber, *J. Chem. Ecol.,* 9, 1363, 1983.

274. **Sonnet, P. E. and Heath, R. R.,** presented at the Symposium on Chemistry and Applications of Insect Pheromone Technology, Annu. Meet. ACS, New York, 1981.

275. **Suzuki, T.,** A facile synthesis of 4.8-dimethyldecanal, aggregation pheromone of flour beetles, and its analogues, *Agric. Biol. Chem.,* 45, 2641, 1981.

276. **Cross, J. H., Byler, R. C., Cassidy, R. F. Jr., Silverstein, R. M., Greenblatt, R. E., Burkholder, W. E., Levinson, A. R., and Levinson, H. C.,** Porapak Q collection of pheromone components and isolation of (Z)- and (E)-14-methyl-8-hexadecenal, sex pheromone components, from the frass of a few species of *Trogoderma* (Coleoptera: Dermestidae), *J. Chem. Ecol.,* 2, 457, 1976.

277. **Silverstein, R. M., Rodin, J. O., and Wood, D. L.,** Methodology for isolation and identification of insect pheromones with reference to studies on Californian five-spined *Ips, J. Econ. Entomol.,* 60, 944, 1967.

278. **Rossi, R.,** Insect pheromones. I. Synthesis of achiral components of insect pheromones, *Synthesis,* p. 817, 1977.

279. **Henrick, C. A.,** The synthesis of insect sex pheromones, *Tetrahedron,* 33, 1845, 1977.

280. **Mori, K.,** Synthetic chemistry of insect pheromones and juvenile hormones, in *Recent Developments in the Chemistry of Natural Carbon Compounds,* Vol. 9, Bognar, R., Bruckner, V., and Szantay, Cs., Eds., Akademiai Kiado, Budapest, 1979, 9.

281. **Mori, K.,** The synthesis of insect pheromones, in *The Total Synthesis of Natural Products,* Vol. 4, ApSimon, J., Ed., John Wiley & Sons, New York, 1981, 1.

282. **Mori, K.,** Synthesis of optically active insect pheromones, *Yuki Kagaku Kyokaishi,* 39, 63, 1981.

283. **Katzenellenbogen, J. A.,** Insect pheromone synthesis: new methodology, *Science,* 194, 139, 1976.

284. **Jacobson, M.,** Methodology for isolation, identification and synthesis of sex pheromones in the Lepidoptera, in *Control of Insect Behavior by Natural Products,* Wood, D. L., Silverstein, R. M., and Nakajima, M., Eds., Academic Press, New York, 1970, 111.

285. **Mori, K.,** Chemistry of natural products, *Kagaku No Ryoiki, Zokan,* p. 155, 1980.

286. **Sonnet, P. E.,** Tabulations of selected methods of syntheses that are frequently employed for insect sex pheromones, emphasizing the literature of 1977—1982, in *Techniques in Pheromone Research,* Hummel, H. E. and Miller, T. E., Eds., Springer-Verlag, Berlin, 1984, 371.

287. **Silverstein, R. M.,** Spectrometric identification of insect sex attractants, *J. Chem. Educ.,* 45, 794, 1968.

288. **Francke, W., Sauerwein, P., Vité, J. P., and Klimetzek, D.,** The pheromone bouquet of *Ips amitinus, Naturwissenschaften,* 67, 147, 1980.

289. **Pearce, G. T., Gore, W. E., and Silverstein, R. M.,** Carbon-13-spectra of some insect pheromones and related compounds of the 6,8-dioxabicyclo[3.2.1]octane system, *J. Magn. Reson.,* 27, 497, 1977.

290. **Francke, W., Reith, W., and Sinwell, V.,** Bestimmung der relativen Konfiguration bei Spiroacetalen durch ^{1}H- und ^{13}C-NMR-Spektroskopie, *Chem. Ber.,* 113, 2686, 1980.

291. **Francke, W., Hindorf, G., and Reith, W.,** Alkyl-1,6-dioxaspiro[4.5]decanes — a new class of pheromones, *Naturwissenschaften,* 66, 618, 1979.

292. **Stenhagen, E., Abrahamsson, S., and McLafferty, F. W., Eds.,** *Registry of Mass Spectral Data,* John Wiley & Sons, New York, 1974.

293. **Ohta, Y., Sakai, T., and Hirose, Y.,** Sesquiterpene hydrocarbons from the oil of cubeb, α-cubebene and β-cubebene, *Tetrahedron Lett.,* p. 6365, 1966.

294. **Beilstein Institut für Literatur der Organischen Chemie, Ed.,** *Beilstein's Handbuch der Organischen Chemie,* 4th ed., Springer-Verlag, Berlin, 1975.

295. **Wheeler, J. W., Happ, G. M., Araujo, J., and Pasteel, J. M.,** γ-Dodecalactone from rove beetles, *Tetrahedron Lett.,* p. 4635, 1972.

296. **Francke, W., Heemann, V., and Heynes, K.,** Flüchtige Inhaltsstoffe von Ambrosiakäfern (Coleoptera: Scolytidae). I., *Z. Naturforsch.,* 29c, 243, 1974.

297. **Suzuki, T.,** Identification of the aggregation pheromone of flour beetles *Tribolium castaneum* and *T. confusum* (Coleoptera: Tenebrionidae), *Agric. Biol. Chem.,* 45, 1357, 1981.

298. **Institute for Spectroscopy, Dortmund, FRG and DMS Scientific Advisory Board, Eds.,** *DMS, Documentation of Molecular Spectroscopy,* Verlag Chemie, Weinheim, and Butterworths, London, 1970.

299. **Francke, W., Levinson, A. R., Jen, T.-L., and Levinson, H. Z.,** Carbonsäure-isopropylester — eine neue Klasse von Insektenpheromonen, *Angew. Chem.,* 91, 843, 1979; *Angew. Chem. Int. Ed. Engl.,* 18, 796, 1979.

300. **Holman, R. T., Lundberg, W. O., Lauer, W. M., and Burr, G. O.,** Spectrophotometric studies of the oxidation of fats. I. Oleic acid, ethyl oleate and elaidic acid, *J. Am. Chem. Soc.,* 67, 1285, 1945.

301. **Bolland, J. L. and Koch, H. P.,** The course of autoxidation reactions in polyisoprenes and allied compounds. IX. The primary thermal oxidation product of ethyl linoleate, *J. Chem. Soc.,* p. 445, 1945.

302. **Wong, J. W., Verigin, V., Oehlschlager, A. C., Borden, J. H., Pierce, H. D. Jr., Pierce, A. M., and Chong, L.,** Isolation and identification of two macrolide pheromones from the frass of *Cryptolestes ferrugineus* (Coleoptera: Cucujidae), *J. Chem. Ecol.,* 9, 451, 1983.

303. **Bradley, J. R., Clower, D. F., and Graves, J. B.,** Field studies of sex attraction in the boll weevil, *J. Econ. Entomol.,* 61, 1457, 1968.

304. **Tumlinson, J. H., Gueldner, R. C., Hardee, D. D., Thompson, A. C., Hedin, P. A., and Minyard, J. P.,** The boll weevil sex attractant, in *Chemicals Controlling Insect Behavior,* Beroza, M., Ed., Academic Press, New York, 1970, 41.

305. **Chang, V. C. S. and Curtis, G. A.,** Pheromone production by the New Guinea surgarcane weevil, *Environ. Entomol.,* 1, 476, 1972.

306. **Phillips, J. K. and Burkholder, W. E.,** Evidence for a male-produced aggregation pheromone in the rice weevil, *J. Econ. Entomol.,* 74, 539, 1981.

307. **Kalo, P.,** Identification of potential sex pheromones in the large pine weevil *Hylobius abietes* L. (Coleoptera, Curculionidae), *Finn. Chem. Lett.,* p. 189, 1979.

308. **Coffelt, J. A., Vick, K. W., Sower, L. L., and McClellan, W. T.,** Sex pheromone of the sweetpotato weevil, *Cylas formicarius elongatulus:* laboratory bioassay and evidence for a multiple components system, *Environ. Entomol.,* 7, 756, 1978.

309. **Hedin, P. A., Payne, J. A., Carpenter, T. L., and Neal, W.,** Sex pheromones of the male and female pecan weevil, *Curculio caryae,* behavioral and chemical studies, *Environ. Entomol.,* 8, 521, 1979.

310. **Mody, N. V., Miles, D. H., Nee, W. W., Hedin, P. A., Thompson, A. C., and Gueldner, R. C.,** Pecan weevil sex attractant: bioassay and chemical studies, *J. Insect Physiol.,* 19, 2063, 1973.

311. **Yun-Tai Qui, and Burkholder, W. E.,** Sex pheromone biology and behavior of the cowpea weevil *Callosobruchus maculatus* (Coleoptera: Bruchidae), *J. Chem. Ecol.,* 8, 527, 1982.

312. **Jacobson, M., Lilly, C. E., and Harding, C.,** Sex attractant of sugar-beet wire worm: identification and biological activity, *Science,* 159, 208, 1968.

313. **Butler, L. I., McDonough, L. M., Onsager, J. A., and Landis, B. J.,** Sex pheromones of the pacific coast wireworm, *Limonius canus,* Elateridae, Coleoptera, *Environ. Entomol.,* 4, 229, 1975.

314. **Voelkel, H.,** Zur Biologie und Bekämpfung des Khraprakäfers, *Trogoderma granarium* Everts., *Ab. Biol. Reichsanst.,* 13, 129, 1924.

315. **Schwarz, M., Jacobson, M., and Cuthbert, F. P.,** Chemical studies of the sex attractant of the banded cucumber beetle Diabrotica balteata, *J. Econ. Entomol.,* 64, 769, 1971.

316. **Matthes, D.,** Sexualstimulation durch Duftfächeln, *Umsch. Wiss. Tech.,* 1970, 112.

317. **Matthes, D.,** Der Meloide Cerocoma schäfferi (L.), *Entomol. Blatter,* 66, 33, 1970.

318. **Klinger, R.,** A species specific sex pheromone on the cuticle of the female *Eusphalerum minutum, Naturwissenschaften,* 65, 597, 1978.

319. **Peschke, K.,** The female sex pheromone of the staphylinid beetle *Aleochara curtula, J. Insect Physiol.,* 24, 197, 1978.

320. **Higgs, M. D. and Evans, D. A.,** Chemical mediators in the oviposition behaviour of the house longhorne beetle, *Hylotrupes bajulus,* Experientia, 34, 46, 1978.

321. **Iwabuchi, K.,** Mating behavior of *Xylotrechus pyrrhoderus* Bates (Coleoptera: Cerambycidae). I. Behavioral sequences and existence of the male sex pheromone, *Appl. Entomol. Zool.,* 17, 494, 1982.

322. **Chuman, T., Kohno, M., Kato, K., and Noguchi, M.,** 4,6-Dimethyl-7-hydroxy-nonan-3-one, a sex pheromone of the cigarette beetle (*Lasioderma serricorne* F.), *Tetrahedron Lett.,* p. 2361, 1979.

323. **Chuman, T., Mochizuki, K., Mori, M., Kohno, M., Kato, K., Nomi, H., and Mori, K.,** Behavioral and electroantennogram responses of male cigarette beetle (Lasioderma serricorne F.) to optically active serricornins, *Agric. Biol. Chem.,* 46, 3109, 1982.

324. **Levinson, H. Z., Levinson, A. R., Francke, W., Mackenroth, W., and Heemann, V.,** The pheromone activity of anhydroserricornin and serricornin for male cigarette beetles (*Lasioderma serricorne* F.), *Naturwissenschaften,* 68, 148, 1981.

325. **Kuwahara, Y., Fukami, H., Howard, R., Ishii, S., Matsumura, F., and Burkholder, W. E.,** Chemical studies on the Anobiidae; sex pheromone of the drugstore beetle, *Stegobium paniceum* (Coleoptera), *Tetrahedron,* 34, 1769, 1978.

326. **Khorramshahi, A. and Burkholder, W. E.,** Behavior of the lesser grain borer *Rhyzopertha dominica* (Coleoptera: Bostrichidae) male-produced aggregation pheromone attracts both sexes, *J. Chem. Ecol.,* 7, 33, 1981.

327. **Halstead, D. G. H.,** Preliminary biological studies on the pheromone produced by male *Acanthoscelides obtectus* (Col., Bruchidae), *J. Stored Prod. Res.,* 9, 109, 1973.

328. **Guss, P. L., Carney, R. L., Sonnet, P. E., and Tumlinson, J. H.,** Stereospecific sex attractant for *Diabrotica cristata* (Coleoptera, Chrysomyelidae), *Environ. Entomol.,* 12, 1296, 1983.

329. **Hedin, P. A., Gueldner, R. C., and Thompson, A. C.,** Utilization of the boll weevil pheromone for insect control, *Pest Management with Insect Sex Attractants,* Beroza, M., Ed., ACS Symp. Ser. No. 23, Am. Chem. Soc., Washington, D.C., 1976, 30.

330. **Hardee, D. D., McKibben, G. H., Gueldner, R. C., Mitchell, E. B., Tumlinson, J. H., and Cross, W. H.,** Boll weevil respond to grandlure, a synthetic pheromone, *J. Econ. Entomol.,* 65, 97, 1972.

331. **Hardee, D. D., McKibben, G. H., Rummel, D. R., Huddleston, P. M., and Coppedge, J. R.,** Response of boll weevils to component ratios and doses of the pheromone, grandlure, *Environ. Entomol.,* 3, 135, 1974.

332. **Huddleston, D. M., Mitchell, E. R., and Wilson, N. M.,** Disruption of boll weevil communication, *J. Econ. Entomol.,* 70, 83, 1977.

333. **Villavaso, E. J. and McGovern, W. L.,** Boll weevil: disruption of pheromonal communication in the laboratory and small field plots, *J. Ga. Entomol. Soc.,* 16, 306, 1981.

334. **Mitchell, E. R., Lloyd, E. P., Hardee, D. D., Cross, W. H., and Davich, T. B.,** In-field traps and insecticides for suppression and elimination of populations of boll weevils, *J. Econ. Entomol.,* 69, 83, 1976.

335. **Gutmann, A., Payne, T. L., Roberts, E. A., Schulte-Elte, K. H., Giersch, W., and Ohloff, G.,** Antennal olfactory response of boll weevil to grandlure and vicinal dimethyl analogs, *J. Chem. Ecol.,* 7, 919, 1981.

336. **Tanaka, K., Ohsawa, K., Honda, H., and Yamamoto, I.,** Copulation release pheromone, erectin, from the azuki bean weevil (*Callosobruchus chinensis* L.), *J. Pest. Sci.,* 6, 75, 1981.

337. **Baker, J. E. and Nelson, D. R.,** Cuticular hydrocarbons of adults of the cowpea weevil, *Callosobruchus maculatus, J. Chem. Ecol.,* 7, 175, 1981.

338. **Booth, D. C., Phillips, T. W., Claesson, A., Silverstein, R. M., Lanier, G. N., and West, J. R.,** Aggregation pheromone components of two species of *Pissodes* weevils (Coleoptera: Curcullionidae): isolation, identification, and field activity, *J. Chem. Ecol.,* 9, 1, 1983.

339. **Sharma, S. P. and Deora, R. K.,** Factors affecting production, release and response to female sex pheromones in *Sitophilus oryzae* (L.) (Coleoptera: Curculionidae), *Indian J. Exp. Biol.,* 18, 463, 1980.

340. **Fukui, H., Matsumura, F., Barak, A. V., and Burkholder, W. E.,** Isolation and identification of a major sex-attracting component of *Attagenus elongatulus* (Casey) (Coleoptera: Dermestidae), *J. Chem. Ecol.,* 3, 539, 1977.

341. **Barak, A. V. and Burkholder, W. E.,** Behavior and pheromone studies with *Attagenus elongatulus* Casey (Coleoptera: Dermestidae), *J. Chem. Ecol.,* 3, 219, 1977.

342. **Levinson, H. Z., Levinson, A. R., Jen, T.-L., Williams, J. L. D., Kahn, G., and Francke, W.,** Production, site, partial composition and olfactory perception of a pheromone in the male hide beetle, *Naturwissenschaften,* 65, 543, 1978.

343. **Shaaya, E.,** Sex pheromone of *Dermestes maculatus* DeGeer (Coleoptera, Dermestidae), *J. Stored Prod. Res.,* 17, 13, 1981.

344. **Levinson, H. Z., Levinson, A. R., and Francke, W.,** Feeding aggregants in the feces of *Dermestes maculatus* DeGeer, *Naturwissenschaften,* 67, 463, 1980.

345. **Levinson, A. R., Levinson, H. Z., and Francke, W.,** Intraspezifische Lockstoffe des Dornspeckkäfers *Dermestes maculatus, Mitt. Dtsch. Ges. Allg. Angew. Entomol.,* 2, 235, 1981.

346. **Pitman, G. B.,** *trans*-Verbenol and α-pinene: their utility in manipulation of the mountain pine beetle, *J. Econ. Entomol.,* 64, 426, 1971.

347. **Yarger, R. G., Silverstein, R. M., and Burkholder, W. E.,** Sex pheromone of the female dermestid beetle *Trogoderma glabrum* (Herbst), *J. Chem. Ecol.,* 1, 323, 1975.

348. **Greenblatt, R. E., Burkholder, W. E., Cross, J. H., Cassidy, R. F., Silverstein, R. M., Levinson, A. R., and Levinson, H. Z.,** Chemical basis for interspecific responses to sex pheromones of *Trogoderma* species (Coleoptera: Dermestidae), *J. Chem. Ecol.,* 3, 337, 1977.

349. **Ikan, R., Bergmann, E. D., Yinon, U., and Schulov, A.,** Identification, synthesis and biological activity of an "ascembling" scent from the beetle *Trogoderma granarium, Nature (London),* 223, 317, 1969.

350. **Yinon, U., Shulov, A., and Ikan, R.,** Olfactory responses of granary beetles towards natural and synthetic fatty acid esters, *J. Insect Physiol.,* 17, 1037, 1971.

351. **Levinson, A. R., Levinson, H. Z., Schwaiger, H., Cassidy, R. F. Jr.,** Olfactory behavior and receptor potentials of the khapra beetle *Trogoderma granarium* induced by the major components of its sex pheromone, certain analogues and fatty acid esters, *J. Chem. Ecol.,* 4, 95, 1978.

352. **Rossi, R., Salvadori, P. A., Carpita, A., and Niccoli, A.,** Chirality influences the biological activity of the sex pheromone of the khapra beetle, *Naturwissenschaften,* 66, 211, 1979.

353. **Levinson, H. Z. and Mori, K.,** The pheromone activity of chiral isomers of trogodermal for male khapra beetles, *Naturwissenschaften,* 67, 148, 1980.

354. **Levinson, H. Z., Levinson, A. R., and Mori, K.,** Olfactory behavior and receptor potentials of two khapra beetle strains induced by enantiomers of trogodermal, *Naturwissenschaften,* 68, 480, 1981.

355. **Rossi, R. and Niccoli, A.,** Relationships between chirality and biological activity. Biological response of *Trogoderma granarium* to optically active synthetic sex attractants, *Naturwissenschaften,* 65, 259, 1978.

356. **Silverstein, R. M. and Young, J. C.,** Insects generally use multicomponent pheromones, in *Pest Management with Insect Attractants*, Beroza, M., Ed., ACS Symp. Ser. No. 23, Am. Chem. Soc., Washington, D.C., 1976, 29.

357. **Cross, J. H., Byler, R. C., Silverstein, R. M., Greenblatt, R. E., Gorman, J. E., and Burkholder, W. E.,** Sex pheromone components and calling behavior of the female dermestid beetle, *Trogoderma variabile* Ballion (Coleoptera: Dermestidae), *J. Chem. Ecol.*, 3, 115, 1977.

358. **Renwick, J. A. A., Vité, J. P., and Billings, R. F.,** Aggregation pheromones in the ambrosia beetle *Platypus flavicornis*, *Naturwissenschaften*, 64, 226, 1977.

359. **Madrid, F., Vité, J. P., and Renwick, J. A. A.,** Evidence of aggregation pheromone in the ambrosia beetle *Platypus flavicornis*, *Z. Angew. Entomol.*, 72, 73, 1972.

360. **Henzell, R. F. and Lauren, D. R.,** Use of sex attractant traps to estimate the development stage of grass grub, *Costelytra zealandica* (White) (Coleoptera: Scarabaeidae) in the soil, *N.Z. J. Agric. Res.*, 20, 75, 1977.

361. **Klein, M. G., Tumlinson, J. H., Ladd, T. L. Jr., and Doolittle, R. E.,** Japanese beetle (Coleoptera: Scarabaeidae): response to synthetic sex attractant plus phenethyl propionate: eugenol, *J. Chem. Ecol.*, 7, 1, 1981.

362. **Ladd, T. L., Klein, M. G., and Tumlinson, J. H.,** Phenethyl propionate + eugenol + geraniol (3:7:3) and Japonilure: a highly effective joint lure for Japanese beetles, *J. Econ. Entomol.*, 74, 665, 1981.

363. **Francke, W. and Heemann, V.,** Das Duftstoff-Bouquet des Großen Waldgärtners *Blastophagus piniperda* (Coleoptera: Scolytidae), *Z. Angew. Entomol.*, 82, 117, 1976.

364. **Hughes, P. R. and Renwick, J. A. A.,** unpublished data, cited in *Naturwissenschaften*, 63, 550, 1976.

365. **Vité, J. P. and Francke, W.,** The aggregation pheromones of bark beetles: progress and problems, *Naturwissenschaften*, 63, 550, 1976.

366. **Rudinsky, J. A.,** Multiple functions of the Douglas-fir beetle pheromone 3-methyl-2-cyclohexen-1-one, *Environ. Entomol.*, 2, 579, 1973.

367. **Renwick, J. A. A.,** Identification of two oxygenated terpenes from the bark beetles *Dendroctonus frontalis* and *Dendroctonus brevicomis*, *Contrib. Boyce Thompson Inst.*, 23, 355, 1967.

368. **Pitman, G. B., Vité, J. P., Kinzer, G. W., and Fentiman, A. F. Jr.,** Specificity of population-aggregating pheromones in *Dendroctonus*, *J. Insect Physiol.*, 15, 363, 1969.

369. **Hughes, P. R.,** *Dendroctonus*; production of pheromones and related compounds in response to host monoterpenes, *Z. Angew. Entomol.*, 73, 294, 1973.

370. **Stewart, T. D., Plummer, E. L., Pearce, G. T., McCandless, L., and Silverstein, R. M.,** presented at 168th ACS Natl. Meet., Atlantic City, N.J., September, 1974.

371. **Hughes, P. R. and Pitman, G. B.,** Method for observing and recording the flight behavior of tethered bark beetles in response to chemical messengers, *Contrib. Boyce Thompson Inst.*, 24, 329, 1970.

372. **Vité, J. P. and Pitman, G. B.,** Insect and host odors in the aggregation of the western pine beetle, *Can. Entomol.*, 101, 113, 1969.

373. **Vité, J. P. and Pitman, G. B.,** Management of western pine beetle populations: use of chemical messengers, *J. Econ. Entomol.*, 63, 1132, 1970.

374. **Bedard, W., Tilden, P. E., Wood, D. L., Silverstein, R. M., Brownlee, R. G., and Rodin, J. O.,** Western pine beetle: field responses to its sex pheromone and a synergistic host terpene, myrcene, *Science*, 164, 1284, 1968.

375. **Vité, J. P. and Pitman, G. B.,** Aggregating behavior of *Dendroctonus brevicomis* in response to synthetic pheromones, *J. Insect Physiol.*, 15, 1617, 1969.

376. **Pitman, G. B.,** Pheromone response in pine bark beetles: influence of host volatiles, *Science*, 166, 905, 1969.

377. **Pitman, G. B. and Vité, J. P.,** Predator-prey response to western pine beetle attractants, *J. Econ. Entomol.*, 64, 402, 1971.

378. **Wood, D. L.,** Selection and colonization of ponderosa pine by bark beetles, in *Insect Plant Relationships*, Vol. 6, van Emden, H. F., Ed., R. Entomol. Soc., London, 1972, 101.

379. **Vité, J. P. and Renwick, J. A. A.,** Inhibition of *Dendroctonus frontalis* response to frontalin by isomers of brevicomin, *Naturwissenschaften*, 58, 418, 1971.

380. **Tilden, P. E., Bedard, W. D., Wood, D. L., and Stubbs, H. A.,** Interruption of response of *Dendroctonus brevicomis* to its attractive pheromone by components of the pheromone, *J. Chem. Ecol.*, 7, 183, 1981.

381. **Byers, J. A. and Wood, D. L.,** Interspecific effects of pheromones on the attraction of the bark beetles, *Dendroctonus brevicomis* and *Ips paraconfusus* in the laboratory, *J. Chem. Ecol.*, 7, 9, 1981.

382. **Wood, D. L., Browne, L. E., Ewing, B., Kindahl, K., Bedard, W. D., Tilden, P. E., Mori, K., Pitman, G. B., and Hughes, P. R.,** Western pine beetle: specificity among enantiomers of male and female components on an attractant pheromone, *Science*, 192, 896, 1976.

383. **Byers, J. A. and Wood, D. L.,** Interspecific inhibition of the response of the bark beetles, *Dendroctonus brevicomis* and *Ips paraconfusus*, to their pheromones in the field, *J. Chem. Ecol.*, 6, 149, 1980.

384. **Renwick, J. A. A., Pitman, G. B., and Vité, J. P.,** 2-Phenylethanol isolated from bark beetles, *Naturwissenschaften,* 63, 198, 1976.

385. **Pitman, G. B., Vité, J. P., Kinzer, G. W., and Fentiman, A. F. Jr.,** Bark beetle attractants: *trans*-Verbenol isolated from *Dendroctonus, Nature (London),* 218, 168, 1968.

386. **Vité, J. P., Islas, S. F., Renwick, J. A. A., Hughes, P. R., and Klieforth, R. A.,** Biochemical and biological variation of southern pine beetle populations in North and Central America, *Z. Angew. Entomol.,* 75, 422, 1974.

387. **Renwick, J. A. A., Hughes, P. R., and Vité, J. P.,** The aggregation pheromone system of a Dendroctonus bark beetle in Guatemala, *J. Insect Physiol.,* 21, 1097, 1975.

388. **Rudinsky, J. A. A.,** Multiple functions of the southern pine beetle pheromone, verbenone, *Environ. Entomol.,* 2, 511, 1973.

389. **Rudinsky, J. A. and Michael, R. R.,** Sound production in Scolytidae: "rivalry" behaviour of male *Dendroctonus* beetles, *J. Insect Physiol.,* 20, 1219, 1974.

390. **Vité, J. P. and Renwick, J. A. A.,** Different diagnosis and isolation of population attractants, *Contrib. Boyce Thompson Inst.,* 24, 323, 1970.

391. **Renwick, J. A. A., Hughes, P. R., and Tanletin, D. R.,** Oxidation products of pinene in the bark beetle *Dendroctonus frontalis, J. Insect Physiol.,* 19, 1735, 1973.

392. **Coster, J. E. and Vité, J. P.,** Effects of feeding and mating on pheromone release in the southern pine beetle, *Ann. Entomol. Soc. Am.,* 65, 263, 1972.

393. **McCarty, F. A., Billings, P. M., Richerson, J. V., Payne, T. L., and Edson, L. J.,** Response of the southern pine beetle to behavioral chemicals in the laboratory, *J. Ga. Entomol. Soc.,* 15, 307, 1980.

394. **Richerson, J. V. and Payne, T. L.,** Effects of bark beetle inhibitors on landing and attack behavior of the southern pine beetle and beetle associates, *Environ. Entomol.,* 8, 360, 1979.

395. **Payne, T. L., Richerson, J. V., Dickens, J. C., West, J. R., Mori, K., Berisford, C. W., Hedden, R. L., Vité, J. P., and Blum, M. S.,** Southern pine beetle: olfactory receptor and behavior discrimination of enantiomers of the attractant pheromone frontalin, *J. Chem. Ecol.,* 8, 873, 1982.

396. **Payne, T. L., Coster, J. E., Richerson, J. V., Edson, L. J., and Hart, E. R.,** Field response of the southern pine beetle to behavioral chemicals, *Environ. Entomol.,* 7, 578, 1978.

397. **Payne, T. L., Coster, J. E., and Johnson, P. C.,** Effects of slow-release formulation of synthetic *endo*- and *exo*-brevicomin on southern pine beetle flight and landing behavior, *J. Chem. Ecol.,* 3, 133, 1977.

398. **Richerson, J. V.,** private communication, cited in *Prog. Chem. Org. Natl. Prod.,* 37, 134, 1979.

399. **Browne, L. E., Wood, D. L., Bedard, W. D., Silverstein, R. M., and West, J. R.,** Quantitative estimates of the western pine beetle attractive pheromone components, *exo*-brevicomin, frontalin and myrcene in nature, *J. Chem. Ecol.,* 5, 397, 1979.

400. **Renwick, J. A. A. and Pitman, G. B.,** An attractant isolated from female Jeffrey pine beetles, *Dendroctonus jeffreyi, Environ. Entomol.,* 8, 40, 1979.

401. **Pitman, G. B. and Vité, J. P.,** Aggregation behavior of *Dendroctonus ponderosae* (Coleoptera: Scolytidae) in response to chemical messengers, *Can. Entomol.,* 101, 143, 1969.

402. **Hughes, P. R.,** Effect of α-pinene exposure on *trans*-verbenol synthesis in *Dendroctonus ponderosae* Hopk., *Naturwissenschaften,* 60, 261, 1973.

403. **Rasmussen, L. A.,** Attraction of mountain pine beetles to small-diameter lodgepole pines baited with *trans*-verbenol and α-pinene, *J. Econ. Entomol.,* 65, 1396, 1972.

404. **Vité, J. P., Pitman, G. B., Fentiman, A. D. Jr., and Kinzer, G. W.,** 3-Methyl-2-cyclohexen-1-ol isolated from *Dendroctonus, Naturwissenschaften,* 59, 469, 1972.

405. **Kinzer, G. W., Fentiman, A. F. Jr., Foltz, R. L., and Rudinsky, J. A.,** Bark beetle attractants: 3-methyl-2-cyclohexen-1-one isolated from *Dendroctonus pseudotsugae, J. Econ. Entomol.,* 64, 970, 1971.

406. **Rudinsky, J. A., Kinzer, G. W., Fentiman, A. F. Jr., and Foltz, R. L.,** *trans*-Verbenol isolated from Douglas-fir beetle. Laboratory and field bioassays in Oregon, *Environ. Entomol.,* 1, 485, 1972.

407. **Rudinsky, J. A., Morgan, M. E., Libbey, L. M., and Putnam, T. B.,** Additional compounds of the Douglas fir beetle (Col., Scolytidae) aggregative pheromone and their possible utility in pest control, *Z. Angew. Entomol.,* 76, 65, 1974.

408. **Plummer, E. L., Stewart, T. E., Byrne, K. J., Gore, W. E., Pearce, G. T., and Silverstein, R. M.,** presented at the 168th ACS Natl. Meet., Atlantic City, N.J., September 1974.

409. **Rudinsky, J. A. and Michael, R. R.,** Sound production in Scolytidae. Chemostimulus of sonic signal by douglas-fir beetle, *Science,* 175, 1386, 1972.

410. **Pitman, G. B. and Vité, J. P.,** Field responses of *Dendroctonus pseudotsugae* (Coleoptera: Scoytidae) to synthetic frontalin, *Ann. Entomol. Soc. Am.,* 63, 661, 1970.

411. **Knopf, J. A. E. and Pitman, G. B.,** Aggregation pheromone for manipulation of the Douglas-fir beetle, *J. Econ. Entomol.,* 65, 723, 1972.

412. **Copony, J. A. and Morris, C. L.,** Southern pine beetle suppression with frontalure and cacodylic acid treatment, *J. Econ. Ecol.,* 65, 754, 1972.

413. **Furniss, M. M., Kline, L. N., Schmitz, R. F., and Rudinsky, J. A.,** Tests of three pheromones to induce or disrupt aggregation of Douglas-fir beetles on live-trees, *Ann. Entomol. Soc. Am.,* 65, 1227, 1972.

414. **Pitman, G. B.,** Further observations on Douglure in a *Dendroctonus pseudotsugae* management system, *Environ. Entomol.,* 2, 109, 1973.

415. **Furniss, M. M., Daterman, G. E., Kline, L. N., McGregor, M. D., Trostle, G. C., Pettinger, L. F., and Rudinsky, J. A.,** Effectiveness of the Douglas-fir beetle antiaggregative pheromones methylcyclo-hexanone at three concentrations and spacings around felled host trees, *Can. Entomol.,* 106, 381, 1974.

416. **Rudinsky, J. A., Sartwell, C. Jr., Graves, T. M., and Morgan, M. E.,** Granular formulation of methylcyclohexenone: an antiaggregative pheromone of the Douglas fir and spruce bark beetles (Col., Scolytidae), *Z. Angew. Entomol.,* 75, 254, 1974.

417. **Pitman, G. B., Hedden, R. L., and Gara, R. J.,** Synergistic effects of ethyl alcohol on the aggregation of *Dendroctonus pseudotsugae* (Col., Scolytidae) in response to pheromones, *Z. Angew. Entomol.,* 78, 203, 1975.

418. **Kline, N. L., Schmitz, R. F., Rudinsky, J. A., and Furniss, M. M.,** Repression of the spruce beetle (Coleoptera) attraction by methylcyclohexenone in Idaho, *Can. Entomol.,* 106, 894, 1976.

419. **Furniss, M. M., Baker, B. H., and Hostetler, B. B.,** Aggregation of spruce beetles (Coleoptera) to seudenol and repression of attraction by methylcyclohexenone in Alaska, *Can. Entomol.,* 108, 1297, 1976.

420. **Baker, B. H., Hostetler, B. B., and Furniss, M. M.,** Response of eastern larch beetle in Alaska to its natural attractant and to Douglas-fir beetle pheromones, *Can. Entomol.,* 109, 289, 1977.

421. **Hall, R. W.,** Attraction of *Dendroctonus valens* (Coleoptera: Scolytidae) to ponderosa pines baited with *Dendroctonus brevicomis* (Coleoptera: Scolytidae) pheromone, *Environ. Entomol.,* 12, 718, 1983.

422. **Borden, J. H., Chong, L., McLean, J. A., Slessor, K. N., and Mori, K.,** *Gnathotrichus sulcatus*; synergistic response to enantiomers of the aggregation pheromone sulcatol, *Science,* 192, 894, 1976.

423. **Cade, S. C., Hrufiord, B. F., and Gara, R. I.,** Identification of a primary attractant for *Gnathotrichus sulcatus* isolated from western hemlock logs, *J. Econ. Entomol.,* 63, 1014, 1970.

424. **Vité, J. P., Bakke, A., and Renwick, J. A. A.,** Pheromones in *Ips* (Coleoptera, Scolytidae): occurrence and production, *Can. Entomol.,* 104, 1967, 1972.

425. **Bakke, A.,** Aggregation pheromone component of the bark beetle *Ips acuminatus, Oikos,* 31, 184, 1978.

426. **Klimetzek, D. and Francke, W.,** Relationships between the enantiomeric composition of α-pinene in host trees and the production of verbenols in *Ips* species, *Experientia,* 36, 1343, 1980.

427. **Hedden, R., Vité, J. P., and Mori, K.,** Synergistic effect of a pheromone and a kairomone on host selection and colonization by *Ips avulsus, Nature (London),* 261, 696, 1976.

428. **Renwick, J. A. A. and Vité, J. P.,** Pheromones and host volatiles that govern aggregation of the six-spined engraver beetle, *Ips calligraphus, J. Insect Physiol.,* 18, 1215, 1972.

429. **Hughes, P. R.,** Myrcene: a precursor of the pheromones in *Ips* beetles, *J. Insect Physiol.,* 20, 1271, 1974.

430. **Vité, J. P., Klimetzek, D., Loskant, G., Hedden, R., and Mori, K.,** Chirality of insect pheromones: response interruption by inactive antipodes, *Naturwissenschaften,* 63, 582, 1976.

431. **Stoakley, J. T., Bakke, A., Renwick, J. A. A., and Vité, J. P.,** The aggregation pheromone system of the larch bark beetle *Ips cembrae, Z. Angew. Entomol.,* 86, 174, 1978.

432. **Rebenstorff, H. and Francke, W.,** Lärchenborkenkäfer. Überwachung mit Lockstoffen, *Allg. Forst-Zeitschr.,* 37, 450, 1982.

433. **Lapis, E. B. and San Valentin, H. O.,** Field response of *Ips* (*Ips calligraphus* Germar) to synthetic aggregation pheromones and other attractants, *Sylvatrop,* 4, 223, 1979.

434. **Young, J. C., Silverstein, R. M., and Birch, M. C.,** Aggregation pheromone of the beetle *Ips confusus.* Isolation and identification, *J. Insect Physiol.,* 19, 2273, 1973.

435. **Birch, M. C., Tilden, P. E., Wood, D. L., Browne, E., Young, J. C., and Silverstein, R. M.,** Biological activity of compounds isolated from air condensates and frass of the bark beetle, *Ips confusus, J. Insect Physiol.,* 23, 1373, 1977.

436. **Bakke, A.,** Aggregation pheromone in the bark beetle, *Ips duplicatus* (Sahlberg), *Norw. J. Entomol.,* 22, 67, 1975.

437. **Vité, J. P. and Renwick, J. A. A.,** Population aggregating pheromone in the bark beetle, *Ips grandicollis, J. Insect Physiol.,* 17, 1699, 1971.

438. **Werner, R. A.,** Response of the beetles, *Ips grandicollis,* to combinations of host and insect produced attractants, *J. Insect Physiol.,* 18, 1403, 1972.

439. **Vité, J. P., Lühl, R., Hughes, P. R., and Renwick, J. A. A.,** Pine beetles of the genus *Dendroctonus*: pest population in Central America, *FAO Plant Protect. Bull.,* 23, 6, 1975.

440. **Wood, D. L., Stark, R. W., Silverstein, R. M., and Rodin, J. O.,** Unique synergistic effect produced by the principal sex attractant compounds of *Ips confusus* (LeConte) [now *I. paraconfusus*] (Coleoptera: Scolytidae), *Nature (London),* 215, 206, 1967.

441. **Furniss, M. M. and Livingston, L. R.,** Inhibition by ipsenol of pine engraver attraction in Northern Idaho, *Environ. Entomol.,* 8, 369, 1979.

442. **Wood, D. L., Browne, L. E., Bedard, W. D., Tilden, P. E., Silverstein, R. M., and Rodin, J. O.,** Response of *Ips confusus* [now *I. paraconfusus*] to synthetic sex pheromones in nature, *Science*, 159, 1373, 1968.

443. **Renwick, J. A. A., Hughes, P. R., and Krull, I. S.,** Selective production of *cis*- and *trans*-verbenol from (−)- and (+)-α-pinene by a bark beetle, *Science*, 191, 199, 1976.

444. **Lanier, G. N., Birch, M. C., Schmitz, R. F., and Furniss, M. M.,** Pheromones of *Ips pini* (Coleoptera: Scolytidae): variation in response among three populations, *Can. Entomol.*, 104, 1917, 1972.

445. **Birch, M. C., Light, D. M., Wood, D. L., Browne, L. E., Silverstein, R. M., Bergot, B. J., Ohloff, G., West, J. R., and Young, J. C.,** Pheromonal attraction and allomonal interruption of *Ips pini* in California by the two enantiomers of ipsdienol, *J. Chem. Ecol.*, 6, 703, 1980.

446. **Birch, M. C.,** Dept. of Entomol., University of California, Davis, cited in *Prog. Chem. Org. Nat. Prod.*, 37, 97, 1979.

447. **Vité, J. P., Bakke, A., and Hughes, P. R.,** Population attractant of the twelve-dentate pine bark beetle *Ips sexdentatus*, *Naturwissenschaften*, 61, 365, 1974.

448. **Schönherr, J., Vité, J. P., and Serez, M.,** Überwachung von *Ips sexdentatus* — Populationen mit synthetischem Lockstoff, *Z. Angew. Entomol.*, 95, 51, 1983.

449. **Bakke, A.,** Spruce bark beetle, *Ips typographus*: pheromone production and field response to synthetic pheromones, *Naturwissenschaften*, 63, 92, 1976.

450. **Bakke, A., Frøjen, P., and Skattebøl, L.,** Field response to a new pheromonal compound isolated from *Ips typographus*, *Naturwissenschaften*, 64, 98, 1977.

451. **Sauerwein, P. and Vité, J. P.,** Die Eignung von Typolure-Formulierungen zur Überwachung und Bekämpfung des Buchdruckers *Ips typographus*, *Mitt. Dtsch. Ges. Allg. Angew. Entomol.*, 1, 189, 1978.

452. **Krawielitzki, S., Klimetzek, D., Bakke, A., Vité, J. P., and Mori, K.,** Field and laboratory response of *Ips typographus* to optically active pure pheromonal components, *Z. Angew. Entomol.*, 83, 300, 1977.

453. **Rudinsky, J. A., Novak, V., and Svihra, P.,** Pheromone and terpene attraction in the bark beetle *Ips typographus* L., *Experientia*, 27, 161, 1971.

454. **Vité, J. P.,** Einsatz von Lockstoffen bei der Borkenkäferbekämpfung, *Allg. Forstz.*, 33, 428, 1978.

455. **Klimetzek, D., Sauerwein, P., Dimitri, L., and Vaupel, O.,** Einsatz von Typolure und Fallen gegen den Buchdrucker, *Allg. Forst-Jagdztg.*, 150, 238, 1979.

456. **Zumr, V.,** Effect of synthetic pheromone pheroprax on the coleopterous predators of the spruce bark beetle, *Z. Angew. Entomol.*, 95, 47, 1983.

457. **Chararas, C. and M'Sadda, K.,** Chemical and sexual attraction in *Orthotomicus erosus*, *C.R. Acad. Sci. Ser. D*, 271, 1904, 1970.

458. **Harring, C. M., Vité, J. P., and Hughes, P. R.,** Ipsenol, the aggregation pheromone of the crooked-toothed fir bark beetle *(Pityokteines curvidens)*, *Naturwissenschaften*, 62, 488, 1975.

459. **Harring, C. M.,** Aggregation pheromones of the European fir engraver beetles *Pityokteines curvidens, P. spinidens* and *P. vorontzovi* and the role of juvenile hormone in pheromone biosynthesis, *Z. Angew. Entomol.*, 85, 281, 1978.

460. **Blight, M. M., Ottridge, A. P., Wadhams, L. J., Wenham, M. J., and King, C. J.,** Response of the European population of *Scolytus multistriatus* to the enantiomers of α-multistriatin, *Naturwissenschaften*, 67, 517, 1980.

461. **Lanier, G. N., Gore, W. E., Pearce, G. T., Peacock, J. W., and Silverstein, R. M.,** Response of the European elm bark beetle, *Scolytus multistriatus* (Coleoptera: Scolytidae), to isomers and components of its pheromone, *J. Chem. Ecol.*, 3, 1, 1977.

462. **Peacock, J. W., Cuthbert, R. A., Gore, W. E., Lanier, G. N., Pearce, G. T., and Silverstein, R. M.,** Collection of Porapack-Q of the aggregation pheromone of *Scolytus multistriatus* (Coleoptera, Scolytidae), *J. Chem. Ecol.*, 1, 149, 1975.

463. **Birch, M. C., Miller, J. C., and Paine, T. D.,** Evaluation of two attempts to trap defined populations of *Scolytus multistriatus*, *J. Chem. Ecol.*, 8, 125, 1982.

464. **Vité, J. P., Lühl, R., Gerken, L. B., and Lanier, G. N.,** Ulmensplintkäfer: Anlockversuche mit synthetischen Pheromonen im Oberrheintal, *Z. Pflanzenkr. Pflanzenschutz*, 83, 166, 1976.

465. **Minks, A. K. and van Deventer, P.,** Phenological observations of elm bark beetles with attractant traps in the Netherlands during 1975 and 1976, *Ned. Bosbouw-Tijdschr.*, 50, 151, 1978.

466. **Arciero, M.,** Use of multilure baited traps in the Californian Dutch Elm Disease Program for survey and detection of *Scolytus multistriatus*, presented at the 25th Annu. Meet. Entomol. Soc. Am., Washington, D.C., November 1977.

467. **O'Callaghan, D. P., Gallagher, E. M., and Lanier, G. N.,** Field evaluation of pheromone-baited trap trees to control elm bark beetles, vectors of Dutch elm disease, *Environ. Entomol.*, 9, 181, 1980.

468. **Blight, M. M., Henderson, N. C., Wadhams, L. J., Fielding, N. J., and King, C. J.,** Field response of elm bark beetles to baits containing 4-methyl-3-heptanone, *Naturwissenschaften*, 69, 554, 1982.

469. **van Deventer, P. and Minks, A. K.,** Enkele waarnemingen over de schorskever, *Scolytus pygmaeus* (F.) (Coleoptera, Scolytidae), *Entomol. Ber. (Amsterdam)*, 37, 138, 1977.

470. **Gerken, B. and Grüne, S.**, Zur biologischen Bedeutung käfereigener Duftstoffe des Großen Ulmensplint-käfers *Scolytus scolytus* F. (Col., Scolytidae), *Mitt. Dtsch. Ges. Allg. Angew. Entomol.*, 1, 38, 1978.

471. **Blight, M. M., King, J. C., Wadhams, L. J., and Wenham, M. J.**, Attraction of *Scolytus scolytus* to the components of multilure, the aggregation pheromone of *Scolytus multistriatus, Experientia*, 34, 1119, 1978.

472. **Blight, M. M., Wadhams, L. J., Wenham, M. J., and King, J. C.**, Field attraction of *Scolytus scolytus* (F.) to the enantiomers of 4-methyl-3-heptanol, the major components of the aggregation pheromone, *Forestry*, 52, 83, 1979.

473. **Wadhams, L. J., Angst, M. E., and Blight, M. M.**, Response of the olfactory receptors of *Scolytus scolytus* (F.) (Coleoptera, Scolytidae) to the stereoisomers of 4-methyl-3-heptanol, *J. Chem. Ecol.*, 8, 477, 1982.

474. **Klimetzek, D., Vité, J. P., and König, E.**, Über das Verhalten mitteleuropäischer *Trypodendron*-Arten gegenüber natürlichen und synthetischen Lockstoffen, *Mitt. Dtsch. Ges. Allg. Angew Entomol.*, 2, 303, 1981.

475. **Francke, W. and Heyns, K.**, Flüchtige Inhaltsstoffe von Ambrosiakäfern (Coleoptera: Scolytidae), II, *Z. Naturforsch.*, 29c, 246, 1974.

476. **Francke, W. and Heemann, V.**, Lockversuche bei *Xyloterus domesticus* (L.) and *X. lineatus* (Oliv.) mit 3-Hydroxy-3-methylbutan-2-on, *Z. Angew. Entomol.*, 75, 67, 1974.

477. **Francke, W.**, Untersuchungen über Aggregationssubstanzen bei *Xyloterus domesticus* L. (Coleoptera: Scolytidae), *Z. Angew. Entomol.*, 74, 319, 1973.

478. **Klimetzek, D., Vité, J. P., and König, E.**, Response of European Trogoderma species to natural and synthetic attractants. Über das Verhalten mitteleuropäischer Trypodendron-Arten gegenüber natürlichen und synthetischen Lockstoffen, *Allg. Forst-Jagdztg-Ztg.*, 152, 64, 1981.

479. **Klimetzek, D., Vité, J. P., and Mori, K.**, Effect and formulation of the population attractant of lumber bark beetle *Trypodendron (= Xyloterus) lineatum, Z. Angew. Entomol.*, 89, 57, 1980.

480. **Borden, J. H., Oehlschlager, A. C., Slessor, K. N., Chong, L., and Pierce, H. D. Jr.**, Field tests of isomers of lineatin, the aggregation pheromone of *Trypodendron lineatum* (Coleoptera: Scolytidae), *Can. Entomol.*, 112, 107, 1980.

481. **Bauer, J. and Vité, J. P.**, Host selection by *Trypodendron lineatum, Naturwissenschaften*, 62, 539, 1975.

482. **Nijholt, W. W. and Schönherr, J.**, Chemical response behavior of scolytids in West Germany and Western Canada, *Can. For. Serv. Bi-Mon. Res. Notes*, 32, 31, 1976.

483. **Vité, J. P. and Bakke, A.**, Synergism between chemical and physical stimuli in host colonization by an Ambrosia beetle, *Naturwissenschaften*, 66, 528, 1979.

484. **Payne, T. L., Klimetzek, D., Kohnle, U., and Mori, K.**, Electrophysiological and field responses of *Trypodendron* spp. to enantiomers of lineatin, *Z. Angew. Entomol.*, 95, 272, 1983.

485. **Suzuki, T. and Sugawara, R.**, Isolation of an aggregation pheromone from the flour beetles *Tribolium castaneum* and *T. confusum* (Tenebrionidae), *Appl. Entomol. Zool.*, 14, 228, 1979.

486. **Faustini, D. L., Burkholder, W. E., and Laub, R. J.**, Sexually dimorphic setiferous sex patch in the male red flour beetle, *Tribolium castaneum* (Herbst) (Coleoptera: Tenebrionidae): site of aggregation pheromone production, *J. Chem. Ecol.*, 7, 465, 1981.

487. **Suzuki, T.**, 4,8-Dimethyldecanal: the aggregation pheromone of the flour beetles, *Tribolium castaneum* and *T. confusum* (Coleoptera: Tenebrionidae), *Agric. Biol. Chem.*, 44, 2519, 1980.

488. **Keville, R. and Kannowski, P. B.**, Sexual excitation by pheromones of the confused flour beetle, *J. Insect Physiol.*, 21, 81, 1975.

489. **Smart, L. E., Martin, A. P., and Cloudsley-Thompson, J. L.**, The response to pheromones of adult and newly emerged mealworm beetles (*Tenebrio molitor* L.) (Col., Tenebrionidae), *Entomol. Mon. Mag.*, 116, 139, 1980.

490. **Wright, R. H.**, After pesticides — what?, *Nature (London)*, 204, 121, 1964.

491. **Bestmann, H. J.**, Probleme der chemischen Informationsübermittlung bei Insekten, *Mitt. Dtsch. Ges. Allg. Angew. Entomol.*, 1, 147, 1978.

492. **Bestmann, H. J.**, Pheromon-Rezeptor-Wechselwirkung bei Insekten, *Mitt. Dtsch. Ges. Allg. Angew. Entomol.*, 2, 242, 1981.

493. **Lie, R. and Bakke, A.**, Practical results from the mass trapping of *Ips typographus* in Scandinavia, in *Management of Insect Pests with Semiochemicals*, Mitchell, E. R., Ed., Plenum Press, New York, 1981, 175.

494. **Roelofs, W. L.**, Insect lures are complex mixtures, *Chem. Eng. News.*, 52, 16, 1974.

495. **Silverstein, R. M. and Young, J. C.**, Insects generally use multicomponent pheromones, in *Pest Management with Insect Sex Attractants*, Beroza, M., Ed., ACS Symp. Ser. 23, Am. Chem. Soc., Washington, D.C., 1976, 1.

496. **Renwick, J. A. A. and Vité, J. P.**, Biology of pheromones, in *Chemie der Pflanzenschutz- und Schäd-lingsbekämpfungsmittel*, Wegler, R., Ed., Vol. 6, Springer-Verlag, Berlin, 1981, 1.

497. **Mustaparta, H.,** personal communication, cited in *Aliphatic Relat. Nat. Prod. Chem.*, 2, 46, 1981.
498. **Levinson, H. Z. and Bar Ilan, A. R.,** Olfactory and tactile behavior of the khapra beetle, *Trogoderma granarium*, with special reference to its assembling scent, *J. Insect Physiol.*, 16, 561, 1970.
499. **Levinson, H. Z. and Bar Ilan, A. R.,** Behavior of the khapra beetle *Trogoderma granarium* towards the assembling scent released by the female, *Experientia*, 26, 846, 1970.
500. **Burkholder, W. E.,** in *Control of Insect Behaviour by Natural Products*, Wood, D. L. et al., Eds., Academic Press, New York, 1970, 1.
501. **Vick, K. W., Burkholder, W. E., and Gorman, J. E.,** Interspecific response of sex pheromones of *Trogoderma* species (Coleoptera: Dermestidae), *Ann. Entomol. Soc. Am.*, 63, 379, 1970.
502. **Chapman, J. A.,** Evidence for the sex attractant in the Elaterid beetle, *Hemicrepidius morio* (LeConte), *Can. Entomol.*, 96, 909, 1964.
503. **Doane, J. F.,** Movement on the soil surface of adult *Ctenicera acripennis destructor* (Brown) and *Hypolithus bicolor* Esch. (Coleoptera: Elateridae), as indicated by funnel pitfall traps, with notes on captures of other arthropods, *Can. Entomol.*, 93, 696, 1961.
504. **Ivashenko, J. J. and Adamenko, E. A.,** Sexual attractant of *Agriotes ligitiosus* (Elateridae), *Zool. Zh. (Moscow)*, 50, 1021, 1971.
505. **Lilly, C. E.,** Response of males of *Limonius californicus* (Mann.) (Coleoptera: Elateridae) to a sex attractant, separable by paper-chromatography, *Can. Entomol.*, 91, 145, 1959.
506. **Lilly, C. E. and McGinnis, A. J.,** Reactions of male click beetles in the laboratory to olfactory pheromones, *Can. Entomol.*, 97, 317, 1965.
507. **Lilly, C. E. and McGinnis, J. A.,** Quantitative responses of males of *Limonius californicus* (Coleoptera: Elateridae) to female sex pheromone, *Can. Entomol.*, 100, 1071, 1968.
508. **Hayes, J. T. and Wheeler, A. G.,** Evidence for a sex attractant in *Hemicrepidus decoloratus* (Coleoptera; Elateridae), *Can. Entomol.*, 100, 207, 1968.
509. **Levinson, H. Z. and Levinson, A. R.,** Trapping of storage insects by sex and food attractants as a tool of integrated control, in *Chemical Ecology: Odour Communication in Animals*, Ritter, F. J., Ed., Elsevier/North Holland, Amsterdam, 1979, 327.
510. **Lanier, G. M. and Burkholder, W. E.,** Pheromones in speciation of Coleoptera, in *Pheromones*, Birch, M. C., Ed., American Elsevier, New York, 1974, 161.
511. **Silverstein, R. M.,** Enantiomeric composition and bio-activity of chiral semiochemicals in insects, in *Chemical Ecology: Odour Communication in Animals*, Ritter, F. J., Ed., Elsevier/North Holland, Amsterdam, 1979, 133.
512. **Brand, J. M., Young, Chr., and Silverstein, R. M.,** *Prog. Chem. Org. Nat. Prod.*, 37, 95, 1979.
513. **Mori, K.,** Optical isomerism and olfaction, *Koryo*, 127, 19, 1980.
514. **Bordon, J. H., Handley, J. R., McLean, J. A., Silverstein, R. M., Chong, Li, Slessor, K. N., Johnston, B. D., and Schuler, H. R.,** Enantiomer-based specificity in pheromone communication by two sympatric *Gnathotrichus* species (Coleoptera: Scolytidae), *J. Chem. Ecol.*, 6, 445, 1980.
515. **Renwick, J. A. A. and Vité, J. P.,** Bark beetle attractants: mechanism of colonization by *Dendroctonus frontalis*, *Nature (London)*, 224, 1222, 1969.
516. **Vité, J. P. and Renwick, J. A. A.,** Insect and host factors in the aggregation of the southern pine beetle, *Contrib. Boyce Thompson Inst.*, 24, 61, 1968.
517. **Dickens, J. C. and Payne, T. L.,** Bark beetle olfaction: pheromone receptor system in *Dendroctonus frontalis*, *J. Insect Physiol.*, 23, 481, 1977.
518. **Renwick, J. A. A. and Vité, J. P.,** Systems of chemical communication in *Dendroctonus*, *Contrib. Boyce Thompson Inst.*, 24, 283, 1970.
519. **Vité, J. P. and Renwick, J. A. A.,** Anwendbarkeit von Borkenkäferpheromonen: Konfiguration und Konsequenzen, *Z. Angew. Entomol.*, 82, 112, 1976.
520. **Dickens, J. C.,** Electrophysiological investigations of olfaction in bark beetles, *Mitt. Schweiz. Entomol. Ges.*, 52, 203, 1979.
521. **Birch, M. C., Light, D. M., and Mori, K.,** Selective inhibition of response of *Ips pini* to its pheromone by the (S) $(-)$-enantiomer of ipsenol, *Nature (London)*, 270, 738, 1977.
522. **Mustaparta, H., Angst, M. E., and Lanier, G. N.,** Receptor discrimination of enantiomers of the aggregation pheromone ipsdienol, in two species of *Ips*, *J. Chem. Ecol.*, 6, 689, 1980.
523. **Mustaparta, H.,** Olfactory receptor specificities for multicomponent chemical signals, in *Recept. Neurotransm., Horm., Pheromones Insects, Proc. Workshop 1979*, Satelle, D. B., Hall, L. M., and Hildebrand, J. G., Eds., Elsevier, Amsterdam, 1980, 283.
524. **Kerck, K.,** Einfluß baum- und käferbürtiger Reize auf das Suchverhalten von *Xyloterus domesticus* L., *Naturwissenschaften*, 65, 452, 1978.
525. **Moeck, H. A.,** Ethanol as the primary attractant for the ambrosia beetle *Trypodendron lineatum* (Coleoptera: Scolytidae), *Can. Entomol.*, 102, 985, 1970.

526. **Tumlinson, J. H., Gueldner, R. C., Hardee, D. D., Thompson, A. D., Hedin, P. A., and Minyard, J. P.,** The boll weevil sex attractant, in *Chemicals Controlling Insect Behaviour,* Beroza, M., Ed., Academic Press, New York, 1970, 41.

527. **Minyard, J. P., Tumlinson, J. H., Hedin, P. A., and Thompson, A. C.,** Constituents of the cotton bud terpene hydrocarbons, *J. Agric. Food Chem.,* 13, 599, 1965.

528. **Hedin, P. A., Thompson, A. C., Gueldner, R. C., and Minyard, J. P.,** Volatile constituents of the boll weevil, *J. Insect Physiol.,* 18, 79, 1972.

529. **Minyard, J. P., Tumlinson, J. H., Thompson, A. C., and Hedin, P. A.,** Constituents of the cotton bud. The carbonyl compounds, *J. Agric. Food Chem.,* 15, 517, 1967.

530. **Hardee, D. D.,** Pheromone production by male boll weevil as affected by food and host factors, *Contrib. Boyce Thompson Inst.,* 24, 315, 1970.

531. **Mitlin, N. and Hedin, P. A.,** Biosynthesis of grandlure, the pheromone of the boll weevil, *Anthonomus grandis,* from acetate, mevalonate, and glucose, *J. Insect Physiol.,* 20, 1825, 1974.

532. **Hedin, P. A.,** A study of factors that control biosynthesis of the compounds which comprise the boll weevil pheromone, *J. Chem. Ecol.,* 3, 279, 1977.

533. **Hughes, P. R. and Renwick, J. A. A.,** Neural and hormonal control of pheromone biosynthesis in the bark beetle, *Ips paraconfusus, Physiol. Entomol.,* 2, 117, 1977.

534. **Hendry, L. B., Piatek, B., Browne, L. E., Wood, D. L., Byers, J. A., Fish, R. H., and Hicks, R. A.,** *In vivo* conversion of a labelled host plant chemical to pheromones in the bark beetle *Ips paraconfusus, Nature (London),* 284, 485, 1980.

535. **Renwick, J. A. A., Hughes, P. R., Pitman, G. B., and Vité, J. P.,** Oxidation products of terpenes identified from *Dendroctonus* and *Ips* bark beetles, *J. Insect Physiol.,* 22, 725, 1976.

536. **Byers, J. A.,** Male-specific conversion of the host plant compound, myrcene, to the pheromone, (+)-ipsdienol, in the bark beetle, *J. Chem. Ecol.,* 8, 363, 1982.

537. **Renwick, J. A. A. and Hughes, R. P.,** Oxidation of unsaturated cyclic hydrocarbons by *Dendroctonus frontalis, Insect Biochem.,* 5, 459, 1975.

538. **Hughes, P. R.,** private communication, cited in *Prog. Chem. Org. Nat. Prod.,* 37, 111, 1979.

539. **Brattsten, L. B., Wilkinson, C. F., and Eisner, T.,** Herbivore-plant interactions: mixed function oxidases and secondary plant substances, *Science,* 196, 1349, 1977.

540. **Brand, J. M., Bracke, J. W., Markovetz, A. J., Wood, D. L., and Browne, L. E.,** Production of verbenol pheromone by a bacterium isolated from bark beetles, *Nature (London),* 254, 136, 1975.

541. **Chararas, C., Riviere, J., Ducauze, C., Rutledge, D., Delpui, G., and Cazelles, M. T.,** Bioconversion of a terpenic compound by bacteria from the digestive tube of *Phloeosinus armatus* (Coleoptera; Scolytidae), *C.R. Acad. Sci. Paris Ser. D,* 291, 299, 1980.

542. **Barras, S. J. and Perry, T.,** Fungal symbionts in the prothoracic mycangium of *Dendroctonus frontalis* (Coleoptera: Scolytidae), *Z. Angew. Entomol.,* 71, 95, 1972.

543. **Barras, S. J. and Taylor, J. J.,** Varietal *Ceratocystis minor* identification from mycangium of *Dendroctonus frontalis, Mycopathol. Mycol. Applic.,* 50, 293, 1973.

544. **Brand, J. M., Bracke, J. W., Britton, L. N., Markovetz, A. J., and Barras, S. J.,** Bark beetle pheromones: production of verbenone by a mycangial fungus of *Dendroctonus frontalis, J. Chem. Ecol.,* 2, 195, 1976.

545. **Brand, J. M., Schultz, J., Barras, S. J., Edson, L. J., Payne, T. L., and Hedden, R. L.,** Bark beetle pheromones: enhancement of *Dendroctonus frontalis* (Coleoptera: Scolytidae) aggregation pheromone by yeast metabolites in laboratory bioassay, *J. Chem. Ecol.,* 3, 657, 1977.

546. **Borden, J. H., Nair, K. K., and Slater, C. E.,** Synthetic juvenile hormone: induction of sex pheromone production in *Ips confusus, Science,* 166, 1626, 1969.

547. **Hackstein, E. and Vité, J. P.,** Pheromonbiosynthese und Reizkette in der Besiedlung von Fichten durch den Buchdrucker *Ips typographus, Mitt. Dtsch. Ges. Allg. Angew. Entomol.,* 1, 185, 1978.

548. **Menon, M.,** Age-dependent effects of synthetic juvenile hormone on pheromone synthesis in adult females of *Tenebrio molitor, Ann. Entomol. Soc. Am.,* 69, 457, 1976.

549. **Hughes, P. R. and Renwick, J. A. A.,** Hormonal and host factors stimulating pheromone synthesis in female western pine beetles, *Dendroctonus brevicomis, Physiol. Entomol.,* 2, 289, 1977.

550. **Gerken, B. and Hughes, P.,** Hormonale Stimulation der Biosynthese geschlechtsspezifischer Duftstoffe bei Borkenkäfern, *Z. Angew. Entomol.,* 82, 108, 1976.

551. **Brand, J. M. and Barras, S. J.,** The major volatile constituents of a Basidiomycete associated with the southern pine beetle, *Lloydia,* 40, 398, 1977.

552. **Hoyt, C. P., Osborne, G. O., and Nulcock, A. P.,** Production of an insect sex attractant by symbiontic bacteria, *Nature (London),* 230, 472, 1971.

553. **Stark, R. W. and Gittins, A. R., Eds.,** *Pest Management of the 21st Century,* Idaho Research Foundation Inc., University Stn., Moscow, Idaho, 1973.

554. **Mitchell, E. R., Ed.,** *Management of Insect Pests with Semiochemicals: Concepts and Practice,* Plenum Press, New York, 1981.

555. **Silverstein, R. M.,** Pheromones: Background and potential for use in insect pest control, *Science,* 213, 1326, 1981.

566. **Roelofs, W. L.,** Pheromones and their chemistry, in *Insect Biology in the Future: "VBW 80'',* Locke, M. and Smith, D. S., Eds., Academic Press, New York, 1980, 583.

557. **Roelofs, W. L.,** Attractive and aggregating pheromones, in *Semiochemicals: Their Role in Pest Control,* Nordlund, D. A., Jones, R. L., and Lewis, W. J., Eds., John Wiley & Sons, New York, 1981, 213.

558. **Nordlund, D. A., Jones, R. L., and Lewis, W. J.,** *Semiochemicals, Their Role in Pest Control,* Nordlund, D. A., Jones, R. L., and Lewis, W. J., Eds., John Wiley & Sons, New York, 1981, 306.

559. **Campion, D. G. and Nesbitt, B. F.,** Lepidopteran sex pheromones and pest management in developing countries, *Trop. Pest Manage.,* 27, 53, 1981.

560. **Kydonieus, A. F. and Beroza, M., Eds.,** *Insect Suppression with Controlled Release Pheromone Systems,* Vols. 1 and 2, CRC Press, Boca Raton, Fla., 1982.

561. **Klassen, W., Ridgway, R. L., and Inscoe, M.,** Chemical attractants in integrated pest management programs, in *Insect Suppression with Controlled Release Pheromone Systems,* Vol. 1, Kydonieus, A. F. and Beroza, M., Eds., CRC Press, Boca Raton, Fla., 1982, 13.

562. **Boneβ, M.,** Die praktische Verwendung von Insektenpheromonen, in *Chemie der Pflanzenschutz- und Schädlingsbekämpfungsmittel,* Vol. 6, Springer-Verlag, Berlin, 1981, 165.

563. **Minks, A. K.,** Present status of insect pheromones in agriculture and forestry, in *Proceedings of the Int. Symp. IOBC/WPRS on Integrated Control in Agriculture and Forestry, Vienna,* IPO, Wageningen, Netherlands, 1980, 127.

564. **Hrdy, J.,** Current status and possibilities of further use of insect pheromones in plant protection, *Agrochemia,* 19, 143, 1979.

565. **Fadeev, Yu. N. and Smetnik, A. I.,** Use of insect pheromones in plant protection in the USSR, *S-hk. Biol.,* 15, 803, 1980.

566. **Roelofs, W. L.,** Pheromones and their chemistry, in *Insect Biology in the Future: "VBW 80'',* Locke, M. and Smith, D. S., Eds., Academic, New York, 1980, 583.

567. **Oloumi-Sadeghi, H., Showers, W. B., and Reed, G. L.,** European corn-borer: lack of synchrony of attraction to sex pheromones and capture in light traps, *J. Econ. Entomol.,* 68, 663, 1975.

568. **Beroza, M., Bierl, B. A., Tardif, J. G. R., Cook, D. A., and Paszek, E. C.,** Activity and persistence of synthetic and natural sex attractants of the gypsy moth in laboratory and field trials, *J. Econ. Entomol.,* 64, 1499, 1971.

569. **Neumark, S. and Teich, I.,** Pink Bollworm: constant-level liquid device for use in trapping moths, *J. Econ. Entomol.,* 66, 298, 1973.

570. **Boneβ, M.,** Freilandversuche zur Bekämpfung der Nonne *L. monacha* mit dem synthetischen Pheromon von *Pectinophora gossypiella, Pflanzensch. Nachr. Bayer,* 28, 155, 1975.

571. **Mitchell, E. R., Webb, J. C., Baumhover, A. H., Hines, R. W., Stanley, J. W., Endris, R. G., Lindquist, D. A., and Masuda, S.,** Evaluation of cylindrical electric grids as pheromone traps for loopers and tobacco hornworms, *Environ. Entomol.,* 1, 365, 1972.

572. **Boneβ, M., Schulze, W., and Skatulla, U.,** Versuche zur Bekämpfung der Nonne *L. monacha* mit dem synthetischen Pheromon Disparlure, *Anz. Schaedlingskd. Pflanz. Umweltschutz,* 47, 119, 1974.

573. **Hardee, D. D., Moody, R., Lowe, J., and Pitts, A.,** Grandlure, in field traps, and insecticides in population management of the boll weevil, *J. Econ. Entomol.,* 68, 502, 1975.

574. **Boness, M.,** Versuche zur Bekämpfung des Fruchtschalenwicklers *Archips podana* mit Pheromonen, *Z. Angew. Entomol.,* 82, 104, 1976.

575. **Renwick, J. A. A. and Vité, J. P.,** in *Chemie der Pflanzenschutz und Schädlingsbekampfungsmittel,* Vol. 6, Wegler, R., Ed., Springer-Verlag, Berlin, 1981, 15.

576. **Vité, J. P. and Gara, R. I.,** Volatile attractants from ponderosa pine attacked by bark beetles (Coleoptera: Scolytidae), *Contrib. Boyce Thompson Inst.,* 21, 251, 1962.

577. **Chapman, J. A.,** Response behavior of scolytid beetles and odour meteorology, *Can. Entomol.,* 99, 1132, 1967.

578. **Creighton, C. S., McFadden, T. L., and Cuthbert, E. R.,** Supplementary data on phenylacetaldehyde. Attractant for Lepidoptera, *J. Econ. Entomol.,* 66, 114, 1973.

579. **Debolt, J. W., Jay, E. L., and Ost, R. W.,** Light traps: effect of modifications on catches of several species of Noctuidae and Arctiidae, *J. Econ. Entomol.,* 68, 186, 1975.

580. **Hardee, D. D., Lindig, O. H., and Davich, T. B.,** Suppression of populations of boll weevils over a large area in west Texas with pheromone traps in 1969, *J. Econ. Entomol.,* 64, 928, 1971.

581. **McLean, J. A. and Borden, J. H.,** Attack by *Gnathotrichus sulcatus* (Coleoptera: Scolytidae) on stumps and felled trees baited with sulcatol and ethanol, *Can. Entomol.,* 109, 675, 1977.

582. **Scott, W. P., Lloyd, E. P., Bryson, J. O., and Davich, T. B.,** Trap plots for suppression of low density overwintered populations of boll weevils, *J. Econ. Entomol.,* 67, 281, 1974.

583. **Lloyd, E. P., Scott, W. P., Shaunak, K. K., Tingle, F. C., and Davich, T. B.,** A modified trapping system for suppressing low-density populations of overwintered boll weevils, *J. Econ. Entomol.,* 65, 1144, 1972.

584. **Beroza, M.,** Insect attractants are taking hold, *Agric. Chem.,* 15, 37, 1960.

585. **Roelofs, W. L.,** Communication disruption by pheromone components, in Proc. Symp. Insect Pheromones and their Application, Nagaoka and Tokyo, December 1976, 123.

586. **Kaae, R. S., Shorey, H. H., Gaston, L. K., and Hummel, H. H.,** Sex pheromones of Lepidoptera: disruption of pheromone communication in *Trichoplusia ni* and *Pectinophora gossypiella* by permeation of the air with nonpheromonal chemicals, *Environ. Entomol.,* 3, 87, 1974.

587. **Daterman, G. E., Daves, G. D., Jr., and Smith, R. G.,** Comparison of sex pheromone *versus* an inhibitor for disruption of pheromone communication in *Rhyacionia buoliana, Environ. Entomol.,* 4, 944, 1975.

588. **Shorey, H. H.,** Application of pheromones for manipulating insect pests of agricultural crops, in Proc. Symp. Insect Pheromones and their Application, Nagoaka and Tokyo, December 1976, 97.

589. **Furniss, M. M., Young, W., McGregor, M. D., Livington, R. L., and Hamel, D. R.,** Effectiveness of controlled-release formulations of MCH for preventing Douglas-fir beetle infestations in felled trees, *Can. Entomol.,* 104, 1063, 1977.

590. **St. Clair, A., Goulding, R. L., and Rudinsky, J. A.,** Controlled-release dispensing system for 3,2-MCH, antiaggregative pheromone of the Douglas-fir beetle (Col.: Scolyt.), *Z. Angew. Entomol.,* 83, 297, 1977.

591. **Knopf, J. A. E. and Pitman, G.,** Aggregation pheromone for manipulation of the Douglas-fir beetle, *J. Econ. Entomol.,* 65, 723, 1972.

592. **Pitman, G.,** personal communication, cited in Roelofs, W. L., *Environ. Lett.,* 8, 41, 1975.

593. Entomol. Soc. Am. Rev. Committee, The pilot boll weevil eradication experiment, *Bull. Entomol. Soc. Am.,* 19, 218, 1973.

594. **Rust, M. K. and Reierson, D. A.,** Using pheromone extract to reduce repellency of blatticides, *J. Econ. Entomol.,* 70, 34, 1977.

595. **Burkholder, W. E. and Boush, G. M.,** Pheromones in stored products. Insect trapping and pathogen dissemination, *EPPO Bull.,* 4, 455, 1974.

596. **Shapas, J., Burkholder, W. E., and Boush, G. M.,** Population suppression of *T. glabrum* by using pheromone luring for protozoan pathogen dissemination, *J. Econ. Entomol.,* 70, 469, 1977.

597. **Silverstein, R. M.,** Complexity, diversity, and specificity of behavior-modifying chemicals: examples mainly from Coleoptera and Hymenoptera, in *Chemical Control of Insect Behavior: Theory and Application,* John Wiley & Sons, New York, 1977, 231.

598. **Wood, D. L.,** Manipulation of forest insect pests, in *Chemical Control of Insect Behavior: Theory and Application,* John Wiley & Sons, New York, 1977.

599. **Wood, D. L. and Bedard, W. D.,** The role of pheromones in the population dynamics of the western pine beetle, in *Proc. 15th Int. Congress Entomol.,* Entomol. Soc. Am., Washington, D.C., 1976.

600. **Wood, D. L. et al.,** Integrated pest management of the western pine beetle: research approach and principal results, in *New Technology of Pest Control,* Huffaker, C. B., Ed., John Wiley & Sons, New York, 1980. York, 1980.

601. **Bedard, W. D., Wood, D. L., and Tilden, P. E.,** Use of behavior modifying chemicals to lower western pine beetle-caused tree mortality and to protect trees, in Current Topics in Forest Entomology, Tech. Rep. WO-8, Water, W. E., Ed., U.S. Forest Service, Washington, D.C., 1979, 159.

602. **Bedard, W. D.,** private communication, U.S. Forestry Service, Berkeley, Calif., cited in *Prog. Chem. Org. Nat. Prod.,* 37, 129, 1979.

603. **Bordon, J. H., Lindgren, B. S., and Chong, L.,** Ethanol and α-pinene as synergists for the aggregation pheromone of two *Gnathotrichus* species, *Can. J. For. Res.,* 10, 290, 1980.

604. **McLean, J. A. and Borden, J. H.,** Survey of *Gnathotrichus sulcatus* (Coleoptera: Scolytidae) in a commercial saw mill with the pheromone, sulcatol, *Can. J. For. Res.,* 5, 586, 1975.

605. **McLean, J. A. and Borden, J. H.,** An operational pheromone-based suppression program for an Ambrosia beetle, *Gnathotrichus sulcatus,* in a commercial sawmill, *J. Econ. Entomol.,* 72, 165, 1979.

606. **McLean, J. A. and Borden, J. H.,** Suppression of *Gnathotrichus sulcatus* with sulcatol-baited traps in a sawmill and notes on the occurrence of *G. retusus* and *Trypodendron lineatum, Can. J. For. Res.,* 7, 348, 1977.

607. **Lanier, G. N., Silverstein, R. M., and Peacock, J. W.,** Attractant pheromone of the European elm bark beetle (*Scolytus multistriatus*); isolation, identification, synthesis and utilization studies, in *Perspectives in Forest Entomology,* Anderson, J. F. and Kaya, H. K., Eds., Academic Press, New York, 1976.

608. **Birch, M. C.,** Dept. of Entomology, University of California, Davis, private communication.

609. **Birch, M. C., Paine, T. D., and Miller, J. C.,** Effectiveness of pheromone mass-trapping of the smaller European elm bark beetle, *Calif. Agric.,* 35, 6, 1981.

610. **Vité, J. P.,** Silviculture and the management of bark beetle pests, *Proc. Tall Timbers Conf. on Ecol. Animal Control by Habit Management,* 3, 155, 1971.

611. **Vité, J. P. and Williamson, D. L.**, *Thanasimus dubius*: prey perception, *J. Insect Physiol.*, 16, 233, 1970.

612. **Nijholt, W. W.**, Ambrosia beetle: a menace to the forest industry, Can. For. Serv. Pacif. For. Res. Cent. Rep. BC-P 22, 1978.

613. **Vité, J. P.**, Mit Lockstoffen gegen Nutzholzborkenkäfer, *Holz-Zentralbl.*, 105, 1507, 1979.

614. **Bakke, A. and Saether, T.**, Granbarkbillen kan fanges i rørfeller, *Skogeieren*, 65, 10, 1978.

615. **Vité, J. P.**, Die Wirkung von Borkenkäferlockstoffen, *Holz-Zentralbl.*, 105, 809, 1979.

616. **O'Sullivan, D. A.**, Pheromone lures help control bark beetles, *Chem. Eng. News*, July 30, 10, 1979.

617. **Vaupel, O., Dimitri, L., and Vité, J. P.**, Untersuchungen über den Einsatz von Lockstoff-beköderten Rohrfallen zur Bekämpfung des Buchdruckers (*Ips typographus* L.), sowie Möglichkeiten der Optimierung von Lockstoffverfahren, *Allg. Forst-Jagdztg.*, 152, 102, 1981.

618. **Riege, L. A.**, On the pheromone system of *Ips typographus*, practical applications, in Proc. EUCHEM Conference on Chemistry of Insects, Sweden, August 1979.

619. **Bakke, A.**, Mass trapping of the spruce bark beetle *Ips typographus* in Norway as part of an integrated control program, in *Insect Suppression with Controlled Release Pheromone Systems*, Vol. 2, Kydonieus, A. F. and Beroza, M., Eds., CRC Press, Boca Raton, Fla., 1982, 17.

620. **Ozols, G. and Bicevskis, M.**, Prospects of the use of *Ips typographus* attractant, *Biol. Akt. Veshchestva Zashch. Rast.*, p. 49, 1979.

621. **Bakke, A. and Kvamme, T.**, Kairomone response in *Thanasinus* predators to pheromone components of *Ips typographus*, *J. Chem. Ecol.*, 7, 305, 1981.

622. **Furniss, M. M.**, private communication, cited in *Prog. Chem. Org. Nat. Prod.*, 37, 135, 1979.

623. **Dyer, E. D. A.**, Spruce bark beetles aggregated by the synthetic pheromone frontalin, *Can. J. For. Res.*, 3, 486, 1973.

624. **Dyer, E. D. A. and Safranyik, L.**, Assessment of the impact of pheromone-baited trees on a spruce beetle population (Coleoptera: Scolytidae), *Can. Entomol.*, 109, 77, 1977.

625. **Lloyd, E. P., McKibben, G. H., Knipling, E. F., Witz, J. H., Hartstack, A. W., Leggett, J. E., and Lockwood, D. F.**, Mass-trapping for detection, suppression and integration with other suppression measures against the boll weevil, in *Management of Insect Pests with Semiochemicals: Concepts and Practice*, Mitchell, E. R., Ed., Plenum Press, New York, 1981, 191.

626. **Hardee, D. D.**, Mass trapping and trap cropping of the boll weevil, *Anthonomus grandis* Boheman, in *Insect Suppression with Controlled Release Pheromone Systems*, Vol. 2, Kydonieus, A. F. and Beroza, M., Eds., CRC Press, Boca Raton, Fla., 1982, 65.

627. **Levinson, H. Z. and Levinson, A. R.**, Schädlingsmanipulation mit Insektistatica, *Mitt. Dtsch. Ges. Allg. Angew. Entomol.*, 2, 228, 1981.

628. **McGovern, T. P., Beroza, M., Ladd, T. L. Jr., Ingangi, J. C., and Jurimas, P. J.**, Phenethyl propionate, a potent new attractant for Japanese beetles, *J. Econ. Entomol.*, 63, 1727, 1970.

629. **McGovern, T. P., Beroza, M., Schwartz, P. H., Hamilton, D. W., Ingangi, J. C., and Ladd, T. L.**, Methyl cyclohexanepropionate and related chemicals as attractants for the Japanese beetles, *J. Econ. Entomol.*, 63, 276, 1970.

630. **Ladd, T. L. Jr. and Klein, M. G.**, Trapping Japanese beetles with synthetic female sex pheromones and food-type lure, in *Insect Suppression with Controlled Release Pheromone Systems*, Vol. 2, Kydonieus, A. F. and Beroza, M., Eds., CRC Press, Boca Raton, Fla., 1982, 17.

631. **Young, J. W., Graves, T. M., Curtis, R., and Furniss, M. M.**, Controlled release formulations of insect growth regulators and pheromones — evaluation methods and field trials results, in *Controlled Release Pesticides*, Scher, H. B., Ed., ACS Symp. Ser. No. 53, Am. Chem. Soc., Washington, D.C., 1977, 184.

632. **Knauf, W., Roth, W., Bestmann, H. J., Stegmeier, R., and Vostrowsky, O.**, Das Verhalten von Pheromonkomponenten in Ködern von Prognosefallen, *Mitt. Dtsch. Ges. Allg. Angew. Entomol.*, 4, 11, 1983.

633. **Kydonieus, A. F., Smith, J. K., and Beroza, M.**, Controlled release of pheromones through multi-layered polymeric dispensers, in *Controlled Release Formulations*, Paul, D. R. and Harras, F. W., Eds., ACS Symp. Ser. No. 53, Am. Chem. Soc., Washington, D.C., 1976.

634. **Herculite Protective Fabrics Corp.**, New York.

635. **Brooks, T. W.**, Controlled vapor release from hollow fibres: theory and application with insect pheromones, in *Controlled Release Technology: Methods, Theory, and Applications*, Vol. 2, Kydonieus, A. F., Ed., CRC Press, Boca Raton, Fla., 1980, 165.

636. **Campion, D. G., Lester, R., and Nesbitt, B. F.**, Controlled release of pheromones, *Pest. Sci.*, 9, 434, 1978.

637. **Roth, W., Heinrich, R., Knauf, W., Bestmann, H. J., Brosche, T., and Vostrowsky, O.**, Formulierung von Pheromonen — Möglichkeiten und Schwierigkeiten, Mitt. Dtsch. Ges. Allg. Angew. Entomol., 2, 279, 1981.

638. **Hall, D. R., Nesbitt, B. F., Marrs, G. J., Green, A., St. J., Campion, D. G., and Critchley, B. R.,** Development of microencapsulated pheromone formulations, in *Insect Pheromone Technology,* Leonhardt, B. A. and Beroza, M., Eds., ACS Symp. Ser. No. 190, Am. Chem. Soc., Washington, D.C., 1981, 131.

639. **Daterman, G. E.,** Monitoring insects with pheromones: trapping objectives and bait formulations, in *Insect Suppression with Controlled Release Pheromone Systems,* Vol. 1, Kydonieus, A. F. and Beroza, M., Eds., CRC Press, Boca Raton, Fla., 1982, 192.

640. **Bierl-Leonhardt, B. A.,** Release rates from formulations and quality control methods, in *Insect Suppression with Controlled Release Pheromone Systems,* Vol. 1, Kydonieus, A. F. and Beroza, M., Eds., CRC Press, Boca Raton, Fla., 1982, 246.

641. Controlled Release Division of Albany International, Albany, N.Y.

INDEX

A

C